Hassan Khalil

LINEAR SYSTEM THEORY

WILSON J. RUGH

Department of Electrical and Computer Engineering
The Johns Hopkins University

PRENTICE HALL, Englewood Cliffs, New Jersey 07632

Library of Congress Cataloging-in-Publication Data

Rugh, Wilson J.
   Linear system theory / Wilson J. Rugh.
      p.   cm.
   Includes bibliographical references and index.
   ISBN 0-13-555038-6
   1. Control theory. 2. Linear systems.   I. Title.
QA402.3.R84   1993
003'.74--dc20   92-10464
                                                    CIP

Acquisitions editor: Pete Janzow
Production editor: Bayani Mendoza de Leon
Cover design: Joe Di Domenico
Prepress buyer: Linda Behrens
Manufacturing buyer: Dave Dickey
Supplements editor: Alice Dworkin
Editorial assistant: Phyllis Morgan

© 1993 by Prentice-Hall, Inc.
A Simon & Schuster Company
Englewood Cliffs, New Jersey 07632

The author and publisher of this book have used their best efforts in preparing this book. These efforts include the development, research, and testing of the theories and programs to determine their effectiveness. The author and publisher make no warranty of any kind, expressed or implied, with regard to these programs or the documentation contained in this book. The author and publisher shall not be liable in any event for incidental or consequential damages in connection with, or arising out of, the furnishing, performance, or use of these programs.

ISBN 0-13-555038-6

90000>

Printed in the United States of America

10 9 8 7 6 5 4 3 2 1

ISBN 0-13-555038-6

9 780135 550380

Prentice-Hall International (UK) Limited, *London*
Prentice-Hall of Australia Pty. Limited, *Sydney*
Prentice-Hall Canada Inc., *Toronto*
Prentice-Hall Hispanoamericana, S.A., *Mexico City*
Prentice-Hall of India Private Limited, *New Delhi*
Prentice-Hall of Japan, Inc., *Tokyo*
Simon & Schuster Asia Pte. Ltd., *Singapore*
Editora Prentice-Hall do Brasil, Ltda., *Rio de Janeiro*

To Terry, David, and Karen

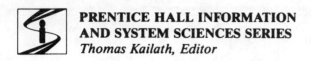

**PRENTICE HALL INFORMATION
AND SYSTEM SCIENCES SERIES**
Thomas Kailath, Editor

CONTENTS

PREFACE

A course on linear system theory at the first-year graduate level typically is a second course on linear state equations for some students, a first course for a few, and somewhere between for the majority. It is the course where students from a variety of backgrounds begin to acquire the tools used in the research literature on linear system and control theory. This book is my notion of what such a course should be. The core material is the theory of time-varying linear systems, with frequent specialization to the time-invariant case. Additional material, included for flexibility in the curriculum, explores refinements and extensions, many confined to time-invariant linear systems.

Motivation for presenting linear system theory in the time-varying context is at least threefold. First, the development provides an excellent review of the time-invariant case, both in the remarkable similarity of the theories and in the perspective provided by specialization. Second, much of the research literature in linear system theory treats the time-varying case — for generality, and because time-varying linear system theory plays an important role in other areas, for example nonlinear system theory. Finally, of course, the theory is directly relevant when a physical system is described by a linear state equation with time-varying coefficients.

Technical development of the material is careful, even rigorous, but not fancy. The presentation is self-contained, and proceeds step-by-step from a modest mathematical base. I have tried to minimize terminology, choose default assumptions that avoid fussy technicalities, and employ a clean, simple notation. These features are intended to maximize clarity and render the theory accessible to beginning graduate students.

Over 250 exercises are included, most applying or extending the theory rather than posing routine calculations. All exercises in Chapter 1 are used in subsequent material. Aside from Chapter 1, results of exercises are used infrequently in the presentation, at least until the later chapters. But linear system theory is not a spectator sport, and the exercises are an integral and important part of the book.

Though the Muse insists that linear systems be described in linear sentences, I demur. However the prose intentionally is lean to not becloud the theory. For those seeking elaboration and congenial discussion, a *Notes* section in each chapter indicates further developments and alternative formulations. These notes are entry points to the literature rather than balanced reviews of the harvest of so many research efforts over the years.

The organization of the book *is* linear, with adjustments as shown on the *Chapter Planning Chart* to provide flexibility in topic selection. Chapter 1 through Chapter 12 contain core material on time-varying linear systems, with Chapter 8 (*Additional Stability Results*) and Chapter 11 (*Minimal Realization*) optional. Depending on preparation of the students, it might be desirable to review the mathematics in Chapter 1 as needed rather than at the outset. Chapter 13 (*Controller and Observer Forms*) contains material necessary for subsequent chapters on the time-invariant case, though results for time-varying systems in Chapter 14 (*State Feedback*) and Chapter 15 (*State Observation*) are independent. Optional topics for time-invariant systems in Chapters 16–19 are the polynomial fraction description, which exhibits the detailed structure of the transfer function representation for multi-input, multi-output systems, and the geometric description of the fine structure of linear state equations.

Acknowledgments

I wrote this book with more than a little help from my friends. Generations of graduate students at Johns Hopkins offered gentle instruction. Brave colleagues taught courses from preliminary versions: Eyad Abed, Aristotle Arapostathis, Pramod Khargonekar, Thordur Runolfsson, Mark Shayman, and Andre Tits. Henk Nijmeijer and Jessy Grizzle reviewed successive versions of Chapters 18 and 19. Elmer Gilbert and Alan Laub ranged over the nearly-final manuscript. I thank all for valuable suggestions, and for pointing out obscurities and errors. Finally, I am grateful to The Johns Hopkins University for an environment where I can freely direct my academic efforts, and to the Air Force Office of Scientific Research for support of research compatible with attention to theoretical foundations.

WJR
Baltimore, Maryland, USA

CHAPTER PLANNING CHART

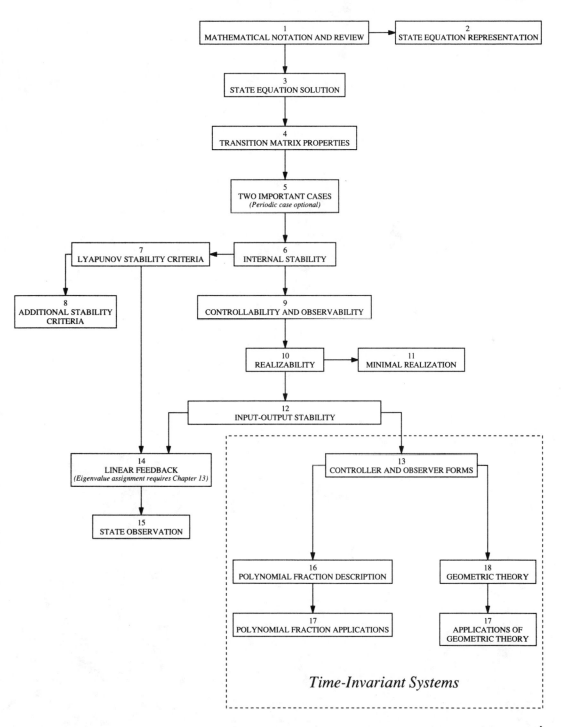

1

MATHEMATICAL NOTATION
AND REVIEW

Throughout this book we use mathematical analysis, linear algebra, and matrix theory at what might be called an advanced undergraduate level. For some topics a review might be beneficial to the typical student, and the best sources for such review are mathematics texts. Here a quick listing of basic notions is provided to set notation and provide reminders. In addition there are exercises that can be solved by reasonably straightforward applications of these notions. Results of exercises in this chapter are used in the sequel, and therefore the exercises should be perused, at least. With minor exceptions, all the mathematical tools in the first 15 chapters are self-contained developments of material reviewed here. In Chapters 16–19 additional mathematical background is introduced for local consumption.

Basic mathematical objects in linear system theory are $n \times 1$ or $1 \times n$ vectors and $m \times n$ matrices with real entries, though on occasion complex entries arise. Typically vectors are in lower-case italics, matrices are in upper-case italics, and scalars (real, or sometimes complex) are represented by Greek letters. Usually the ith-entry in a vector x is denoted x_i, and the i,j-entry in a matrix A is written a_{ij} or $[A]_{ij}$. These notations are not completely consistent, if for no other reason than scalars can be viewed as special cases of vectors, and vectors can be viewed as special cases of matrices. Moreover, notational conventions are abandoned when they collide with strong tradition.

With the usual definition of addition and scalar multiplication, the set of all $n \times 1$ vectors and, more generally, the set of all $m \times n$ matrices, can be viewed as vector spaces over the real (or complex) field. In the real case the vector space of $n \times 1$ vectors is written as $R^{n \times 1}$, or simply R^n, and a vector space of matrices is written as $R^{m \times n}$. The default throughout is the real case. When matrices or vectors with complex entries ($i = \sqrt{-1}$) are at issue, special mention will be made. It is useful for some of the later chapters to review the axioms for a field and a vector space, though for most of the book technical developments are phrased in the language of matrix algebra.

Vectors

Two $n \times 1$ vectors x_a and x_b are called *linearly independent* if no nontrivial linear combination of x_a and x_b gives the zero vector. This means that if $\alpha_1 x_a + \alpha_2 x_b = 0$, then both scalars α_1 and α_2 are zero. Of course the definition extends to a linear combination of any number of vectors. A set of n linearly independent $n \times 1$ vectors forms a *basis* for the vector space of all $n \times 1$ vectors. The set of all linear combinations of a specified set of vectors is a vector space called the *span* of the set of vectors. For example *span* { x_a, x_b, x_c } is a 3-dimensional subspace of R^n, if x_a, x_b, and x_c are linearly independent $n \times 1$ vectors.

Without exception we use the *Euclidean norm* for $n \times 1$ vectors, defined as follows. Writing a vector and its *transpose* in the form

$$x = \begin{bmatrix} x_1 \\ x_2 \\ \vdots \\ x_n \end{bmatrix}, \quad x^T = \begin{bmatrix} x_1 & x_2 & \cdots & x_n \end{bmatrix}$$

let

$$\|x\| = \sqrt{x^T x} = \left[\sum_{i=1}^{n} x_i^2 \right]^{1/2} \tag{1}$$

Elementary inequalities relating the Euclidean norm of a vector to the absolute values of entries are

$$\max_{1 \leq i \leq n} |x_i| \leq \|x\| \leq \sqrt{n} \max_{1 \leq i \leq n} |x_i|$$

As any norm must, the Euclidean norm has the following properties for arbitrary $n \times 1$ vectors x and y, and any scalar α :

$$\|x\| \geq 0$$
$$\|x\| = 0 \quad \text{if and only if } x = 0$$
$$\|\alpha x\| = |\alpha| \|x\|$$
$$\|x + y\| \leq \|x\| + \|y\| \tag{2}$$

The last of these is called the *triangle inequality*. Also the *Cauchy-Schwarz inequality* in terms of the Euclidean norm is

$$|x^T y| \leq \|x\| \|y\| \tag{3}$$

If x is complex, then the transpose of x must be replaced by *conjugate transpose*, also known as *Hermitian transpose* and thus written x^H, throughout the above discussion of norms. This notation is correctly construed as complex conjugate in the scalar case, though we use overbar to denote complex conjugate when transpose is not desired.

Matrices

For matrices there are several standard concepts and special notations used in the sequel. The $m \times n$ matrix with all entries zero is written as $0_{m \times n}$, or simply 0 when dimensional emphasis is not needed. For square matrices, $m = n$, the zero matrix sometimes is written as 0_n, while the identity matrix is written similarly as I_n or I. We reserve the notation e_k for the kth-column or k^{th}-row, depending on context, of the identity matrix.

The notions of addition and multiplication for conformable matrices are presumed to be familiar. Of course the multiplication operation is more interesting, in part because it is not *commutative* in general. That is, AB and BA are not always the same. Similar to the vector case, the transpose of a matrix A with entries a_{ij} is the matrix A^T with i,j-entry given by a_{ji}. A useful fact is $(AB)^T = B^T A^T$.

For a square $n \times n$ matrix A, the *trace* is the sum of the diagonal entries, written

$$\text{tr } A = \sum_{i=1}^{n} a_{ii} \tag{4}$$

If B also is $n \times n$, then

$$\text{tr } [AB] = \text{tr } [BA]$$

A familiar scalar-valued function of a square matrix A is the *determinant*. The determinant of A can be evaluated via the *Laplace expansion* described as follows. Let c_{ij} denote the *cofactor* corresponding to the entry a_{ij}. Recall that c_{ij} is $(-1)^{i+j}$ times the determinant of the $(n-1) \times (n-1)$ matrix that results when the ith-row and j^{th}-column of A are deleted. Then for any fixed i, $1 \le i \le n$,

$$\det A = \sum_{j=1}^{n} a_{ij} c_{ij}$$

This is the expansion of the determinant along the ith-row. A similar formula holds for the expansion along a column. Aside from being a useful representation for the determinant, recursive use of this expression provides a method for computing the determinant of a matrix from the fact that the determinant of a scalar is simply the scalar itself. Since this procedure expresses the determinant as a sum of products of entries of the matrix, the determinant viewed as a function of the matrix entries is continuously differentiable any number of times. Finally if B also is $n \times n$, then

$$\det (AB) = (\det A)(\det B) = \det (BA) \tag{5}$$

The matrix A has an inverse, written A^{-1}, if and only if $det\ A \ne 0$. One formula for A^{-1} that occurs often is based on the cofactors of A. The *adjugate* of A, written $adj\ A$, is the matrix with i,j-entry given by the cofactor c_{ji}. In other words, $adj\ A$ is the transpose of the matrix of cofactors. Then

$$A^{-1} = \frac{\text{adj } A}{\det A}$$

a standard, collapsed way of writing the product of the scalar $1/(\det A)$ and the matrix *adj A*. The inverse of a product of square, invertible matrices is given by

$$(AB)^{-1} = B^{-1}A^{-1}$$

If A is $n \times n$ and p is a nonzero $n \times 1$ vector such that for some scalar λ,

$$Ap = \lambda p \tag{6}$$

then p is an *eigenvector* corresponding to the *eigenvalue* λ. Of course p must be presumed nonzero, for if $p = 0$ then this equation is satisfied for any λ. Also any nonzero scalar multiple of an eigenvector is another eigenvector. We must be a bit careful here, because a real matrix can have complex eigenvalues and eigenvectors, though the eigenvalues must occur in conjugate pairs, and corresponding conjugate eigenvectors can be assumed. In other words if $Ap = \lambda p$, then $A\bar{p} = \bar{\lambda}\bar{p}$. These notions can be refined by viewing (6) as the definition of a *right eigenvector*. Then it is natural to define a *left eigenvector* for A as a nonzero $1 \times n$ vector q such that $qA = \lambda q$ for some eigenvalue λ.

The n eigenvalues of A are precisely the n roots of the *characteristic polynomial* of A, given by $\det(sI_n - A)$. Since the roots of a polynomial are continuous functions of the coefficients of the polynomial, the eigenvalues of a matrix are continuous functions of the matrix entries. The *Cayley-Hamilton theorem* states that if

$$\det(sI_n - A) = s^n + a_{n-1}s^{n-1} + \cdots + a_0$$

then

$$A^n + a_{n-1}A^{n-1} + \cdots + a_1A + a_0I_n = 0_n$$

Our main application of this result is to write A^{n+k}, for integer $k \geq 0$, as a linear combination of $I, A, \ldots, A^{n-1}$.

A *similarity transformation* of the type $T^{-1}AT$, where A and invertible T are $n \times n$, occurs frequently. It is a simple exercise to show that $T^{-1}AT$ and A have the same set of eigenvalues. If A has distinct eigenvalues, and T has as columns a corresponding set of (linearly independent) eigenvectors for A, then $T^{-1}AT$ is a diagonal matrix, with the eigenvalues of A as the diagonal entries. Therefore this computation can lead to a matrix with complex entries.

1.1 Example The characteristic polynomial of

$$A = \begin{bmatrix} 0 & -2 \\ 2 & -2 \end{bmatrix} \tag{7}$$

is

$$\det(\lambda I - A) = \det \begin{bmatrix} \lambda & 2 \\ -2 & \lambda+2 \end{bmatrix}$$

$$= (\lambda + 1 + i\sqrt{3})(\lambda + 1 - i\sqrt{3})$$

Therefore A has eigenvalues

$$\lambda_a = -1 + i\sqrt{3}, \quad \lambda_b = -1 - i\sqrt{3}$$

Setting up (6) to compute a right eigenvector p^a corresponding to λ_a gives the linear equation

$$\begin{bmatrix} 0 & -2 \\ 2 & -2 \end{bmatrix} \begin{bmatrix} p_1^a \\ p_2^a \end{bmatrix} = \begin{bmatrix} (-1+i\sqrt{3})p_1^a \\ (-1+i\sqrt{3})p_2^a \end{bmatrix}$$

One nonzero solution is

$$p^a = \begin{bmatrix} 2 \\ 1-i\sqrt{3} \end{bmatrix} \tag{8}$$

A similar calculation gives an eigenvector corresponding to λ_b that is simply the complex conjugate of p^a. Then the invertible matrix

$$T = \begin{bmatrix} 2 & 2 \\ 1-i\sqrt{3} & 1+i\sqrt{3} \end{bmatrix}$$

yields the diagonal form

$$T^{-1}AT = \begin{bmatrix} -1+i\sqrt{3} & 0 \\ 0 & -1-i\sqrt{3} \end{bmatrix}$$

□ □ □

We often use the basic solvability conditions for a linear equation

$$Ax = b \tag{9}$$

where A is a given $m \times n$ matrix, and b is a given $m \times 1$ vector. The *range space* or *image* of A is the vector space (subspace of R^m) spanned by the columns of A. The *null space* or *kernel* of A is the vector space of all $n \times 1$ vectors x such that $Ax = 0$. The linear equation (9) has a solution if and only if b is in the range space of A, or, more subtly, if and only if $b^T y = 0$ for all y in the null space of A^T. Of course if $m = n$ and A is invertible, then there is a unique solution for any given b; namely $x = A^{-1}b$. The *rank* of an $m \times n$ matrix A is equivalently the dimension of the range space of A as a vector subspace of R^n, the number of linearly independent column vectors in the matrix, or the number of linearly independent row vectors. An important inequality involving an $m \times n$ matrix A and an $n \times p$ matrix B is

$$\text{rank } A + \text{rank } B - n \leq \text{rank } (AB) \leq \min \{ \text{rank } A, \text{rank } B \}$$

For many calculations it is convenient to make use of partitioned vectors and matrices. Standard computations can be expressed in terms of operations on the partitions, when the partitions are conformable. For example, with all partitions square and of the same dimension,

$$\begin{bmatrix} A_1 & A_2 \\ 0 & A_4 \end{bmatrix} + \begin{bmatrix} B_1 & B_2 \\ B_3 & 0 \end{bmatrix} = \begin{bmatrix} A_1+B_1 & A_2+B_2 \\ B_3 & A_4 \end{bmatrix}$$

$$\begin{bmatrix} A_1 & A_2 \\ 0 & A_4 \end{bmatrix} \begin{bmatrix} B_1 & B_2 \\ B_3 & 0 \end{bmatrix} = \begin{bmatrix} A_1B_1+A_2B_3 & A_1B_2 \\ A_4B_3 & 0 \end{bmatrix}$$

If x is an $n \times 1$ vector and A is an $m \times n$ matrix partitioned by rows,

$$\begin{bmatrix} A_1 \\ \vdots \\ A_m \end{bmatrix} x = \begin{bmatrix} A_1x \\ \vdots \\ A_mx \end{bmatrix}$$

If A is partitioned by columns, and z is $m \times 1$,

$$z^T \begin{bmatrix} A_1 & \cdots & A_n \end{bmatrix} = \begin{bmatrix} z^TA_1 & \cdots & z^TA_n \end{bmatrix}$$

A useful feature of partitioned square matrices with square partitions as diagonal blocks is

$$\det \begin{bmatrix} A_{11} & A_{12} \\ 0 & A_{22} \end{bmatrix} = (\det A_{11})(\det A_{22})$$

When in doubt about a specific partitioned calculation, always pause and carefully check a simple yet nontrivial example.

The *induced norm* of an $m \times n$ matrix A can be defined in terms of a constrained maximization problem. Let

$$\|A\| = \max_{\|x\|=1} \|Ax\| \tag{10}$$

where notation is several ways abused. First, the same symbol is used for the induced norm of a matrix as for the norm of a vector. Second, the norms appearing on the right side of (10) are the Euclidean norms of the vectors x and Ax, and Ax is $m \times 1$ while x is $n \times 1$. Finally, that the indicated maximum actually is attained for some unity-norm x is something that needs to be proved. Alternately it can be shown (Exercise 1.6) that the norm of A induced by the Euclidean norm is equal to the (nonnegative) square root of the largest eigenvalue of A^TA, or of AA^T. While induced norms corresponding to other vector norms can be defined, only this so-called *spectral norm* for matrices is used in the sequel.

1.2 Example   If λ_1 and λ_2 are real numbers, then the spectral norm of

$$A = \begin{bmatrix} \lambda_1 & 1 \\ 0 & \lambda_2 \end{bmatrix} \tag{11}$$

is given by (10) as

$$\|A\| = \max_{\sqrt{x_1^2+x_2^2}=1} \sqrt{(\lambda_1 x_1 + x_2)^2 + \lambda_2^2 x_2^2}$$

To elude this constrained maximization problem, we compute $\|A\|$ by computing the eigenvalues of $A^T A$. The characteristic polynomial of $A^T A$ is

$$\det(\lambda I - A^T A) = \det \begin{bmatrix} \lambda - \lambda_1^2 & -\lambda_1 \\ -\lambda_1 & \lambda - \lambda_2^2 - 1 \end{bmatrix}$$

$$= \lambda^2 - (1 + \lambda_1^2 + \lambda_2^2)\lambda + \lambda_1^2 \lambda_2^2$$

The roots of this quadratic are given by

$$\lambda = \frac{1 + \lambda_1^2 + \lambda_2^2 \pm \sqrt{(1 + \lambda_1^2 + \lambda_2^2)^2 - 4\lambda_1^2 \lambda_2^2}}{2}$$

The radical can be rewritten so that its positivity is obvious. Then the largest root is obtained by choosing the plus sign, and a little algebra gives

$$\|A\| = \tfrac{1}{2} \left[\sqrt{(\lambda_1 + \lambda_2)^2 + 1} + \sqrt{(\lambda_1 - \lambda_2)^2 + 1} \right]$$

□□□

The induced norm of an $m \times n$ matrix satisfies the axioms of a norm on $R^{m \times n}$, and additional properties as well. In particular $\|A^T\| = \|A\|$, a neat instance of which is that the induced norm $\|x^T\|$ of the $1 \times n$ matrix x^T is the square root of the largest eigenvalue of $x^T x$, or of xx^T. Choosing the more obvious of the two configurations immediately gives $\|x^T\| = \|x\|$. Also $\|Ax\| \le \|A\| \|x\|$ for any $n \times 1$ vector x (Exercise 1.1), and for conformable A and B,

$$\|AB\| \le \|A\| \|B\| \tag{12}$$

(Exercise 1.2). If A is $m \times n$, then inequalities relating $\|A\|$ to absolute values of the entries of A are

$$\max_{i,j} |a_{ij}| \le \|A\| \le \sqrt{mn} \max_{i,j} |a_{ij}|$$

When complex matrices are involved, all transposes in the above discussion of the induced norm should be replaced by Hermitian transposes.

Quadratic Forms

For a specified $n \times n$ matrix Q and any $n \times 1$ vector x, both with real entries, the product $x^T Q x$ is called a *quadratic form* in x. Without loss of generality Q can be taken as *symmetric, $Q = Q^T$*, in the study of quadratic forms. To verify this, multiply out a typical case to show that

$$x^T \left[\frac{Q + Q^T}{2} \right] x = x^T Q x \tag{13}$$

for all x. It follows that any Q can be replaced by $(Q + Q^T)/2$, which clearly is symmetric. A symmetric matrix Q is called *positive semidefinite* if $x^T Q x \geq 0$ for all x. It is called *positive definite* if it is positive semidefinite, and $x^T Q x = 0$ implies $x = 0$. Negative definiteness and semidefiniteness are defined in terms of positive definiteness and positive semidefiniteness of $-Q$. Often the short-hand notations $Q > 0$ and $Q \geq 0$ are used to denote positive definiteness, and positive semidefiniteness, respectively. Of course $Q_a \geq Q_b$ simply means that $Q_a - Q_b$ is positive semidefinite.

All eigenvalues of a symmetric matrix must be real. It follows that positive definiteness is equivalent to all eigenvalues positive, and positive semidefiniteness is equivalent to all eigenvalues nonnegative. An important inequality for a symmetric $n \times n$ matrix Q is the *Rayleigh-Ritz inequality*, which states that for any real $n \times 1$ vector x,

$$\lambda_{\min} x^T x \leq x^T Q x \leq \lambda_{\max} x^T x \tag{14}$$

where $\lambda_{\min}$ and $\lambda_{\max}$ denote the smallest and largest eigenvalues of Q. See Exercise 1.5 for the spectral norm of Q. If we assume $Q \geq 0$, then $\|Q\| = \lambda_{\max}$ and the trace is bounded by

$$\|Q\| \leq \text{tr } Q \leq n \|Q\|$$

Tests for definiteness properties can be based on sign properties of various submatrix determinants. These tests are difficult to state in a fashion that is both precise and economical, so a careful prescription will be given. Suppose Q is a real, $n \times n$, symmetric matrix with entries q_{ij}. For integers $p = 1, \ldots, n$ and $1 \leq i_1 < i_2 < \cdots < i_p \leq n$, the scalars

$$Q(i_1, i_2, \ldots, i_p) = \det \begin{bmatrix} q_{i_1 i_1} & q_{i_1 i_2} & \cdots & q_{i_1 i_p} \\ q_{i_2 i_1} & q_{i_2 i_2} & \cdots & q_{i_2 i_p} \\ \vdots & \vdots & \vdots & \vdots \\ q_{i_p i_1} & a_{i_p i_2} & \cdots & q_{i_p i_p} \end{bmatrix} \tag{15}$$

are called *principal minors* of Q. The scalars $Q(1, 2, \ldots, p)$, $p = 1, 2, \ldots, n$, which simply are the determinants of the upper left $p \times p$ submatrices of Q,

$$Q(1) = q_{11}, \quad Q(1,2) = \det \begin{bmatrix} q_{11} & q_{12} \\ q_{21} & q_{22} \end{bmatrix}, \quad Q(1,2,3) = \det \begin{bmatrix} q_{11} & q_{12} & q_{13} \\ q_{21} & q_{22} & q_{23} \\ q_{31} & q_{32} & q_{33} \end{bmatrix}, \cdots$$

are called *leading principal minors*.

1.3 Theorem   The symmetric matrix Q is positive definite if and only if

$$Q(1, 2, \ldots, p) > 0, \quad p = 1, 2, \ldots, n$$

It is negative definite if and only if

$$(-1)^p \, Q(1, 2, \ldots, p) > 0, \quad p = 1, 2, \ldots, n$$

The test for semidefiniteness is much more complicated since all principal minors are involved, not just the leading principal minors.

1.4 Theorem   The symmetric matrix Q is positive semidefinite if and only if

$$Q(i_1, i_2, \ldots, i_p) \geq 0, \quad \begin{cases} 1 \leq i_1 < i_2 < \cdots < i_p \leq n \\ p = 1, 2, \ldots, n \end{cases}$$

It is negative semidefinite if and only if

$$(-1)^p \, Q(i_1, i_2, \ldots, i_p) \geq 0, \quad \begin{cases} 1 \leq i_1 < i_2 < \cdots < i_p \leq n \\ p = 1, 2, \ldots, n \end{cases}$$

1.5 Example   The symmetric matrix

$$Q = \begin{bmatrix} q_{11} & q_{12} \\ q_{12} & q_{22} \end{bmatrix} \tag{16}$$

is positive definite if and only if $q_{11} > 0$ and $q_{11}q_{22} - q_{12}^2 > 0$. It is positive semidefinite if and only if $q_{11} \geq 0$, $q_{22} \geq 0$, and $q_{11}q_{22} - q_{12}^2 \geq 0$.
□□□

If Q has complex entries but is *Hermitian*, that is $Q = Q^H$ where again ^{H} denotes Hermitian (conjugate) transpose, then a quadratic form is defined as $x^H Q x$. This is a real quantity, and the various definitions and definiteness tests above apply.

Matrix Calculus

Often the vectors and matrices in these chapters have entries that are functions of time. With only one or two exceptions, the entries are at least continuous functions, and often they are continuously differentiable. For convenience of discussion assume the latter. Standard notation is used for various intervals of time, for example, $t \in [t_0, t_1)$ means $t_0 \leq t < t_1$. To avoid silliness we assume always that the right endpoint of an interval is greater than the left endpoint. If no interval is specified, the default is $(-\infty, \infty)$.

The sophisticated mathematical view is to treat matrices whose entries are functions of time as matrix-valued functions of a real variable. For example, $x(t)$ would denote a function with domain a time interval, and range R^n. However this framework is not needed for our purposes, and actually can be confusing because of conventional interpretations of matrix concepts and calculations in linear system theory.

In mathematics a norm, for example $\|x(t)\|$, always denotes a real number. However this 'function space' viewpoint is less useful for our purposes than interpreting $\|x(t)\|$ 'pointwise in time.' That is, $\|x(t)\|$ is viewed as the real-valued function of t that gives the Euclidean norm of $x(t)$ at each value of t. Namely,

$$\|x(t)\| = \sqrt{x^T(t)x(t)}$$

Also we say that a square matrix function $M(t)$ is invertible for all t if for any value of t the inverse matrix $M^{-1}(t)$ exists. This is completely different from invertibility of the mapping $M(t)$ with domain R and range $R^{n \times n}$, even when $n = 1$. Other algebraic constructs are handled in a similar pointwise-in-time fashion. For example at each t an $n \times n$ matrix $A(t)$ has eigenvalues $\lambda_1(t), \ldots, \lambda_n(t)$, and an induced norm $\|A(t)\|$, all of which are viewed as scalar functions of time. If $Q(t)$ is a symmetric $n \times n$ matrix at each t, then $Q(t) > 0$ means that at each value of t the matrix is positive definite. Sometimes this viewpoint is said to treat matrices 'parameterized' by t rather than 'matrix functions' of t. However we retain the latter terminology.

Confusion also can arise in the rules of 'matrix calculus.' In general, matrix calculations are set up to be consistent with scalar calculus in the following sense. If the matrix expression is written out in scalar terms, the usual scalar calculations performed, and the result repacked into matrix form, then we should get the same result as is given by the rules of matrix calculus. This principle leads to the conclusion that differentiation and integration of matrices should be defined entry-wise. Thus the i,j-entry of

$$\int_0^t A(\sigma)\, d\sigma, \qquad \frac{d}{dt}A(t)$$

is, respectively,

$$\int_0^t a_{ij}(\sigma)\, d\sigma, \qquad \frac{d}{dt}a_{ij}(t)$$

Using these facts it is easy to verify that the product rule holds for differentiation of matrices. That is, with overdot denoting differentiation with respect to time,

$$\frac{d}{dt}[A(t)B(t)] = \dot{A}(t)B(t) + A(t)\dot{B}(t)$$

The fundamental theorem of calculus applies in the matrix case,

$$\frac{d}{dt}\int_0^t A(\sigma)\,d\sigma = A(t)$$

and also the *Leibniz rule:*

$$\frac{d}{dt}\int_{f(t)}^{g(t)} A(t,\,\sigma)\,d\sigma = A(t,\,g(t))\,\dot{g}(t) - A(t,\,f(t))\,\dot{f}(t)$$

$$+ \int_{f(t)}^{g(t)} \frac{\partial}{\partial t} A(t,\,\sigma)\,d\sigma \tag{17}$$

However we must be careful about the generalization of certain familiar calculations from the scalar case — particularly those having the appearance of a chain rule. For example if $A(t)$ is square the product rule gives

$$\frac{d}{dt}A^2(t) = \dot{A}(t)A(t) + A(t)\dot{A}(t)$$

This is not in general the same thing as $2A(t)\dot{A}(t)$, since $A(t)$ and its derivative need not commute. (The diligent might want to figure out why the appearance of a chain rule is false.) Of course in any suspicious case the way to verify a matrix-calculus rule is to write out the scalar form, compute, and repack.

In view of the interpretations of norm and integration, a particularly useful inequality for an $n \times 1$ vector function $x(t)$ follows from the triangle inequality applied to approximating sums for the integral:

$$\left\| \int_{t_o}^t x(\sigma)\,d\sigma \right\| \le \left| \int_{t_o}^t \|x(\sigma)\|\,d\sigma \right| \tag{18}$$

Often we apply this when $t \ge t_o$, in which case the absolute value signs on the right side can be erased.

Convergence

Familiarity with basic notions of convergence for sequences or series of real numbers is assumed at the outset. A brief review of some more general notions is provided here, though it is appropriate to note that the only explicit use of this material is in discussing existence and uniqueness of solutions to linear state equations.

An infinite sequence of $n \times 1$ vectors is written as $\{x_k\}_{k=0}^{\infty}$, where the subscript notation in this context denotes different vectors rather than entries of a vector. A vector $\hat{x}$ is called the *limit* of the sequence if for any given $\varepsilon > 0$ there exists a positive integer, written $K(\varepsilon)$ to indicate that the integer depends on ε, such that

$$\|\hat{x} - x_k\| < \varepsilon , \quad k > K(\varepsilon) \tag{19}$$

If such a limit exists, the sequence is said to *converge to* $\hat{x}$, written $\lim_{k \to \infty} x_k = \hat{x}$. Notice that the use of the norm converts the question of convergence for a sequence of vectors $\{x_k\}_{k=0}^{\infty}$ to a vector $\hat{x}$ into a question of convergence of the sequence of scalars $\{\|\hat{x} - x_k\|\}_{k=0}^{\infty}$ to zero.

More often we are interested in sequences of vector functions of time, denoted $\{x_k(t)\}_{k=0}^{\infty}$, and defined on some interval, say $[t_0, t_1]$. Such a sequence is said to converge (pointwise) on the interval if there exists a vector function $\hat{x}(t)$ such that for any $t_a \in [t_0, t_1]$ the sequence of vectors $\{x_k(t_a)\}_{k=0}^{\infty}$ converges to the vector $\hat{x}(t_a)$. In this case, given an ε, the K can depend on both ε and t_a. The sequence of functions *converges uniformly* on $[t_0, t_1]$ if there exists a function $\hat{x}(t)$ such that given $\varepsilon > 0$ there exists a positive integer $K(\varepsilon)$ such that for any t_a in the interval,

$$\|\hat{x}(t_a) - x_k(t_a)\| < \varepsilon , \quad k > K(\varepsilon)$$

The distinction is that, given $\varepsilon > 0$, the same $K(\varepsilon)$ can be used for any value of t_a to show convergence of the vector sequence $\{x_k(t_a)\}_{k=0}^{\infty}$.

For an infinite series of vector functions, written

$$\sum_{j=0}^{\infty} x_j(t) \tag{20}$$

with each $x_j(t)$ defined on $[t_0, t_1]$, convergence is defined in terms of the sequence of *partial sums*

$$s_k(t) = \sum_{j=0}^{k} x_j(t)$$

The series converges (pointwise) to the function $\hat{x}(t)$ if for each $t_a \in [t_0, t_1]$,

$$\lim_{k \to \infty} \|\hat{x}(t_a) - s_k(t_a)\| = 0$$

The series (20) is said to *converge uniformly to* $\hat{x}(t)$ on $[t_0, t_1]$ if the sequence of partial sums converges uniformly to $\hat{x}(t)$ on $[t_0, t_1]$. Namely, given an $\varepsilon > 0$ there must exist a positive integer $K(\varepsilon)$ such that for any $t \in [t_0, t_1]$,

$$\|\hat{x}(t) - \sum_{j=0}^{k} x_j(t)\| < \varepsilon , \quad k > K(\varepsilon)$$

While infinite series important in this book converge pointwise for $t \in (-\infty, \infty)$, the emphasis is on showing uniform convergence on arbitrary but finite intervals of the form $[t_0, t_1]$. This permits the use of special properties of uniformly convergent series with regard to continuity and differentiation.

1.6 Theorem   If (20) is an infinite series of continuous vector functions on $[t_0, t_1]$ that converges uniformly to $\hat{x}(t)$ on $[t_0, t_1]$, then $\hat{x}(t)$ is continuous for $t \in [t_0, t_1]$.

It is an inconvenient fact that term-by-term differentiation of a uniformly convergent series of functions does not always yield the derivative of the sum. Another uniform convergence analysis is required.

1.7 Theorem   Suppose (20) is an infinite series of continuously-differentiable functions on $[t_0, t_1]$ that converges uniformly to $\hat{x}(t)$ on $[t_0, t_1]$. If the series

$$\sum_{j=0}^{\infty} \frac{d}{dt}x_j(t) \tag{21}$$

converges uniformly on $[t_0, t_1]$, it converges to $(d/dt)\hat{x}(t)$.

The infinite series (20) is said to *converge absolutely* if the series of real functions

$$\sum_{j=0}^{\infty} \|x_j(t)\|$$

converges on the interval. The key property of an absolutely convergent series is that terms in the series can be reordered without changing the fact of convergence.

The specific convergence test applied in developing solutions of linear state equations is the *Weierstrass M-Test*, which can be stated as follows.

1.8 Theorem   If the infinite series of positive real numbers

$$\sum_{j=0}^{\infty} \alpha_j \tag{22}$$

converges, and if $\|x_j(t)\| \le \alpha_j$ for all $t \in [t_0, t_1]$ and all j, then the series (20) converges uniformly and absolutely on $[t_0, t_1]$.

For the special case of *power series* in t, a basic fact is that if a power series with vector coefficients,

$$\sum_{j=0}^{\infty} x_j t^j$$

converges on an interval, it converges uniformly and absolutely on that interval. A vector function $f(t)$ is called *analytic* on a time interval if for any point t_a in the interval it can be represented by the power series

$$\sum_{j=0}^{\infty} \frac{d^j}{dt^j} f(t) \bigg|_{t=t_a} \frac{(t-t_a)^j}{j!} \tag{23}$$

that converges on some subinterval containing t_a. That is, $f(t)$ is analytic on an interval if it has a convergent *Taylor series* representation at each point in the interval. Thus $f(t)$ is analytic at t_a if and only if it has derivatives of any order at t_a, and these derivatives satisfy a certain growth condition. (Sometimes the term *real analytic* is used to distinguish analytic functions of a real variable from analytic functions of a complex variable. Except for Laplace transforms, functions of a complex variable do not arise in the sequel, and we use the simpler terminology.)

Similar definitions of convergence properties for sequences and series of $m \times n$ matrix functions of time can be made using the induced norm for matrices. It is not difficult to show that these matrix or vector convergence notions are equivalent to applying the corresponding notion to the scalar sequence formed by each particular entry of the matrix or vector sequence.

Laplace Transform

Since we use the Laplace transform only for functions that are sums of terms of the form $t^k e^{\alpha t}$, only the most basic features need review. If $F(t)$ is an $m \times n$ matrix of such functions defined for $t \in [0, \infty)$, the Laplace transform is defined as the $m \times n$ matrix function of the complex variable s given by

$$\mathsf{F}(s) = \int_0^{\infty} F(t) e^{-st} \, dt \tag{24}$$

Often this operation is written in the format $\mathsf{F}(s) = \mathbf{L}[F(t)]$. (For much of the book, Laplace transforms are represented in Helvetica font to distinguish, yet connect, the corresponding time function in Italic font.)

Because of the exponential nature of each entry of $F(t)$, there is always a half-plane of convergence of the form Re $[s] > \alpha$ for the integral in (24). Also easy calculations show that each entry of $\mathsf{F}(s)$ is a *strictly proper* rational function — a ratio of two polynomials in s where the degree of the denominator polynomial is strictly greater than the degree of the numerator polynomial. A convenient method of computing the matrix $F(t)$ from such a transform $\mathsf{F}(s)$ is entry-by-entry partial fraction expansion and table-lookup.

Our applications require only a few properties of the Laplace transform. These include linearity, and the derivative and integral relations

$$\mathbf{L}[\dot{F}(t)] = s\mathbf{L}[F(t)] - F(0)$$

$$\mathbf{L}\left[\int_0^{\infty} F(\sigma) \, d\sigma\right] = \frac{1}{s} \mathbf{L}[F(t)]$$

Recall that in certain applications to linear systems, the evaluation of $F(t)$ in the derivative property should be interpreted as an evaluation at $t = 0^-$. The *convolution property*

$$\mathbf{L}\left[\int_0^\infty F(t-\sigma)G(\sigma)\,d\sigma\right] = \mathbf{L}[F(t)]\,\mathbf{L}[G(t)] \qquad (25)$$

is used, and also the *Initial Value theorem* and *Final Value theorem,* which state that if the indicated limits exist, then

$$\lim_{t\to 0} F(t) = \lim_{s\to\infty} s\mathsf{F}(s)$$

$$\lim_{t\to\infty} F(t) = \lim_{s\to 0} s\mathsf{F}(s)$$

EXERCISES

Exercise 1.1   For an $m \times n$ matrix A, prove from the definition in (10) that the spectral norm is given by

$$\|A\| = \max_{x\neq 0} \frac{\|Ax\|}{\|x\|}$$

Conclude that for any $n \times 1$ vector x,

$$\|Ax\| \leq \|A\|\,\|x\|$$

Exercise 1.2   Using the conclusion in Exercise 1.1, prove that for conformable matrices A and B,

$$\|AB\| \leq \|A\|\,\|B\|$$

If A is invertible, show that

$$\|A^{-1}\| \geq \frac{1}{\|A\|}$$

Exercise 1.3   For a partitioned matrix

$$A = \begin{bmatrix} A_{11} & A_{12} \\ A_{21} & A_{22} \end{bmatrix}$$

show that $\|A_{ij}\| \leq \|A\|$ for $i,\,j = 1,\,2$.

Exercise 1.4   If A is an $n \times n$ matrix, show that for all $n \times 1$ vectors x

$$|x^T Ax| \leq \|A\|\,\|x\|^2, \quad x^T Ax \geq -\|A\|\,\|x\|^2$$

Show that for any eigenvalue λ of A,

$$|\lambda| \leq \|A\|$$

(In words, the *spectral radius* of A is no larger than the spectral norm of A.)

Exercise 1.5   If Q is a symmetric $n \times n$ matrix, prove that the spectral norm is given by

$$\|Q\| = \max_{\|x\| = 1} |x^T Q x| = \max_{1 \leq i \leq n} |\lambda_i|$$

where $\lambda_1, \ldots, \lambda_n$ are the eigenvalues of Q.

Exercise 1.6   Show that the spectral norm of an $m \times n$ matrix A is given by

$$\|A\| = \left[\max_{\|x\| = 1} x^T A^T A x \right]^{1/2}$$

Conclude from the Rayleigh-Ritz inequality that $\|A\|$ is given by the nonnegative square root of the largest eigenvalue of $A^T A$.

Exercise 1.7   If A is an invertible $n \times n$ matrix, prove that

$$\|A^{-1}\| \leq \frac{\|A\|^{n-1}}{|\det A|}$$

(*Hint*: Work with the symmetric matrix $A^T A$, and Exercise 1.5.)

Exercise 1.8   Show that the spectral norm of an $m \times n$ matrix A is given by

$$\|A\| = \max_{\|x\|, \|y\| = 1} |y^T A x|$$

Exercise 1.9   If $A(t)$ is a continuous, $n \times n$ matrix function of t, show that its eigenvalues $\lambda_1(t), \ldots, \lambda_n(t)$ and the spectral norm $\|A(t)\|$ are continuous functions of t. Show by example that continuous differentiability of $A(t)$ does not imply continuous differentiability of the eigenvalues or the spectral norm. (*Hint*: The composition of continuous functions is a continuous function.)

Exercise 1.10   If Q is an $n \times n$ symmetric matrix, and $\varepsilon_1, \varepsilon_2$ are such that

$$0 < \varepsilon_1 I \leq Q \leq \varepsilon_2 I$$

show that

$$\frac{1}{\varepsilon_2} I \leq Q^{-1} \leq \frac{1}{\varepsilon_1} I$$

Exercise 1.11   Suppose $W(t)$ is an $n \times n$ time-dependent matrix such that $W(t) - \varepsilon I$ is symmetric and positive semidefinite for all t, where $\varepsilon > 0$. Show there exists a $\gamma > 0$ such that $\det W(t) \geq \gamma$ for all t.

Exercise 1.12   If $A(t)$ is a continuously-differentiable $n \times n$ matrix that is invertible at each t, show that

$$\frac{d}{dt} A^{-1}(t) = -A^{-1}(t)\dot{A}(t)A^{-1}(t)$$

Exercise 1.13   If $x(t)$ is an $n \times 1$ differentiable function of t, and $\|x(t)\|$ also is a differentiable function of t, prove that

$$\left| \frac{d}{dt} \|x(t)\| \right| \leq \left\| \frac{d}{dt} x(t) \right\| , \quad \text{for all } t$$

Show necessity of the assumption that $\|x(t)\|$ is differentiable by considering the scalar case $x(t) = t$.

Exercise 1.14   Suppose that $A(t)$ is $m \times n$, and such that there is no finite constant α for which

$$\int_0^t \|A(\sigma)\| \, d\sigma \le \alpha, \quad t \ge 0$$

Show that there is at least one entry of $A(t)$, say $a_{ij}(t)$, that has the same property. That is, there is no finite β for which

$$\int_0^t |a_{ij}(\sigma)| \, d\sigma \le \beta, \quad t \ge 0$$

Exercise 1.15   Suppose $M(t)$ is an $n \times n$ matrix that is invertible for each t. Show that if there is a finite constant α such that $\|M^{-1}(t)\| \le \alpha$ for all t, then there is a positive constant β such that $|\det M(t)| \ge \beta$ for all t.

Exercise 1.16   Suppose $Q(t)$ is $n \times n$, symmetric, and positive semidefinite for all t. If $t_b \ge t_a$ and

$$\int_{t_a}^{t_b} Q(\sigma) \, d\sigma \le \varepsilon I$$

show that

$$\int_{t_a}^{t_b} \|Q(\sigma)\| \, d\sigma \le n\varepsilon$$

(*Hint*: Use Exercise 1.5.)

NOTES

Note 1.1   Standard references for matrix analysis are

F.R. Gantmacher, *Theory of Matrices,* (two volumes), Chelsea Publishing, New York, 1959

R.A. Horn, C.A. Johnson, *Matrix Analysis,* Cambridge University Press, Cambridge, 1985

G. Strang, *Linear Algebra and Its Applications,* Third Edition, Harcourt, Brace, Janovich, San Diego, 1988

All three go well beyond what we need. In particular, the second reference contains an extensive treatment of induced norms. The compact reviews of linear algebra and matrix algebra in texts on linear systems also are valuable. For example consult the appropriate sections in the books

R.W. Brockett, *Finite Dimensional Linear Systems,* John Wiley, New York, 1970

D.F. Delchamps, *State Space and Input-Output Linear Systems,* Springer-Verlag, New York, 1988

T. Kailath, *Linear Systems,* Prentice Hall, Englewood Cliffs, New Jersey, 1980

L.A. Zadeh, C.A. Desoer, *Linear System Theory,* McGraw-Hill, New York, 1963

Note 1.2 Matrix theory and linear algebra provide effective computational tools in addition to a mathematical language for linear system theory. Several commercial packages are available that provide convenient computational environments. A basic reference for matrix computation is

G.H. Golub, C.F. Van Loan, *Matrix Computations,* Second Edition, Johns Hopkins University Press, Baltimore, 1989

Numerical aspects of the theory of time-invariant linear systems are covered in

P.H. Petkov, N.N. Christov, M.M. Konstantinov, *Computational Methods for Linear Control Systems,* Prentice Hall, New York, 1991

Note 1.3 Various induced norms for matrices can be defined corresponding to various vector norms. For a specific purpose there may be one induced norm that is most suitable, but from a theoretical perspective, any choice will do in most circumstances. For economy we use the spectral norm, ignoring all others.

A fundamental construct related to the spectral norm, but not explicitly used in this book, is the following. The nonnegative square roots of the eigenvalues of $A^T A$ are called the *singular values* of A. (The spectral norm of A is then the largest singular value of A.) The *singular value decomposition* of A is based on the existence of orthogonal matrices U and V, ($U^{-1} = U^T$ and $V^{-1} = V^T$) such that $U^T A V$ displays the singular values of A on the quasi-diagonal, with all other entries zero. Singular values and the corresponding decomposition have theoretical implications in linear system theory, and are central to numerical computation. See the citations in Note 1.2, the paper

V.C. Klema, A.J. Laub, "The singular value decomposition: its computation and some applications," *IEEE Transactions on Automatic Control,* Vol. 25, No. 2, pp. 164–176, 1980

or Chapter 19 of

R.A. DeCarlo, *Linear Systems,* Prentice Hall, New York, 1989

Note 1.4 The growth condition that an infinitely-differentiable function of a real variable must satisfy to be an analytic function is proved in Section 15.7 of

W. Fulks, *Advanced Calculus,* Third Edition, John Wiley, New York, 1978

Basic material on convergence and uniform convergence of series of functions are treated in this text, and many, many others.

Note 1.5 Linear algebraic notions associated to a time-dependent matrix, for example range space and rank structure, can be delicate to work out, and can depend on smoothness assumptions on the time-dependence. For examples related to linear system theory, see

L. Weiss, P.L. Falb, "Dolezal's theorem, linear algebra with continuously parametrized elements, and time-varying systems," *Mathematical Systems Theory,* Vol. 3, No. 1, pp. 67–75, 1969

L.M. Silverman, R.S. Bucy, "Generalizations of a theorem of Dolezal," *Mathematical Systems Theory,* Vol. 4, No. 4, pp. 334–339, 1970

2

STATE EQUATION REPRESENTATION

The basic representation for linear systems is the *linear state equation,* customarily written in the standard form

$$\dot{x}(t) = A(t)x(t) + B(t)u(t)$$

$$y(t) = C(t)x(t) + D(t)u(t) \tag{1}$$

where the overdot denotes differentiation with respect to time t. The $n \times 1$ vector function of time, $x(t)$, is called the *state vector,* and its components, $x_1(t), \ldots, x_n(t)$, are the *state variables.* The *input signal* is the $m \times 1$ function $u(t)$, and $y(t)$ is the $p \times 1$ *output signal.* We assume throughout that p, $m \le n$ — a sensible formulation in terms of independence considerations on the components of the vector input and output signals.

Default assumptions on the coefficient matrices in (1) are that the entries of $A(t)$ $(n \times n)$, $B(t)$ $(n \times m)$, $C(t)$ $(p \times n)$, and $D(t)$ $(p \times m)$ are continuous, real-valued functions defined for all $t \in (-\infty, \infty)$. Standard terminology is that (1) is *time invariant* if these coefficient matrices are constant. The linear state equation is called *time varying* if any entry of any coefficient matrix varies with time.

Mathematical hypotheses weaker than continuity can be adopted as the default setting. The resulting theory changes little, except in sophistication of the mathematics that must be employed. The assumptions adopted here are intended to balance engineering generality against simplicity of the required mathematical tools. There are isolated instances when complex-valued coefficient matrices arise, namely when certain special forms for state equations obtained by a change of state variables are considered. However such exceptions to the standing assumptions are noted locally.

The input signal $u(t)$ is assumed to be piecewise continuous. Piecewise continuity is adopted so that at one place in Chapter 5 an input signal can be pieced together on subintervals of time, leaving jump discontinuities at the boundaries of adjacent subintervals. Aside from this construction and occasional mention of impulse (generalized function) inputs, the input signal can be regarded as a continuous function of time.

Typically in engineering problems there is a fixed initial time t_o, and properties of the solution $x(t)$ of a linear state equation for given initial state $x(t_o) = x_o$ and input signal $u(t)$, specified for $t \in [t_o, \infty)$, are of interest for $t \geq t_o$. However from a mathematical viewpoint there are occasions when solutions "backward in time" are of interest, and this is the reason that the interval of definition of the coefficient matrices in the state equation is $(-\infty, \infty)$. That is, the input signal might be defined for all $t \in (-\infty, \infty)$, and the solution $x(t)$ for $t < t_o$ as well as $t \geq t_o$ can be sought. Of course if the state equation is defined and of interest only in a smaller interval, say $t \in [0, \infty)$, the domain of definition of the coefficient matrices can be extended to $(-\infty, \infty)$ simply by setting, for example, $A(t) = A(0)$ for $t < 0$, and the default set-up above is attained.

The fundamental theoretical issues for the class of linear state equations just introduced are the existence and uniqueness of solutions. Consideration of these issues is postponed to Chapter 2, while we provide motivation for the state equation representation. In fact linear state equations of the form (1) can arise in many ways. Sometimes a time-varying linear state equation results directly from a physical model of interest. More generally the classical n^{th}-order, linear differential equations from mathematical physics can be placed in state-equation form. Also a time-varying linear state equation arises as the linearization of a nonlinear state equation about a particular solution of interest. Of course the advantage of describing physical systems in the standard format (1) is that system properties can be characterized in terms of properties of the coefficient matrices. Thus the study of (1) can bring out the common properties of diverse physical settings.

Examples

We begin with a collection of simple examples that illustrate the genesis of time-varying linear state equations. Relying also on previous exposure to linear systems, the universal should emerge from the particular.

2.1 Example   Suppose a rocket ascends from the surface of the Earth propelled by a thrust force due to an ejection of mass. As shown in Figure 2.2, let $h(t)$ be the altitude of the rocket at time t, and $v(t)$ be the (vertical) velocity at time t, both with initial values zero at $t = 0$. Also, let $m(t)$ be the mass of the rocket at time t. Acceleration due to gravity is denoted by the constant g, and the thrust force is the product $v_e u_o$, where v_e is the assumed-constant relative exhaust velocity, and u_o is the assumed-constant rate of change of mass. Note $v_e < 0$ since the exhaust velocity direction is opposite $v(t)$, and $u_o < 0$ since the mass of the rocket decreases.

Because of the time-variable mass of the rocket, the equations of motion must be based on consideration of both the rocket mass and the expelled mass. Attention to basic physics (see Note 2.1) leads to the force equation

$$m(t)\dot{v}(t) = -m(t)g + v_e u_o \qquad (2)$$

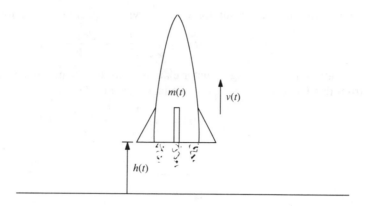

2.2 Figure   A rocket ascends, with altitude $h(t)$ and velocity $v(t)$.

Vertical velocity is the rate of change of altitude, so an additional differential equation describing the system is

$$\dot{h}(t) = v(t)$$

Finally the rocket mass variation is given by $\dot{m}(t) = u_o$, which gives, by integration,

$$m(t) = m_o + u_o t$$

where m_o is the initial mass of the rocket. Let $x_1(t) = h(t)$ and $x_2(t) = v(t)$ be the state variables, and suppose altitude also is the output. A linear state equation description that is valid until the mass supply is exhausted is

$$\dot{x}(t) = \begin{bmatrix} 0 & 1 \\ 0 & 0 \end{bmatrix} x(t) + \begin{bmatrix} 0 \\ -g + v_e u_o / (m_o + u_o t) \end{bmatrix}, \quad x(0) = 0 \tag{3}$$

$$y(t) = [\, 1 \quad 0\,] x(t)$$

Here the input signal has a fixed form, so the input term is written as a forcing function. This should be viewed as a time-invariant linear state equation with a time-varying forcing function, not a time-varying linear state equation. We return to this system in Example 2.6, and consider a variable rate of mass expulsion.

2.3 Example   Time-varying versions of the basic linear circuit elements can be devised in simple ways. A time-varying resistor exhibits the voltage/current characteristic

$$v(t) = r(t) i(t)$$

where $r(t)$ is a fixed time function. For example if $r(t)$ is a sinusoid, then this is the basis for some modulation schemes in communication systems. A time-varying capacitor exhibits a time-varying charge/voltage characteristic, $q(t) = c(t) v(t)$. Here $c(t)$ is a fixed time function describing, for example, the variation in plate spacing of a parallel-plate capacitor. Since current is the instantaneous rate of change of charge, the voltage/current relationship for a time-varying capacitor has the form

$$i(t) = c(t) \frac{dv(t)}{dt} + \frac{dc(t)}{dt} v(t) \tag{4}$$

Similarly a time-varying inductor exhibits a time-varying flux/current characteristic, and from this fact can be derived the voltage/current relation

$$v(t) = l(t) \frac{di(t)}{dt} + \frac{dl(t)}{dt} i(t)$$

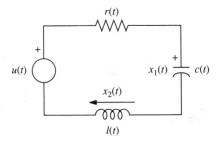

2.4 Figure   A series connection of time-variable circuit elements.

Consider the series circuit shown in Figure 2.4, which includes one of each of these circuit elements, with a voltage source providing the input signal $u(t)$. Suppose the output signal $y(t)$ is the voltage across the resistor. Following a standard prescription, we choose as state variables the voltage $x_1(t)$ across the capacitor, and the current $x_2(t)$ through the inductor (which also is the current through the entire series circuit). Then Kirchhoff's voltage law for the circuit gives

$$\dot{x}_2(t) = \frac{-1}{l(t)} x_1(t) - \frac{1}{l(t)} [\, r(t) + \dot{l}(t)\,]x_2(t) + \frac{1}{l(t)} u(t) \tag{5}$$

Another equation describing the circuit (a trivial application of Kirchhoff's current law) is (4), which in the present context is written in the form

$$\dot{x}_1(t) = -\frac{\dot{c}(t)}{c(t)} x_1(t) + \frac{1}{c(t)} x_2(t)$$

The output equation is

$$y(t) = r(t)x_2(t)$$

This yields a linear state equation description of the circuit with coefficients

$$A(t) = \begin{bmatrix} \dfrac{-\dot{c}(t)}{c(t)} & \dfrac{1}{c(t)} \\[2ex] -\dfrac{1}{l(t)} & -\dfrac{r(t)+\dot{l}(t)}{l(t)} \end{bmatrix}, \quad B(t) = \begin{bmatrix} 0 \\[1ex] \dfrac{1}{l(t)} \end{bmatrix}, \quad C(t) = \begin{bmatrix} 0 & r(t) \end{bmatrix}$$

2.5 Example   Consider an nth-order linear differential equation in the dependent variable $y(t)$, with forcing function $b_0(t)u(t)$,

$$\frac{d^n y(t)}{dt^n} + a_{n-1}(t)\frac{d^{n-1}y(t)}{dt^{n-1}} + \cdots + a_0(t)y(t) = b_0(t)u(t) \tag{6}$$

defined for $t \geq t_o$, with initial conditions

$$y(t_o), \frac{dy}{dt}(t_o), \ldots, \frac{d^{n-1}y}{dt^{n-1}}(t_o)$$

A simple device can be used to recast this differential equation into the form of a linear state equation with input $u(t)$ and output $y(t)$. Though it seems an arbitrary choice, it is convenient to define state variables (entries in the state vector) by

$$x_1(t) = y(t)$$

$$x_2(t) = \frac{dy(t)}{dt}$$

$$\vdots$$

$$x_n(t) = \frac{d^{n-1}y(t)}{dt^{n-1}}$$

That is, the output and its first $n-1$ derivatives are defined as state variables. Then

$$\dot{x}_1(t) = x_2(t)$$

$$\dot{x}_2(t) = x_3(t)$$

$$\vdots$$

$$\dot{x}_{n-1}(t) = x_n(t) \tag{7}$$

and, according to the differential equation,

$$\dot{x}_n(t) = -a_0(t)x_1(t) - a_1(t)x_2(t) - \cdots - a_{n-1}(t)x_n(t) + b_0(t)u(t)$$

Writing these equations in vector-matrix form, with the obvious definition of the state vector $x(t)$, gives a time-varying linear state equation,

$$\dot{x}(t) = \begin{bmatrix} 0 & 1 & \cdots & 0 \\ \vdots & \vdots & \vdots & \vdots \\ 0 & 0 & \cdots & 1 \\ -a_0(t) & -a_1(t) & \cdots & -a_{n-1}(t) \end{bmatrix} x(t) + \begin{bmatrix} 0 \\ \vdots \\ 0 \\ b_0(t) \end{bmatrix} u(t) \tag{8}$$

The output equation can be written as

$$y(t) = \begin{bmatrix} 1 & 0 & \cdots & 0 \end{bmatrix} x(t)$$

and the initial conditions on the output and its derivatives form the initial state

$$x(t_o) = \begin{bmatrix} y(t_o) \\ \dfrac{dy}{dt}(t_o) \\ \vdots \\ \dfrac{d^{n-1}y}{dt^{n-1}}(t_o) \end{bmatrix}$$

Linearization

A linear state equation (1) is useful as an approximation to a nonlinear state equation in the following sense. Consider

$$\dot{x}(t) = f(x(t), u(t), t), \quad x(t_o) = x_o \tag{9}$$

where the state $x(t)$ is an $n \times 1$ vector, and $u(t)$ is an $m \times 1$ vector input. Written in scalar terms, the ith component equation has the form

$$\dot{x}_i(t) = f_i(x_1(t), \ldots, x_n(t); u_1(t), \ldots, u_m(t); t), \quad x_i(t_o) = x_{i0}$$

for $i = 1, \ldots, n$. Suppose (9) has been solved for a particular input signal called the *nominal input* $\tilde{u}(t)$ and a particular initial state called the *nominal initial state* $\tilde{x}_o$ to obtain a unique nominal solution, often called a *nominal trajectory* $\tilde{x}(t)$. Of interest is the behavior of the nonlinear state equation for an input and initial state that are 'close' to the nominal values. That is, consider $u(t) = \tilde{u}(t) + u_\delta(t)$ and $x_o = \tilde{x}_o + x_{o\delta}$, where $\|x_{o\delta}\|$ and $\|u_\delta(t)\|$ are appropriately small for $t \geq t_o$. We assume that the corresponding solution remains close to $\tilde{x}(t)$, at each t, and write $x(t) = \tilde{x}(t) + x_\delta(t)$. Of course this is not always the case, though we will not pursue further the analysis of this assumption. In terms of the nonlinear state equation description, these notations are related according to

$$\frac{d}{dt}\tilde{x}(t) + \frac{d}{dt}x_\delta(t) = f(\tilde{x}(t)+x_\delta(t), \tilde{u}(t)+u_\delta(t), t),$$

$$\tilde{x}(t_o) + x_\delta(t_o) = \tilde{x}_o + x_{o\delta} \tag{10}$$

Assuming derivatives exist, we can expand the right side using Taylor series about $\tilde{x}(t)$ and $\tilde{u}(t)$, and then retain only the terms through order 1. This should provide a reasonable approximation since $u_\delta(t)$ and $x_\delta(t)$ are assumed to be small for all t. Note that the expansion describes the behavior of the function $f(x, u, t)$ with respect to arguments x and u; there is no expansion in terms of the third argument t. For the ith component, retaining terms through first order, and momentarily dropping most t-arguments for simplicity, we can write

$$f_i(\tilde{x} + x_\delta, \tilde{u} + u_\delta, t) \cong f_i(\tilde{x}, \tilde{u}, t) + \frac{\partial f_i}{\partial x_1}(\tilde{x}, \tilde{u}, t)x_{\delta 1} + \cdots + \frac{\partial f_i}{\partial x_n}(\tilde{x}, \tilde{u}, t)x_{\delta n}$$

$$+ \frac{\partial f_i}{\partial u_1}(\tilde{x}, \tilde{u}, t)u_{\delta 1} + \cdots + \frac{\partial f_i}{\partial u_m}(\tilde{x}, \tilde{u}, t)u_{\delta m} \tag{11}$$

Performing this expansion for $i = 1, \ldots, n$ and arranging into vector-matrix form gives

$$\frac{d}{dt}\tilde{x}(t) + \frac{d}{dt}x_\delta(t) \cong f(\tilde{x}(t), \tilde{u}(t), t) + \frac{\partial f}{\partial x}(\tilde{x}(t), \tilde{u}(t), t)\ x_\delta(t)$$

$$+ \frac{\partial f}{\partial u}(\tilde{x}(t), \tilde{u}(t), t)\ u_\delta(t)$$

where the notation $\partial f/\partial x$ denotes the *Jacobian*, a matrix with i,j-entry $\partial f_i/\partial x_j$. Since

$$\frac{d}{dt}\tilde{x}(t) = f(\tilde{x}(t), \tilde{u}(t), t), \quad \tilde{x}(t_o) = \tilde{x}_o$$

the relation between $x_\delta(t)$ and $u_\delta(t)$ is approximately described by a time-varying linear state equation of the form

$$\dot{x}_\delta(t) = A(t)x_\delta(t) + B(t)u_\delta(t), \quad x_\delta(t_o) = x_o - \tilde{x}_o$$

where $A(t)$ and $B(t)$ are the matrices of partial derivatives evaluated using the nominal trajectory data, namely

$$A(t) = \frac{\partial f}{\partial x}(\tilde{x}(t), \tilde{u}(t), t), \quad B(t) = \frac{\partial f}{\partial u}(\tilde{x}(t), \tilde{u}(t), t)$$

If there is a nonlinear output equation, say

$$y(t) = h(x(t), u(t), t)$$

the function $h(x, u, t)$ can be expanded about $x = \tilde{x}(t)$ and $u = \tilde{u}(t)$ in a similar fashion to give, after dropping higher-order terms, the approximate description

$$y_\delta(t) = C(t)x_\delta(t) + D(t)u_\delta(t)$$

Here the deviation output is given by $y_\delta(t) = y(t) - \tilde{y}(t)$, with $\tilde{y}(t) = h(\tilde{x}(t), \tilde{u}(t), t)$, and

$$C(t) = \frac{\partial h}{\partial x}(\tilde{x}(t), \tilde{u}(t), t), \quad D(t) = \frac{\partial h}{\partial u}(\tilde{x}(t), \tilde{u}(t), t)$$

In this development a nominal solution is assumed to exist for all $t \geq t_o$, and be unique so that the linearization makes sense as an approximation. Also the nominal solution is assumed to be given in the computation of the linearization. Determining an appropriate nominal trajectory often is a difficult problem, though physical insight can be helpful.

2.6 Example   Consider the behavior of the rocket in Example 2.1 when the rate of mass expulsion can be varied with time: $u(t) = \dot{m}(t)$, in place of a constant u_o. The velocity and altitude considerations remain the same, leading to

$$\dot{h}(t) = v(t)$$

$$\dot{v}(t) = -g + \frac{v_e}{m(t)} u(t) \tag{12}$$

In addition the rocket mass $m(t)$ is described by

$$\dot{m}(t) = u(t)$$

Therefore $m(t)$ is regarded as another state variable, with $u(t)$ as the input signal. Setting

$$x_1(t) = h(t), \quad x_2(t) = v(t), \quad x_3(t) = m(t)$$

we obtain

$$\begin{bmatrix} \dot{x}_1(t) \\ \dot{x}_2(t) \\ \dot{x}_3(t) \end{bmatrix} = \begin{bmatrix} x_2(t) \\ -g + v_e u(t)/x_3(t) \\ u(t) \end{bmatrix}$$

$$y(t) = x_1(t) \tag{13}$$

This is a nonlinear state equation description of the system, and we consider linearization about a nominal trajectory corresponding to the constant nominal input $\tilde{u}(t) = u_o < 0$. The nominal trajectory is not difficult to compute by integrating in turn the differential equations for $x_3(t)$, $x_2(t)$, and $x_1(t)$. This calculation, equivalent to solving the linear state equation (3) in Example 2.1, gives

$$\tilde{x}_1(t) = \frac{-g}{2} t^2 + \frac{m_o v_e}{u_o} \left[\left(1 + \frac{u_o}{m_o} t \right) \ln\left(1 + \frac{u_o}{m_o} t \right) - \frac{u_o}{m_o} t \right]$$

$$\tilde{x}_2(t) = -gt + v_e \ln\left(1 + \frac{u_o}{m_o} t \right)$$

$$\tilde{x}_3(t) = m_o + u_o t \tag{14}$$

As before, these expressions are valid until the available mass is exhausted.

To compute the linearized state equation about this nominal trajectory, the partial derivatives needed are

$$\frac{\partial f(x, u)}{\partial x} = \begin{bmatrix} 0 & 1 & 0 \\ 0 & 0 & -v_e u/x_3^2 \\ 0 & 0 & 0 \end{bmatrix}, \quad \frac{\partial f(x, u)}{\partial u} = \begin{bmatrix} 0 \\ v_e/x_3 \\ 1 \end{bmatrix}$$

Evaluating these derivatives at the nominal data, the linearized state equation in terms of the deviation variables $x_\delta(t) = x(t) - \tilde{x}(t)$ and $u_\delta(t) = u(t) - u_o$ is

$$\dot{x}_\delta(t) = \begin{bmatrix} 0 & 1 & 0 \\ 0 & 0 & \dfrac{-v_e u_o}{(m_o + u_o t)^2} \\ 0 & 0 & 0 \end{bmatrix} x_\delta(t) + \begin{bmatrix} 0 \\ \dfrac{v_e}{m_o + u_o t} \\ 1 \end{bmatrix} u_\delta(t) \qquad (15)$$

(Here $u_\delta(t)$ can be positive or negative, representing deviations from the negative constant value u_o.) The initial conditions for the deviation state variables are given by

$$x_\delta(0) = x(0) - \begin{bmatrix} 0 \\ 0 \\ m_o \end{bmatrix}$$

Of course the nominal output is simply $\tilde{y}(t) = \tilde{x}_1(t)$, and the linearized output equation is

$$y_\delta(t) = \begin{bmatrix} 1 & 0 & 0 \end{bmatrix} x_\delta(t)$$

2.7 Example   An Earth satellite of unit mass can be modeled as a point mass moving in a plane while attracted to the origin of the plane by an inverse square law force. It is convenient to choose polar coordinates, with $r(t)$ the radius from the origin to the mass, and $\theta(t)$ the angle from an appropriate axis. Assuming the satellite can apply force $u_1(t)$ in the radial direction and $u_2(t)$ in the tangential direction, as shown in Figure 2.8, the equations of motion have the form

$$\ddot{r}(t) = r(t)\dot{\theta}^2(t) - \frac{\beta}{r^2(t)} + u_1(t)$$

$$\ddot{\theta}(t) = -\frac{2\dot{r}(t)\dot{\theta}(t)}{r(t)} + \frac{u_2(t)}{r(t)} \qquad (16)$$

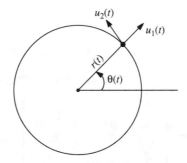

2.8 Figure   A unit point mass in gravitational orbit.

where β is a constant. When the thrust forces are identically zero, solutions can be ellipses, parabolas, or hyperbolas, describing orbital motion in the first instance, and escape trajectories of the satellite in the others. The simplest orbit is a circle, where $r(t)$ and $\dot{\theta}(t)$ are constant. Specifically it is easy to verify that for the nominal input $\tilde{u}_1(t) = \tilde{u}_2(t) = 0$, $t \geq 0$, and nominal initial conditions

$$r(0) = r_o, \quad \dot{r}(0) = 0$$

$$\theta(0) = \theta_o, \quad \dot{\theta}(0) = \omega_o$$

where $\omega_o = (\beta/r_o^3)^{1/2}$, a nominal trajectory is

$$\tilde{r}(t) = r_o, \quad \tilde{\theta}(t) = \omega_o t + \theta_o$$

To determine a state equation representation, let

$$x_1(t) = r(t), \quad x_2(t) = \dot{r}(t), \quad x_3(t) = \theta(t), \quad x_4(t) = \dot{\theta}(t)$$

so that the equations of motion are described by

$$
\begin{bmatrix} \dot{x}_1(t) \\ \dot{x}_2(t) \\ \dot{x}_3(t) \\ \dot{x}_4(t) \end{bmatrix}
=
\begin{bmatrix}
x_2(t) \\
x_1(t)x_4^2(t) - \dfrac{\beta}{x_1^2(t)} + u_1(t) \\
x_4(t) \\
\dfrac{-2x_2(t)x_4(t)}{x_1(t)} + \dfrac{u_2(t)}{x_1(t)}
\end{bmatrix}
\tag{17}
$$

The nominal data is then

$$
\tilde{u}(t) = \begin{bmatrix} \tilde{u}_1(t) \\ \tilde{u}_2(t) \end{bmatrix} = 0, \quad
\tilde{x}(t) = \begin{bmatrix} r_o \\ 0 \\ \omega_o t + \theta_o \\ \omega_o \end{bmatrix}, \quad
\tilde{x}(0) = \begin{bmatrix} r_o \\ 0 \\ \theta_o \\ \omega_o \end{bmatrix}
$$

With the deviation variables

$$x_\delta(t) = x(t) - \tilde{x}(t), \quad u_\delta(t) = u(t)$$

the corresponding linearized state equation is computed to be

$$
\dot{x}_\delta(t) = \begin{bmatrix}
0 & 1 & 0 & 0 \\
3\omega_o^2 & 0 & 0 & 2r_o\omega_o \\
0 & 0 & 0 & 1 \\
0 & -2\omega_o/r_o & 0 & 0
\end{bmatrix} x_\delta(t)
+
\begin{bmatrix}
0 & 0 \\
1 & 0 \\
0 & 0 \\
0 & 1/r_o
\end{bmatrix} u_\delta(t)
\tag{18}
$$

Of course the outputs are given by

$$
\begin{bmatrix} r_\delta(t) \\ \theta_\delta(t) \end{bmatrix} = \begin{bmatrix} 1 & 0 & 0 & 0 \\ 0 & 0 & 1 & 0 \end{bmatrix} x_\delta(t)
$$

where $r_\delta(t) = r(t) - r_o$, and $\theta_\delta(t) = \theta(t) - \omega_o t - \theta_o$. For a circular orbit the linearized state equation about the time-varying nominal solution is a time-invariant linear state equation — an unusual occurrence. If a nominal trajectory corresponding to an elliptical orbit is considered, a linearized state equation with periodic coefficients is obtained.
□□□

In a fashion closely related to linearization, time-varying linear state equations provide descriptions of the parameter sensitivity of solutions of nonlinear state equations. As a simple illustration consider an unforced nonlinear state equation of dimension n, including a scalar parameter that enters both the right side of the state equation and the initial state. A solution of the state equation also depends on the parameter, so we adopt the notation

$$
\dot{x}(t, \alpha) = f(x(t, \alpha), \alpha) , \quad x(0, \alpha) = x_o(\alpha) \tag{19}
$$

Suppose that the function $f(x, \alpha)$ is continuously differentiable both in x and α, and that a solution $x(t, \alpha_o)$, $t \geq 0$, exists for a nominal value α_o of the parameter. Then a standard result in the theory of differential equations is that a solution $x(t, \alpha)$ exists and is continuously differentiable in both t and α, for α close to α_o. The issue of interest is the effect of changes in α on such solutions.

We can differentiate (19) with respect to α and write

$$
\frac{\partial}{\partial \alpha} \dot{x}(t, \alpha) = \frac{\partial f}{\partial x}(x(t, \alpha), \alpha)\frac{\partial}{\partial \alpha} x(t, \alpha) + \frac{\partial f}{\partial \alpha}(x(t, \alpha), \alpha) ,
$$

$$
\frac{\partial}{\partial \alpha} x(0, \alpha) = \frac{\partial}{\partial \alpha} x_o(\alpha) \tag{20}
$$

To simplify notation denote derivatives with respect to α, evaluated at α_o, by

$$
z(t) = \frac{\partial x}{\partial \alpha}(t, \alpha_o) , \quad g(t) = \frac{\partial f}{\partial \alpha}(x(t, \alpha_o), \alpha_o)
$$

and let

$$
A(t) = \frac{\partial f}{\partial x}(x(t, \alpha_o), \alpha_o)
$$

Then since

$$
\frac{\partial^2}{\partial \alpha \partial t} x(t, \alpha) = \frac{\partial^2}{\partial t \partial \alpha} x(t, \alpha)
$$

we can write (20) for $\alpha = \alpha_o$ as

$$
\dot{z}(t) = A(t)z(t) + g(t) , \quad z(0) = \frac{\partial x_o}{\partial \alpha}(\alpha_o) \tag{21}
$$

The solution $z(t)$ of this forced linear state equation describes the dependence of the solution of (19) on the parameter α, at least for $|\alpha - \alpha_o|$ small. If in a particular instance $\|z(t)\|$ remains small for $t \geq 0$, then the solution of the nonlinear state equation is relatively insensitive to changes in α near α_o.

State Equation Implementation

In a reversal of the discussion so far, we briefly note that a linear state equation can be implemented directly in electronic hardware. One implementation is based on electronic devices called *operational amplifiers* that can be arranged to produce on electrical signals the three underlying operations in a linear state equation.

The first operation is the (signed) sum of scalar functions of time, diagramed in Figure 2.9(a). The second is integration, which conveniently represents the relationship between a scalar function of time, its derivative, and an initial value. This is shown in Figure 2.9(b). The third operation is multiplication of a scalar signal by a time-varying coefficient, as represented in Figure 2.9(c). The basic building blocks shown in Figure 2.9 can be connected together as prescribed by a given linear state equation. The resulting diagram, called a *state variable diagram,* is very close to a hardware layout for electronic implementation. From a theoretical perspective such a diagram sometimes reveals structural features of the linear state equation that are not apparent from the coefficient matrices.

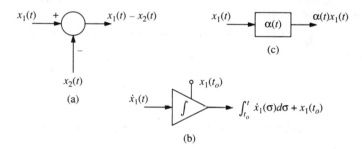

2.9 Figure   The elements of a state variable diagram.

2.10 Example   The linear state equation (8) in Example 2.5 can be represented by the state variable diagram shown in Figure 2.11.

EXERCISES

Exercise 2.1   Rewrite the nth-order linear differential equation

$$y^{(n)}(t) + a_{n-1}(t)y^{(n-1)}(t) + \cdots + a_0(t)y(t) = b_0(t)u(t) + b_1(t)u^{(1)}(t)$$

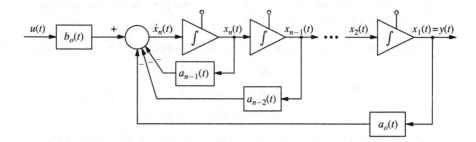

2.11 Figure   A state variable diagram for Example 2.5.

as a dimension-n linear state equation,

$$\dot{x}(t) = A(t)x(t) + B(t)u(t)$$

$$y(t) = C(t)x(t) + D(t)u(t)$$

(*Hint*: let $x_n(t) = y^{(n-1)}(t) - b_1(t)u(t)$.)

Exercise 2.2   Define state variables such that the nth-order differential equation

$$y^{(n)}(t) + a_{n-1}t^{-1}y^{(n-1)}(t) + a_{n-2}t^{-2}y^{(n-2)}(t) +$$

$$\cdots + a_1 t^{-n+1}y^{(1)}(t) + a_0 t^{-n}y(t) = 0$$

can be written as a linear state equation

$$\dot{x}(t) = t^{-1} Ax(t)$$

where A is a constant $n \times n$ matrix.

Exercise 2.3   For the differential equation

$$\ddot{y}(t) + (4/3)y^3(t) = -(1/3)u(t)$$

use a simple trigonometry identity to help find a nominal solution corresponding to $\tilde{u}(t) = \sin(3t)$, $y(0) = 0$, $\dot{y}(0) = 1$. Determine a linearized state equation that describes the behavior about this nominal.

Exercise 2.4   Linearize the nonlinear state equation

$$\dot{x}_1(t) = \frac{-1}{x_2^2(t)}$$

$$\dot{x}_2(t) = u(t)x_1(t)$$

about the nominal trajectory arising from $\tilde{x}_1(0) = \tilde{x}_2(0) = 1$, and $\tilde{u}(t) = 0$, for all $t \geq 0$.

Exercise 2.5   For the nonlinear state equation

$$\begin{bmatrix} \dot{x}_1(t) \\ \dot{x}_2(t) \end{bmatrix} = \begin{bmatrix} x_2(t) - 2x_1(t)x_2(t) \\ -x_1(t) + x_1^2(t) + x_2^2(t) + u(t) \end{bmatrix}$$

with constant nominal input $\tilde{u}(t) = 0$, compute the possible constant nominal solutions and the corresponding linearized state equations.

Exercise 2.6   The Euler equations for the angular velocities of a rigid body are

$$I_1\dot{\omega}_1(t) = (I_2 - I_3)\omega_2(t)\omega_3(t) + u_1(t)$$

$$I_2\dot{\omega}_2(t) = (I_3 - I_1)\omega_1(t)\omega_3(t) + u_2(t)$$

$$I_3\dot{\omega}_3(t) = (I_1 - I_2)\omega_1(t)\omega_2(t) + u_3(t)$$

Here $\omega_1(t)$, $\omega_2(t)$, and $\omega_3(t)$ are the angular velocities in a body-fixed coordinate system coinciding with the principal axes, $u_1(t)$, $u_2(t)$, and $u_3(t)$ are the applied torques, and I_1, I_2, and I_3 are the principal moments of inertia. For $I_1 = I_2$, a symmetrical body, linearize the equations about the nominal solution

$$\tilde{u}_1(t) = \tilde{u}_2(t) = \tilde{u}_3(t) = 0 , \quad \tilde{\omega}_1(0) = 0, \; \tilde{\omega}_2(0) = 1, \; \tilde{\omega}_3(0) = \omega_o$$

$$\tilde{\omega}_1(t) = \sin\left[\omega_o \frac{(I-I_3)}{I} t\right], \quad \tilde{\omega}_2(t) = \cos\left[\omega_o \frac{(I-I_3)}{I} t\right], \quad \tilde{\omega}_3(t) = \omega_o$$

where $I = I_1 = I_2$.

Exercise 2.7   Consider a single-input, single-output, time-invariant linear state equation

$$\dot{x}(t) = Ax(t) + bu(t) , \quad x(0) = x_o$$

$$y(t) = cx(t)$$

If the nominal input is a nonzero constant, $u(t) = \tilde{u}$, under what conditions does there exist a constant nominal solution $\tilde{x}(t) = x_o$ for some x_o. (The condition is more subtle than assuming A is invertible.)   Under what conditions is the corresponding nominal output zero?   Under what conditions do there exist constant nominal solutions that satisfy $\tilde{y} = \tilde{u}$ for all $\tilde{u}$?

Exercise 2.8   A time-invariant linear state equation

$$\dot{x}(t) = Ax(t) + Bu(t)$$

$$y(t) = Cx(t)$$

with $p = m$ is said to have *identity dc-gain* if for any given $m \times 1$ vector $\tilde{u}$ there exists an $n \times 1$ vector $\tilde{x}$ such that

$$A\tilde{x} + B\tilde{u} = 0 , \quad C\tilde{x} = \tilde{u}$$

That is, given any constant input there is a constant nominal solution with output identical to input. Show that given any $m \times n$ matrix K there exists an $m \times m$ matrix N such that

$$\dot{x}(t) = (A + BK)x(t) + BNu(t)$$

$$y(t) = Cx(t)$$

has identity dc-gain if

$$\begin{bmatrix} A & B \\ C & 0 \end{bmatrix}$$

is invertible. Under this condition show that if $(A + BK)$ is invertible, then $C(A + BK)^{-1}B$ is invertible.

Exercise 2.9   Consider a so-called bilinear state equation

$$\dot{x}(t) = Ax(t) + Dx(t)u(t) + bu(t) , \quad x(0) = x_o$$

$$y(t) = cx(t)$$

where A, D are $n \times n$, b is $n \times 1$, c is $1 \times n$, and all are constant matrices. Under what condition does this state equation have a constant nominal solution for a constant nominal input $u(t) = \tilde{u}$? If A is invertible, show that there exists a constant nominal solution if $|\tilde{u}|$ is 'sufficiently small.' What is the linearized state equation about such a nominal solution?

Exercise 2.10   For the nonlinear state equation

$$\dot{x}(t) = \begin{bmatrix} -x_2(t) + u(t) \\ x_1(t) - 2x_2(t) \\ x_1(t)u(t) - 2x_2(t)u(t) \end{bmatrix}$$

$$y(t) = x_3(t)$$

show that for every constant nominal input $\tilde{u}(t) = \tilde{u}$, $t \geq 0$, there exists a constant nominal trajectory $\tilde{x}(t) = \tilde{x}$, $t \geq 0$. What is the nominal output $\tilde{y}$ in terms of $\tilde{u}$? Explain. Linearize the state equation about an arbitrary constant nominal. If $\tilde{u} = 0$ and $x_\delta(0) = 0$, what is the response $y_\delta(t)$ of the linearized state equation for any $u_\delta(t)$? (Solution of the linear state equation is not needed.)

Exercise 2.11   Consider the nonlinear state equation

$$\dot{x}(t) = \begin{bmatrix} u(t) \\ u(t)x_1(t) - x_3(t) \\ x_2(t) - 2x_3(t) \end{bmatrix}$$

$$y(t) = x_2(t) - 2x_3(t)$$

with nominal initial state

$$\tilde{x}(0) = \begin{bmatrix} 0 \\ -3 \\ -2 \end{bmatrix}$$

and constant nominal input $\tilde{u}(t) = 1$. Show that the nominal output is $\tilde{y}(t) = 1$. Linearize the state equation about the nominal solution. Is there anything unusual about this example?

Exercise 2.12   For the nonlinear state equation

$$\dot{x}(t) = \begin{bmatrix} x_1(t) + u(t) \\ 2x_2(t) + u(t) \\ 3x_3(t) + x_1^2(t) - 4x_1(t)x_2(t) + 4x_2^2(t) \end{bmatrix}$$

$$y(t) = x_3(t)$$

determine the constant nominal solution corresponding to any given constant nominal input $u(t) = \tilde{u}$. Linearize the state equation about such a nominal. Show that if $x_\delta(0) = 0$, then $y_\delta(t)$ is zero regardless of $u_\delta(t)$.

Exercise 2.13   For the time-invariant linear state equation

$$\dot{x}(t) = Ax(t) + Bu(t) , \quad x(0) = x_o$$

suppose A is invertible, and $u(t)$ is continuously differentiable. Let

$$q(t) = -A^{-1}Bu(t)$$

and derive a state equation description for $z(t) = x(t) - q(t)$. Interpret this description in terms of deviation from an 'instantaneous constant nominal.'

NOTES

Note 2.1   Developing an appropriate mathematical model for a physical system often is difficult, and always it is the most important step in system analysis and design. The examples offered here are not intended to substantiate this claim — they serve only to motivate. Most engineering models begin with elementary physics. Since the laws of physics presumably do not change with time, the appearance of a time-varying differential equation is because of special circumstances in the physical system, or because of a particular formulation. The electrical circuit with time-varying elements in Example 2.2 is a case of the former, and the linearized state equation for the rocket in Example 2.6 is a case of the latter. Specifically in Example 2.6 where the rocket thrust is time variable, a time-invariant nonlinear state equation is obtained with $m(t)$ as a state variable. This leads to a linear time-varying state equation as an approximation via linearization about a constant-thrust nominal trajectory. Introductory details on the physics of variable-mass systems, including the ubiquitous rocket example, can be found in many elementary physics books, for example

R. Resnick, D. Halliday, *Physics*, Part I, Third Edition, John Wiley, New York, 1977

J.P. McKelvey, H. Grotch, *Physics for Science and Engineering*, Harper & Row, New York, 1978

Elementary physical properties of time-varying electrical circuit elements are discussed in

L.O. Chua, C.A. Desoer, E.S. Kuh, *Linear and Nonlinear Circuits*, McGraw-Hill, New York, 1987

The dynamics of central-force motion, such as a satellite in a gravitational field, are treated in many books on mechanics. See, for example,

B.H. Karnopp, *Introduction to Dynamics*, Addison-Wesley, Reading, Massachusetts, 1974

Elliptical nominal trajectories for Example 2.7 are much more complicated than the circular case.

Note 2.2   For the mathematically inclined, precise axiomatic formulations of 'system' and 'state' are available in the literature. Starting from these axioms the linear state equation description must be unpacked from complicated definitions. See for example

L.A. Zadeh, C.A. Desoer, *Linear System Theory*, McGraw-Hill, New York, 1963

E.D. Sontag, *Mathematical Control Theory*, Springer-Verlag, New York, 1990

Note 2.3   The *direct transmission term* $D(t)u(t)$ in the standard linear state equation causes a dilemma. It should be included on grounds that a theory of linear systems ought to encompass the identity system where $D(t)$ is unity, $C(t)$ is zero, and $A(t)$ and $B(t)$ are anything, or nothing. Also it should be included because physical systems with nonzero $D(t)$ do arise. In many topics, stability, realization, and so on, the direct transmission term is a side issue in the theoretical development and causes no problem. But in other topics, feedback and the polynomial fraction description are examples, a direct transmission complicates the situation. The decision in this book is to simplify matters by occasionally invoking a zero-$D(t)$ assumption.

Note 2.4   More general types of linear state equations also can be considered. A linear state equation where $\dot{x}(t)$ on the left side is multiplied by an $n \times n$ matrix that is singular for at least some values of t is called a *singular state equation* or *descriptor state equation*. To pursue this topic consult

F.L. Lewis, ''A survey of linear singular systems,'' *Circuits, Systems, and Signal Processing,* Vol. 5, pp. 3–36, 1986

Linear state equations that include derivatives of the input signal on the right side are discussed from an advanced viewpoint in

M. Fliess, ''Some basic structural properties of generalized linear systems,'' *Systems & Control Letters,* Vol. 15, No. 5, pp. 391–396, 1990

Finally, the notion of specifying inputs and outputs can be abandoned altogether, and a system can be viewed as a relationship among exogenous time signals. See the papers

J.C. Willems, ''From time series to linear systems,'' *Automatica,* Vol. 22, pp. 561–580 (Part I), pp. 675–694 (Part II), 1986

J.C. Willems, ''Paradigms and puzzles in the theory of dynamical systems,'' *IEEE Transactions on Automatic Control,* Vol. 36, No. 3, pp. 259–294, 1991

Note 2.5   Our informal treatment of linearization of nonlinear state equations provides only a glimpse of the topic. More advanced considerations can be found in the book by Sontag cited in Note 2.2, and in

C.A. Desoer, M. Vidyasagar, *Feedback Systems: Input-Output Properties,* Academic Press, New York, 1975

Note 2.6   The use of state variable diagrams to represent special structural features of linear state equations is typical in earlier references, in part because of the legacy of analog computers. See Section 4.9 of the book by Zadeh and Desoer cited in Note 2.2. Also consult Section 2.1 of

T. Kailath, *Linear Systems,* Prentice Hall, New York, 1980

where the idea of using integrators to represent a differential equation is attributed to Lord Kelvin.

Note 2.7   Can linear system theory contribute to the social sciences? A harsh assessment is entertainingly delivered in

D.J. Berlinski, *On Systems Analysis,* MIT Press, Cambridge, 1976

Those contemplating grand applications of linear system theory to the social, political, or biological sciences might ponder Berlinski's deconstruction.

3

STATE EQUATION SOLUTION

The basic questions of existence and uniqueness of solutions are first addressed for linear state equations unencumbered by inputs and outputs. That is, we consider

$$\dot{x}(t) = A(t)x(t), \quad x(t_o) = x_o \tag{1}$$

where the initial time t_o and initial state x_o are given. The $n \times n$ matrix function $A(t)$ is assumed to be continuous, and defined for all t. By definition a solution is a continuously-differentiable, $n \times 1$ function $x(t)$ that satisfies (1) for all t, though at the outset only solutions for $t \geq t_o$ are considered. Among other things this avoids absolute-value signs in certain inequalities, as mentioned in Chapter 1. A general contraction mapping approach that applies to both linear and nonlinear state equations is typical in mathematics references dealing with existence of solutions, though a more specialized method is used here. One reason is simplicity, but more importantly the calculations provide a good warm-up for developments in the sequel.

An alternative is simply to guess a solution to (1), and verify the guess by substitution into the state equation. This is unscientific, though perhaps reasonable for the very special case of constant $A(t)$ and $n = 1$. (What is your guess?) But the form of the general solution of (1) is too intricate to be guessed without guidance, and our development provides that and more. Requisite mathematical tools are the notions of convergence reviewed in Chapter 1.

When the basic existence question is answered, we show that for a given t_o and x_o there is precisely one solution of (1). Then linear state equations with nonzero input signals are considered, and the important result is that, under our default hypotheses, there exists a unique continuously-differentiable solution for any specified initial time, initial state, and input signal. We conclude the chapter with a review of standard terminology associated with properties of state equation solutions.

Existence

Given t_o, x_o, and an arbitrary time $T > 0$, we will construct a sequence of $n \times 1$ vector functions $\{x_k(t)\}_{k=0}^{\infty}$, defined on the interval $[t_o, t_o+T]$, that can be interpreted as a sequence of 'approximate' solutions of (1). Then we prove that the sequence converges uniformly and absolutely on $[t_o, t_o+T]$, and that the limit function is continuously differentiable and satisfies (1). This settles existence of a solution of (1) with specified t_o and x_o, and also leads to a representation for solutions.

The sequence of approximating functions on $[t_o, t_o+T]$ is defined in an iterative fashion by

$$x_0(t) = x_o$$

$$x_1(t) = x_o + \int_{t_o}^{t} A(\sigma_1) x_0(\sigma_1) \, d\sigma_1$$

$$x_2(t) = x_o + \int_{t_o}^{t} A(\sigma_1) x_1(\sigma_1) \, d\sigma_1$$

$$\vdots$$

$$x_k(t) = x_o + \int_{t_o}^{t} A(\sigma_1) x_{k-1}(\sigma_1) \, d\sigma_1 \tag{2}$$

(Of course the subscripts in (2) denote different $n \times 1$ functions, not entries in a vector.) This iterative prescription can be compiled to write $x_k(t)$ as a sum of terms involving iterated integrals of $A(t)$,

$$x_k(t) = x_o + \int_{t_o}^{t} A(\sigma_1) x_o \, d\sigma_1 + \int_{t_o}^{t} A(\sigma_1) \int_{t_o}^{\sigma_1} A(\sigma_2) x_o \, d\sigma_2 d\sigma_1$$

$$+ \cdots + \int_{t_o}^{t} A(\sigma_1) \int_{t_o}^{\sigma_1} A(\sigma_2) \cdots \int_{t_o}^{\sigma_{k-1}} A(\sigma_k) x_o \, d\sigma_k \cdots d\sigma_1 \tag{3}$$

For the convergence analysis it is more convenient to write each vector function in (2) as a 'telescoping' sum:

$$x_k(t) = x_0(t) + \sum_{j=0}^{k-1} [x_{j+1}(t) - x_j(t)], \quad k = 0, 1, \cdots \tag{4}$$

Then the sequence of partial sums of the infinite series of $n \times 1$ vector functions

$$x_0(t) + \sum_{j=0}^{\infty} [x_{j+1}(t) - x_j(t)] \tag{5}$$

is precisely the sequence $\{x_k(t)\}_{k=0}^{\infty}$. Therefore convergence properties of the infinite series (5) are equivalent to convergence properties of the sequence, and the advantage is that a straightforward convergence argument applies to the series.

Let

$$\alpha = \max_{t_o \le t \le t_o+T} \|A(t)\|$$

$$\beta = \int_{t_o}^{t_o+T} \|A(\sigma_1)x_o\| \, d\sigma_1 \tag{6}$$

where α and β are guaranteed to be finite since $A(t)$ is continuous and the time interval is finite. Then, addressing the terms in (5),

$$\|x_1(t) - x_0(t)\| = \left\| \int_{t_o}^{t} A(\sigma)x_o \, d\sigma \right\|$$

$$\le \int_{t_o}^{t} \|A(\sigma)x_o\| \, d\sigma \le \beta , \quad t \in [t_o, t_o+T]$$

Next,

$$\|x_2(t) - x_1(t)\| = \left\| \int_{t_o}^{t} A(\sigma_1)x_1(\sigma_1) - A(\sigma_1)x_0(\sigma_1) \, d\sigma_1 \right\|$$

$$\le \int_{t_o}^{t} \|A(\sigma_1)\| \, \|x_1(\sigma_1) - x_0(\sigma_1)\| \, d\sigma_1$$

$$\le \int_{t_o}^{t} \alpha\beta \, d\sigma_1 = \beta\alpha \, (t - t_o) , \quad t \in [t_o, t_o+T]$$

It is easy to show that in general

$$\|x_{j+1}(t) - x_j(t)\| = \left\| \int_{t_o}^{t} A(\sigma_1)x_j(\sigma_1) - A(\sigma_1)x_{j-1}(\sigma_1) \, d\sigma_1 \right\|$$

$$\le \int_{t_o}^{t} \|A(\sigma_1)\| \, \|x_j(\sigma_1) - x_{j-1}(\sigma_1)\| \, d\sigma_1$$

$$\le \beta \, \frac{\alpha^j (t - t_o)^j}{j!} , \quad t \in [t_o, t_o+T] , \quad j = 0, 1, \cdots \tag{7}$$

These bounds are all we need to apply the *Weierstrass M-Test* reviewed in Theorem 1.8. The terms in the infinite series (5) are bounded for $t \in [t_o, t_o+T]$ according to

$$\|x_0(t)\| = \|x_o\| , \quad \|x_{j+1}(t) - x_j(t)\| \le \beta \frac{\alpha^j T^j}{j!} , \quad j = 0, 1, \cdots$$

and the series of bounds

$$\|x_o\| + \sum_{j=0}^{\infty} \beta \frac{\alpha^j T^j}{j!}$$

converges to $\|x_o\| + \beta e^{\alpha T}$. Therefore the infinite series (5) converges uniformly and absolutely on the interval $[t_o, t_o+T]$. Since each term in the series is continuous on the interval, the limit function, denoted $x(t)$, is continuous on the interval by Theorem 1.6. Again these properties carry over to the sequence $\{x_k(t)\}_{k=0}^{\infty}$ whose terms are the partial sums of the series (5).

From (3), letting $k \to \infty$, the limit of the sequence (2) can be written as the infinite series expression

$$x(t) = x_o + \int_{t_o}^{t} A(\sigma_1)x_o \, d\sigma_1 + \int_{t_o}^{t} A(\sigma_1)\int_{t_o}^{\sigma_1} A(\sigma_2)x_o \, d\sigma_2 d\sigma_1$$

$$+ \cdots + \int_{t_o}^{t} A(\sigma_1) \int_{t_o}^{\sigma_1} A(\sigma_2) \cdots \int_{t_o}^{\sigma_{k-1}} A(\sigma_k)x_o \, d\sigma_k \cdots d\sigma_1 + \cdots \qquad (8)$$

The last step is to show that this limit $x(t)$ is continuously differentiable, and that it satisfies the linear state equation (1). Evaluating (8) at $t = t_o$ yields $x(t_o) = x_o$. Next, term-by-term differentiation of the series on the right side of (8) gives

$$0 + A(t)x_o + A(t) \int_{t_o}^{t} A(\sigma_2)x_o \, d\sigma_2$$

$$+ \cdots + A(t) \int_{t_o}^{t} A(\sigma_2) \cdots \int_{t_o}^{\sigma_{k-1}} A(\sigma_k)x_o \, d\sigma_k \cdots d\sigma_2 + \cdots \qquad (9)$$

The kth partial sum of this series is the kth partial sum of the series $A(t)x(t)$ — compare the right side of (8) with (9) — and uniform convergence of (9) on $[t_o, t_o+T]$ follows. Thus by Theorem 1.7 this term-by-term differentiation yields the derivative of $x(t)$, and the derivative is $A(t)x(t)$. Because solutions are required by definition to be continuously differentiable, we explicitly note that terms in the series (9) are continuous. Therefore by Theorem 1.6 the derivative of $x(t)$ is continuous, and we have shown that, indeed, (8) is a solution of (1).

This same development works for $t \in [t_o-T, t_o]$, though absolute values must be used in various inequality strings.

It is convenient to rewrite the $n \times 1$ vector series in (8) by factoring x_o out the right side of each term to obtain

$$x(t) = \left[I + \int_{t_o}^{t} A(\sigma_1) \, d\sigma_1 + \int_{t_o}^{t} A(\sigma_1)\int_{t_o}^{\sigma_1} A(\sigma_2) \, d\sigma_2 d\sigma_1 \right.$$

$$\left. + \cdots + \int_{t_o}^{t} A(\sigma_1) \int_{t_o}^{\sigma_1} A(\sigma_2) \cdots \int_{t_o}^{\sigma_{k-1}} A(\sigma_k) \, d\sigma_k \cdots d\sigma_1 + \cdots \right] x_o \qquad (10)$$

Denoting the $n \times n$ matrix series in square brackets by $\Phi(t, t_o)$, the solution just constructed can be written in terms of this *transition matrix* as

$$x(t) = \Phi(t,\ t_o)x_o \tag{11}$$

Since for any x_o the $n \times 1$ vector series $\Phi(t,\ t_o)x_o$ in (8) converges absolutely and uniformly for $t \in [t_o{-}T,\ t_o{+}T]$, where $T > 0$ is arbitrary, it follows that the $n \times n$ matrix series $\Phi(t,\ t_o)$ converges absolutely and uniformly on the same interval. Simply choose $x_o = e_j$, the jth-column of I_n, to prove the convergence properties of the jth-column of $\Phi(t,\ t_o)$.

It is convenient for some purposes to view the transition matrix as a function of two variables, written as $\Phi(t,\ \tau)$, defined by the *Peano-Baker series*

$$\Phi(t,\ \tau) = I + \int_{\tau}^{t} A(\sigma_1)\, d\sigma_1 + \int_{\tau}^{t} A(\sigma_1) \int_{\tau}^{\sigma_1} A(\sigma_2)\, d\sigma_2\, d\sigma_1$$

$$+ \int_{\tau}^{t} A(\sigma_1) \int_{\tau}^{\sigma_1} A(\sigma_2) \int_{\tau}^{\sigma_2} A(\sigma_3)\, d\sigma_3\, d\sigma_2\, d\sigma_1 + \cdots \tag{12}$$

Though we have established convergence properties for fixed τ, it takes a little more work to show the series (12) converges uniformly and absolutely for $t,\ \tau \in [-T,\ T]$, where $T > 0$ is arbitrary. See Exercise 3.10.

By slightly modifying the analysis, it can be shown that the various series considered above converge for any value of t in the whole interval $(-\infty,\ \infty)$. The restriction to finite (though arbitrary) intervals is made to acquire the property of uniform convergence, which implies convenient rules for application of differential and integral calculus.

3.1 Example   For a scalar, time-invariant linear state equation, where we write $A(t) = a$, the approximating sequence in (2) generates

$$x_0(t) = x_o$$

$$x_1(t) = x_o + ax_o\frac{(t{-}t_o)}{1!}$$

$$x_2(t) = x_o + ax_o\frac{(t{-}t_o)}{1!} + a^2 x_o\frac{(t{-}t_o)^2}{2!}$$

and so on. The general term in the sequence is

$$x_k(t) = \left[1 + a\frac{(t{-}t_o)}{1!} + \cdots + a^k\frac{(t{-}t_o)^k}{k!} \right] x_o$$

and the limit of the sequence is the presumably familiar solution

$$x(t) = e^{a(t-t_o)}\, x_o \tag{13}$$

Thus the transition matrix in this case is simply a scalar exponential.

Uniqueness

We next verify that the solution just obtained for the linear state equation (1) with specified t_o and x_o is the only solution. The *Gronwall inequality* is the main tool. Generalizations of this inequality are presented in the exercises, some for use in the sequel.

3.2 Lemma   Suppose that $\phi(t)$ and $v(t)$ are continuous functions defined for $t \geq t_o$ with $v(t)$ nonnegative, and suppose ψ is a constant. Then the implicit inequality

$$\phi(t) \leq \psi + \int_{t_o}^{t} v(\sigma)\phi(\sigma)\, d\sigma , \quad t \geq t_o \tag{14}$$

implies the explicit inequality

$$\phi(t) \leq \psi\, e^{\int_{t_o}^{t} v(\sigma)\, d\sigma} , \quad t \geq t_o \tag{15}$$

Proof   Write the right side of (14) as

$$r(t) = \psi + \int_{t_o}^{t} v(\sigma)\phi(\sigma)\, d\sigma$$

to simplify notation. Then

$$\dot{r}(t) = v(t)\phi(t)$$

and (14) implies, since $v(t)$ is nonnegative,

$$\dot{r}(t) = v(t)\phi(t) \leq v(t)r(t) \tag{16}$$

Multiply both sides of (16) by the positive function

$$e^{-\int_{t_o}^{t} v(\sigma)\, d\sigma}$$

to obtain

$$\frac{d}{dt}\left[r(t)\, e^{-\int_{t_o}^{t} v(\sigma)\, d\sigma} \right] \leq 0 , \quad t \geq t_o$$

Integrating both sides from t_o to any $t \geq t_o$ gives

$$r(t)\, e^{-\int_{t_o}^{t} v(\sigma)\, d\sigma} - \psi \leq 0 , \quad t \geq t_o$$

and this yields (15).
□□□

A proof that there is only one solution of the linear state equation (1) can be accomplished by showing that any two solutions necessarily are identical. Given t_o and

x_o, suppose $x_a(t)$ and $x_b(t)$ both are (continuously-differentiable) solutions of (1) for $t \geq t_o$. Then

$$z(t) = x_a(t) - x_b(t)$$

satisfies

$$\dot{z}(t) = A(t)z(t), \quad z(t_o) = 0 \tag{17}$$

and the objective is to show that (17) implies $z(t) = 0$ for all $t \geq t_o$. (Zero clearly is a solution of (17), but we need to show that it is the only solution in order to elude a vicious circle.)

Integrating both sides of (17) from t_o to any $t \geq t_o$ and taking the norms of both sides of the result yields the inequality

$$\|z(t)\| \leq \int_{t_o}^{t} \|A(\sigma)\| \, \|z(\sigma)\| \, d\sigma$$

Applying Lemma 3.2 (with $\psi = 0$) to this inequality gives immediately that $\|z(t)\| = 0$ for all $t \geq t_o$.

On using a similar demonstration for $t < t_o$, uniqueness of solutions for all t is established. Then the development can be summarized as a result that even the jaded must admit is remarkable, in view of the possible complicated nature of the entries of $A(t)$.

3.3 Theorem   For any t_o and x_o the linear state equation (1), with $A(t)$ continuous, has the unique, continuously-differentiable solution

$$x(t) = \Phi(t, t_o)x_o$$

The transition matrix $\Phi(t, \tau)$ is given by the Peano-Baker series (12) that converges absolutely and uniformly for $t, \tau \in [-T, T]$, where $T > 0$ is arbitrary.

3.4 Example   The properties of existence and uniqueness of solutions that are defined for all t in an arbitrary interval quickly evaporate when nonlinear state equations are considered. Easy substitution verifies that the scalar state equation

$$\dot{x}(t) = x^{1/3}(t), \quad x(0) = 0 \tag{18}$$

has two distinct solutions, $x(t) = (2t/3)^{3/2}$ and $x(t) = 0$, both defined for all t. The scalar state equation

$$\dot{x}(t) = 1 + x^2(t), \quad x(0) = 0 \tag{19}$$

has the solution $x(t) = \tan t$, but only on the time interval $t \in (-\pi/2, \pi/2)$. Specifically this solution blows up at $t = \pm\pi/2$ and there is no continuously differentiable function that satisfies the state equation on any larger interval. Thus we see that Theorem 3.3 is an important foundation for a reasoned theory, and not simply mathematical decoration.
□□□

The Peano-Baker series is a basic theoretical tool for ascertaining properties of linear state equation solutions. We concede that computation of solutions via the Peano-Baker series is a frightening prospect, though calm calculation is profitable in the simplest cases.

3.5 Example   For

$$A(t) = \begin{bmatrix} 0 & t \\ 0 & 0 \end{bmatrix} \tag{20}$$

the Peano-Baker series (12) is

$$\Phi(t,\ \tau) = \begin{bmatrix} 1 & 0 \\ 0 & 1 \end{bmatrix} + \int_\tau^t \begin{bmatrix} 0 & \sigma_1 \\ 0 & 0 \end{bmatrix} d\sigma_1 + \int_\tau^t \begin{bmatrix} 0 & \sigma_1 \\ 0 & 0 \end{bmatrix} \int_\tau^{\sigma_1} \begin{bmatrix} 0 & \sigma_2 \\ 0 & 0 \end{bmatrix} d\sigma_2\ d\sigma_1 + \cdots$$

It is straightforward to verify that all terms in the series beyond the second are zero, and thus

$$\Phi(t,\ \tau) = \begin{bmatrix} 1 & (t^2 - \tau^2)/2 \\ 0 & 1 \end{bmatrix} \tag{21}$$

3.6 Example   For a diagonal $A(t)$, the Peano-Baker series (12) simplifies greatly. Each term of the series is a diagonal matrix, and therefore $\Phi(t,\ \tau)$ is diagonal. The kth-diagonal entry of $\Phi(t,\ \tau)$ has the form

$$\phi_{kk}(t,\ \tau) = 1 + \int_\tau^t a_{kk}(\sigma_1)\ d\sigma_1 + \int_\tau^t a_{kk}(\sigma_1) \int_\tau^{\sigma_1} a_{kk}(\sigma_2)\ d\sigma_2 d\sigma_1 + \cdots$$

where $a_{kk}(t)$ is the kth-diagonal entry of $A(t)$. This expression can be simplified by proving that

$$\int_\tau^t a_{kk}(\sigma_1) \int_\tau^{\sigma_1} a_{kk}(\sigma_2) \cdots \int_\tau^{\sigma_j} a_{kk}(\sigma_{j+1})\ d\sigma_{j+1} \cdots d\sigma_1 = \frac{1}{(j+1)!} \left[\int_\tau^t a_{kk}(\sigma)\ d\sigma \right]^{j+1}$$

To verify this identity note that for any fixed value of τ the two sides agree at $t = \tau$, and the derivatives of the two sides with respect to t (Leibniz rule on the left, chain rule on the right) are identical. Therefore

$$\phi_{kk}(t,\ \tau) = e^{\int_\tau^t a_{kk}(\sigma)\ d\sigma} \tag{22}$$

and consequently $\Phi(t,\ \tau)$ can be written in explicit terms of diagonal entries in $A(t)$.

Complete Solution

The standard approach to considering existence and uniqueness of solutions of

$$\dot{x}(t) = A(t)x(t) + B(t)u(t),\quad x(t_o) = x_o \tag{23}$$

with given t_o, x_o and continuous $u(t)$, $t \geq t_o$, involves using properties of the transition matrix that are discussed in Chapter 4. However the guess-and-verify approach some-times is successful, so in Exercise 3.1 the reader is invited to verify by direct differen-tiation that a solution of (23) is

$$x(t) = \Phi(t, t_o)x_o + \int_{t_o}^{t} \Phi(t, \sigma)B(\sigma)u(\sigma)\, d\sigma, \quad t \geq t_o \tag{24}$$

A little thought shows that this solution is unique since the difference $z(t)$ between any two solutions of (23) must satisfy (17). Thus $z(t)$ must be identically zero.

Taking account of an output equation,

$$y(t) = C(t)x(t) + D(t)u(t) \tag{25}$$

(24) leads to

$$y(t) = C(t)\Phi(t, t_o)x_o + \int_{t_o}^{t} C(t)\Phi(t, \sigma)B(\sigma)u(\sigma)\, d\sigma + D(t)u(t) \tag{26}$$

Under the default assumptions of continuous input signal and continuous state-equation coefficients, $x(t)$ in (24) is continuously differentiable, while $y(t)$ in (26) is continuous. If the assumption on the input signal is relaxed to piecewise continuity, then $x(t)$ is continuous (an exception to our default of continuously differentiable solu-tions) and $y(t)$ is piecewise continuous (continuous if $D(t)$ is zero).

The solution formulas for both $x(t)$ and $y(t)$ comprise two independent com-ponents. The first depends only on the initial state, while the second depends only on the input signal. Adopting an entrenched converse terminology, we call the response component due to the initial state the *zero-input response,* and the component due to the input signal the *zero-state response.* Then the *complete solution* of the linear state equation is the sum of the zero-input and zero-state responses.

The complete solution can be used in conjunction with the general solution of unforced scalar state equations embedded in Example 3.6 to divide and conquer the transition matrix computation in some higher-dimensional cases.

3.7 Example   To compute the transition matrix for

$$A(t) = \begin{bmatrix} 1 & 0 \\ 1 & a(t) \end{bmatrix}$$

write the corresponding pair of scalar equations

$$\dot{x}_1(t) = x_1(t), \quad x_1(t_o) = x_{1o}$$

$$\dot{x}_2(t) = a(t)x_2(t) + x_1(t), \quad x_2(t_o) = x_{2o}$$

From Example 3.1 we have

$$x_1(t) = e^{t-t_o}x_{1o}$$

Then the second scalar equation can be written as a forced scalar state equation $(B(t)u(t) = e^{t-t_o}x_{1o})$

$$\dot{x}_2(t) = a(t)x_2(t) + e^{t-t_o}x_{1o} , \quad x_2(t_o) = x_{2o}$$

The transition matrix for scalar $a(t)$ is computed in Example 3.6, and applying (24) gives

$$x_2(t) = e^{\int_{t_o}^{t} a(\sigma)\,d\sigma} x_{2o} + \int_{t_o}^{t} e^{\int_{\sigma}^{t} a(\tau)\,d\tau} e^{\sigma-t_o}x_{1o}\,d\sigma$$

Repacking into matrix notation yields

$$x(t) = \begin{bmatrix} e^{t-t_o} & 0 \\ \int_{t_o}^{t} \exp[\sigma-t_o+\int_{\sigma}^{t} a(\tau)\,d\tau]\,d\sigma & e^{\int_{t_o}^{t} a(\sigma)\,d\sigma} \end{bmatrix} x_o$$

from which we immediately ascertain $\Phi_A(t, t_o)$.
□ □ □

We close with a few observations on the response properties of the standard linear state equation that are based on the complete solution formulas (24) and (26). Computing the zero-input solution $x(t)$ for the initial state $x_o = e_i$, the ith-column of I_n, at the initial time t_o yields the ith-column of $\Phi(t, t_o)$. Repeating this for the obvious set of n initial states provides the whole matrix function of t, $\Phi(t, t_o)$. However if t_o changes, then the computation in general must be repeated. This can be contrasted with the presumably familiar case of constant A, where knowledge of the transition matrix for any one value of t_o completely determines $\Phi(t, t_o)$ for any other value of t_o. (See Chapter 5.)

Assuming a scalar input for simplicity, the zero-state response for the output with unit impulse input $u(t) = \delta(t - t_o)$ is, from (26),

$$y(t) = C(t)\Phi(t, t_o)B(t_o) + D(t_o)\delta(t - t_o) \tag{27}$$

(We assume that all the effect of the impulse is included under the integral sign in (26). Alternatively we assume that the initial time is t_o^-, and the impulse occurs at time t_o.) Unfortunately the zero-state response to a single impulse occurring at t_o in general provides quite limited information about the response to other inputs. Specifically it is clear from (26) that the zero-state response involves the dependence of the transition matrix on its second argument. Again this can be contrasted with the time-invariant case, where the zero-state response to a single impulse characterizes the zero-state response to all input signals. (Chapter 5, again.)

Finally we review terminology introduced in Chapter 2 from the viewpoint of the complete solution. The state equation (23), (25) is called *linear* because the right sides of both (23) and (25) are linear in the variables $x(t)$ and $u(t)$. Also the solution components in $x(t)$ and $y(t)$ exhibit a linearity property in the following way. The zero-state response is linear in the input signal $u(t)$, and the zero-input response is linear in the initial state x_o. A linear state equation exhibits *causal* input-output behavior because the response $y(t)$ at any $t_a \geq t_o$ does not depend on input values for $t > t_a$. Recall that the response 'waveshape' depends on the initial time in general. More precisely let $y_o(t)$, $t \geq t_o$, be the output signal corresponding to the initial state $x(t_o) = x_o$ and input $u(t)$, $t \geq t_o$. For a new initial time $t_a > t_o$, let $y_a(t)$, $t \geq t_a$, be the output signal corresponding to the same initial state $x(t_a) = x_o$ and the shifted input $u(t - t_a)$, $t \geq t_a$. Then $y_o(t - t_a)$ and $y_a(t)$ in general are not identical. This again is in contrast to the time-invariant case.

EXERCISES

Exercise 3.1   By direct differentiation show that

$$x(t) = \Phi(t, t_o)x_o + \int_{t_o}^{t} \Phi(t, \sigma)B(\sigma)u(\sigma) \, d\sigma$$

is a solution of

$$\dot{x}(t) = A(t)x(t) + B(t)u(t) , \quad x(t_o) = x_o$$

Exercise 3.2   Use term-by-term differentiation of the Peano-Baker series to prove that

$$\frac{\partial}{\partial \tau} \Phi(t, \tau) = - \Phi(t, \tau)A(\tau)$$

Exercise 3.3   If the transition matrix for $A(t)$ is $\Phi_A(t, \tau)$, for what matrix $F(t)$ is $\Phi_F(t, \tau) = \Phi_A^T(-\tau, -t)$?

Exercise 3.4   Show that the inequality

$$\phi(t) \leq \psi(t) + \int_{t_o}^{t} v(\sigma)\phi(\sigma) \, d\sigma , \quad t \geq t_o$$

where $\phi(t)$, $\psi(t)$, $v(t)$ are real, continuous functions with $v(t) \geq 0$ for all $t \geq t_o$, implies

$$\phi(t) \leq \psi(t) + \int_{t_o}^{t} v(\sigma)\psi(\sigma) \, e^{\int_{\sigma}^{t} v(\tau)d\tau} \, d\sigma , \quad t \geq t_o$$

This is called the *Gronwall-Bellman* inequality. (*Hint*: Let

$$r(t) = \int_{t_o}^{t} v(\sigma)\phi(\sigma) \, d\sigma$$

and work with $\dot{r}(t) - v(t)r(t) \leq v(t)\psi(t)$.)

Exercise 3.5   Using the Gronwall-Bellman inequality, show that with the additional assumption that $\psi(t)$ is continuously differentiable,

$$\phi(t) \le \psi(t) + \int_{t_o}^{t} v(\sigma)\phi(\sigma)\,d\sigma\,, \quad t \ge t_o$$

implies

$$\phi(t) \le \psi(t_o)\,e^{\int_{t_o}^{t} v(\sigma)d\sigma} + \int_{t_o}^{t} e^{\int_{\sigma}^{t} v(\tau)d\tau}\,\frac{d}{d\sigma}\,\psi(\sigma)\,d\sigma\,, \quad t \ge t_o$$

Exercise 3.6 Prove the following variation on the Gronwall-Bellman inequality. Suppose ψ is a constant, and $\phi(t)$, $w(t)$, and $v(t)$ are continuous functions with $v(t) \ge 0$ for all $t \ge t_o$. Then

$$\phi(t) \le \psi + \int_{t_o}^{t} w(\sigma) + v(\sigma)\phi(\sigma)\,d\sigma\,, \quad t \ge t_o$$

implies

$$\phi(t) \le \psi\,e^{\int_{t_o}^{t} v(\sigma)d\sigma} + \int_{t_o}^{t} w(\sigma)\,e^{\int_{\sigma}^{t} v(\tau)d\tau}\,d\sigma\,, \quad t \ge t_o$$

Exercise 3.7 Devise an alternate uniqueness proof for linear state equations as follows. Show that if

$$\dot{z}(t) = A(t)z(t)\,, \quad z(t_o) = 0$$

then there is a continuous scalar function $a(t)$ such that

$$\frac{d}{dt}\,\|z(t)\|^2 \le a(t)\|z(t)\|^2$$

Then use an argument similar to one in the proof of Lemma 3.2 to conclude that $z(t) = 0$ for all $t \ge t_o$.

Exercise 3.8 Consider the 'integrodifferential state equation'

$$\dot{x}(t) = A(t)x(t) + \int_{t_o}^{t} E(t,\sigma)x(\sigma)\,d\sigma + B(t)u(t)\,, \quad x(t_o) = x_o$$

where $A(t)$, $E(t,\sigma)$, and $B(t)$ are $n \times n$, $n \times n$, and $n \times m$ continuous matrix functions, respectively. Given x_o, t_o, and a continuous $m \times 1$ input signal $u(t)$ defined for $t \ge t_o$, show that there is at most one (continuously-differentiable) solution. (*Hint:* Consider the equivalent integral equation and rewrite the double-integral term.)

Exercise 3.9 Show that a solution of the matrix differential equation (all matrices $n \times n$)

$$\dot{X}(t) = A_1(t)X(t) + X(t)A_2^T(t) + F(t)\,, \quad X(t_o) = X_o$$

is given by

$$X(t) = \Phi_1(t,\,t_o)X_o\Phi_2^T(t,\,t_o) + \int_{t_o}^{t} \Phi_1(t,\,\sigma)F(\sigma)\Phi_2^T(t,\,\sigma)\,d\sigma$$

For a given $F(t)$ and X_o, is this solution unique?

Exercise 3.10   Use an estimate of

$$\left\| \sum_{j=k+1}^{\infty} \int_{\tau}^{t} A(\sigma_1)\int_{\tau}^{\sigma_1} A(\sigma_2) \cdots \int_{\tau}^{\sigma_{j-1}} A(\sigma_j)\, d\sigma_j \cdots d\sigma_1 \right\|$$

and the definition of uniform convergence of a series to show that the Peano-Baker series converges uniformly to $\Phi(t, \tau)$ for t, $\tau \in [-T, T]$, where $T > 0$ is arbitrary. *Hint*:

$$\frac{\alpha^{k+j}}{(k+j)!} \le \frac{\alpha^k}{k!}\,\frac{\alpha^j}{k^j}$$

Exercise 3.11   For a continuous $n \times n$ matrix function $A(t)$, establish existence of an $n \times n$, continuously-differentiable solution $X(t)$ to the matrix differential equation

$$\dot{X}(t) = A(t)X(t) , \quad X(t_o) = X_o$$

by constructing a suitable sequence of approximate solutions, and showing uniform and absolute convergence on finite intervals of the form $[t_o-T, t_o+T]$.

Exercise 3.12   Consider a linear state equation with specified forcing function

$$\dot{x}(t) = A(t)x(t) + f(t)$$

and specified *two-point boundary conditions*

$$H_o x(t_o) + H_f x(t_f) = h$$

on $x(t)$. Here H_o and H_f are $n \times n$ matrices, h is an $n \times 1$ vector, and $t_f > t_o$. Under what hypotheses does there exist a solution $x(t)$ of the state equation that satisfies the boundary conditions? Under what hypotheses does there exist a unique solution satisfying the boundary conditions? Supposing a solution exists, outline a strategy for computing it under the assumption that you can compute the transition matrix for $A(t)$.

NOTES

Note 3.1   In this chapter we are retracing particular aspects of the classical mathematical topic of ordinary differential equations. Any academic library contains several shelf-feet of reference material. To see the depth and breadth of the subject, consult for instance

P. Hartman, *Ordinary Differential Equations,* Second Edition, Birkhauser, Boston, 1982

The following two books treat the subject at a less-advanced level, and are oriented toward engineering. The first is more introductory than the second.

R.K. Miller, A.N. Michel, *Ordinary Differential Equations,* Academic Press, New York, 1982

D.L. Lukes, *Differential Equations: Classical to Controlled,* Academic Press, New York, 1982

Note 3.2   The default continuity assumptions on linear state equations — adopted to keep technical detail simple — can be weakened without changing the form of the theory. (However, some proofs must be changed.) For example the entries of $A(t)$ might be only piecewise continuous because of switching in the physical system being modeled. In this situation our requirement of continuous-differentiability on solutions is too restrictive, and a continuous $x(t)$ can satisfy the state equation everywhere except for isolated values of t. The books by Hartman and Lukes cited in Note 3.1 treat more general formulations.

Note 3.3 The transition matrix for $A(t)$ can be defined without explicitly involving the Peano-Baker series. This is done by considering the solution of the linear state equation for n linearly independent initial states. Arranging the n solutions as the columns of an $n \times n$ matrix $X(t)$, called a *fundamental matrix,* it can be shown that $\Phi(t, t_o) = X(t)X^{-1}(t_o)$. See, for example, the book by Miller and Michel cited above, or

L.A. Zadeh, C.A. Desoer, *Linear System Theory,* McGraw-Hill, New York, 1963

Use of the Peano-Baker series to define the transition matrix, and develop properties, was emphasized for the system theory community in

R.W. Brockett, *Finite Dimensional Linear Systems,* John Wiley, New York, 1970

Note 3.4 Exercise 3.12 introduces the notion of boundary-value problems in differential equations — an important topic that we do not pursue. For both basic theory and numerical approaches, consult

U.M. Ascher, R.M.M. Mattheij, R.D. Russell, *Numerical Solution of Boundary Value Problems for Ordinary Differential Equations,* Prentice-Hall, Englewood Cliffs, New Jersey, 1988

Note 3.5 Our focus in the next two chapters is on developing theoretical properties of transition matrices. These properties aside there are many commercial simulation packages containing effective, efficient numerical algorithms for solving linear state equations. Via the prosaic device of computing solutions for various initial states, say $e_1, \ldots, e_n$, any of these packages can provide a numerical solution for the transition matrix as a function of one argument. Of course the complete solution of a linear state equation with specified initial state and specified input signal can be calculated and displayed by these simulation packages — often at the click of a mouse in luxurious, colorful window environments.

4

TRANSITION MATRIX
PROPERTIES

Properties of linear state equations rest on properties of transition matrices, and the complicated form of the Peano-Baker series

$$\Phi(t,\ \tau) = I + \int_{\tau}^{t} A(\sigma_1)\ d\sigma_1 + \int_{\tau}^{t} A(\sigma_1) \int_{\tau}^{\sigma_1} A(\sigma_2)\ d\sigma_2 d\sigma_1$$

$$+ \int_{\tau}^{t} A(\sigma_1) \int_{\tau}^{\sigma_1} A(\sigma_2) \int_{\tau}^{\sigma_2} A(\sigma_3)\ d\sigma_3 d\sigma_2 d\sigma_1 + \cdots \qquad (1)$$

tends to mask marvelous features that can be gleaned from careful study. After pointing out two important special cases, general properties of $\Phi(t,\ \tau)$ (holding for any continuous $A(t)$) are developed in this chapter. Further properties in the special cases of constant and periodic $A(t)$ are discussed in Chapter 5.

Two Special Cases

Before developing a list of properties, it might help to connect the general form of the transition matrix to a simpler, perhaps-familiar case. If $A(t) = A$, a constant matrix, then a typical term in the Peano-Baker series becomes

$$\int_{\tau}^{t} A(\sigma_1) \int_{\tau}^{\sigma_1} A(\sigma_2) \int_{\tau}^{\sigma_2} \cdots \int_{\tau}^{\sigma_{k-1}} A(\sigma_k)\ d\sigma_k \cdots d\sigma_1$$

$$= A^k \int_{\tau}^{t} \int_{\tau}^{\sigma_1} \int_{\tau}^{\sigma_2} \cdots \int_{\tau}^{\sigma_{k-1}} 1\ d\sigma_k \cdots d\sigma_1$$

$$= \frac{A^k (t - \tau)^k}{k!}$$

With this observation our first property inherits a convergence proof from the treatment of Peano-Baker series in Chapter 3. However, to emphasize the importance of the time-invariant case, we specialize the general convergence analysis and present the proof again.

4.1 Property   If $A(t) = A$, an $n \times n$ constant matrix, then the transition matrix is

$$\Phi(t, \tau) = e^{A(t - \tau)}$$

where the *matrix exponential* is defined by the power series

$$e^{At} = \sum_{k=0}^{\infty} \frac{1}{k!} A^k t^k \tag{2}$$

that converges uniformly and absolutely on $[-T, T]$, where $T > 0$ is arbitrary.

Proof   On any time interval $[-T, T]$, the matrix functions in the series (2) are bounded according to

$$\frac{\|A^k t^k\|}{k!} \leq \frac{\|A\|^k (2T)^k}{k!}, \quad k = 0, 1, \cdots$$

Since the bounding series of real numbers converges,

$$e^{\|A\| 2T} = \sum_{k=0}^{\infty} \frac{\|A\|^k (2T)^k}{k!}$$

we have from the Weierstrass M-test that the series in (2) converges uniformly and absolutely on $[-T, T]$.
□□□

Because of the convergence properties of the defining power series (2), the matrix exponential e^{At} is analytic on any finite interval. Thus the zero-input solution of a time-invariant linear state equation is analytic on any finite interval.

Properties of the transition matrix in the general case will suggest that $\Phi(t, \tau)$ is as close to being an exponential, without actually being an exponential, as could be hoped. A formula for $\Phi(t, \tau)$ that involves another special class of $A(t)$-matrices supports this prediction, and provides a generalization of the diagonal-$A(t)$ case considered in Example 3.6.

4.2 Property   If for every value of τ and t,

$$A(t) \int_{\tau}^{t} A(\sigma) \, d\sigma = \int_{\tau}^{t} A(\sigma) \, d\sigma \, A(t) \tag{3}$$

then

$$\Phi(t, \tau) = e^{\int_{\tau}^{t} A(\sigma) \, d\sigma} = \sum_{k=0}^{\infty} \frac{1}{k!} \left[\int_{\tau}^{t} A(\sigma) \, d\sigma \right]^k \tag{4}$$

Proof   Our strategy, motivated by Example 3.6, is to show that the commutativity condition (3) implies, for any nonnegative integer j,

$$\int_\tau^t A(\gamma) \left[\int_\tau^\gamma A(\sigma)\, d\sigma \right]^j d\gamma = \frac{1}{j+1} \left[\int_\tau^t A(\sigma)\, d\sigma \right]^{j+1} \tag{5}$$

Then using this identity repeatedly on a general term of the Peano-Baker series (from the right, for $j = 1, 2, \cdots$) gives

$$\int_\tau^t A(\sigma_1) \int_\tau^{\sigma_1} A(\sigma_2) \int_\tau^{\sigma_2} \cdots \left[\int_\tau^{\sigma_{k-2}} A(\sigma_{k-1}) \int_\tau^{\sigma_{k-1}} A(\sigma_k)\, d\sigma_k\, d\sigma_{k-1} \right] d\sigma_{k-2} \cdots d\sigma_1$$

$$= \int_\tau^t A(\sigma_1) \int_\tau^{\sigma_1} A(\sigma_2) \int_\tau^{\sigma_2} \cdots \int_\tau^{\sigma_{k-3}} A(\sigma_{k-2}) \frac{1}{2} \left[\int_\tau^{\sigma_{k-2}} A(\sigma)\, d\sigma \right]^2 d\sigma_{k-2} \cdots d\sigma_1$$

$$= \frac{1}{2} \int_\tau^t A(\sigma_1) \int_\tau^{\sigma_1} A(\sigma_2) \int_\tau^{\sigma_2} \cdots \int_\tau^{\sigma_{k-4}} A(\sigma_{k-3}) \frac{1}{3} \left[\int_\tau^{\sigma_{k-3}} A(\sigma)\, d\sigma \right]^3 d\sigma_{k-3} \cdots d\sigma_1$$

and so on, until we obtain

$$\frac{1}{k!} \left[\int_\tau^t A(\sigma)\, d\sigma \right]^k$$

Of course this is the corresponding general term of the exponential series in (4).

To show (5), first note that it holds at $t = \tau$, for any fixed value of τ. Before continuing, we emphasize again that the tempting chain rule calculation generally is not valid for matrix calculus. However the product rule and Leibniz rule for differentiation are valid, and differentiating the left side of (5) with respect to t gives

$$\frac{\partial}{\partial t} \left[\int_\tau^t A(\gamma) \left[\int_\tau^\gamma A(\sigma)\, d\sigma \right]^j d\gamma \right] = A(t) \left[\int_\tau^t A(\sigma)\, d\sigma \right]^j \tag{6}$$

Differentiating the right side of (5) gives

$$\frac{\partial}{\partial t} \frac{1}{j+1} \left[\int_\tau^t A(\sigma)\, d\sigma \right]^{j+1} = \frac{1}{j+1} \left\{ A(t) \int_\tau^t A(\sigma_2)\, d\sigma_2 \cdots \int_\tau^t A(\sigma_{j+1})\, d\sigma_{j+1} \right.$$

$$+ \int_\tau^t A(\sigma_1)\, d\sigma_1\, A(t) \int_\tau^t A(\sigma_3)\, d\sigma_3 \cdots \int_\tau^t A(\sigma_{j+1})\, d\sigma_{j+1}$$

$$+ \cdots + \int_\tau^t A(\sigma_1)\, d\sigma_1 \cdots \int_\tau^t A(\sigma_j)\, d\sigma_j\, A(t) \left. \right\}$$

$$= A(t) \left[\int_\tau^t A(\sigma)\, d\sigma \right]^j$$

where, in the last step, (3) has been used repeatedly to rewrite each of the $j + 1$ terms in the same form. Therefore we have that the left and right sides of (5) are continuously differentiable, have identical derivatives for all t, and agree at $t = \tau$. Thus the left and right sides of (5) are identical functions of t for any value of τ, and the proof is complete.
□□□

For $n = 1$, where $A(t)$ always commutes with its integral, the 'transition scalar'

$$e^{\int_\tau^t A(\sigma)\, d\sigma}$$

often appears in elementary mathematics courses as an integrating factor in solving linear differential equations. We first encountered this exponential in the proof Lemma 3.2, and then again in Example 3.6.

4.3 Example   For

$$A(t) = \begin{bmatrix} a(t) & a(t) \\ 0 & 0 \end{bmatrix} \tag{7}$$

where $a(t)$ is a continuous scalar function, it is easy to check that the commutativity condition (3) is satisfied. Since

$$\int_\tau^t A(\sigma)\, d\sigma = \begin{bmatrix} \int_\tau^t a(\sigma)\, d\sigma & \int_\tau^t a(\sigma)\, d\sigma \\ 0 & 0 \end{bmatrix}$$

the exponential series (4) is not difficult to sum, giving

$$\Phi(t,\ \tau) = \begin{bmatrix} \exp[\ \int_\tau^t a(\sigma)\, d\sigma\ ] & \exp[\ \int_\tau^t a(\sigma)\, d\sigma\ ] - 1 \\ 0 & 1 \end{bmatrix} \tag{8}$$

If $a(t)$ is a constant, say $a(t) = 2$, then

$$\Phi(t,\ \tau) = e^{A(t-\tau)} = \begin{bmatrix} e^{2(t-\tau)} & e^{2(t-\tau)} - 1 \\ 0 & 1 \end{bmatrix}$$

General Properties

While vector linear differential equations — linear state equations — have been the sole topic so far, it proves useful to also consider matrix differential equations. Given $A(t)$, an $n \times n$ continuous matrix function, consider

$$\frac{d}{dt} X(t) = A(t)X(t), \quad X(t_o) = X_o \tag{9}$$

where $X(t)$ is $n \times n$. Of course this matrix differential equation can be viewed column-by-column, yielding a set of n linear state equations. But a direct matrix representation of the solution is of interest. So with the observation that the column-by-column interpretation trivially yields existence and uniqueness of solutions, the following property is straightforward to verify by differentiation.

4.4 Property   Given X_o and t_o, the linear $n \times n$ matrix differential equation (9) has the unique, continuously-differentiable solution

$$X(t) = \Phi_A(t, t_o)X_o \tag{10}$$

Notice that when the initial condition matrix is the identity, $X_o = I$, then the unique solution is $X(t) = \Phi_A(t, t_o)$. This characterization of the transition matrix is often useful.

Property 4.4, as well as the solution of the linear state equation

$$\dot{x}(t) = A(t)x(t) , \quad x(t_o) = x_o \tag{11}$$

focus on the behavior of the transition matrix $\Phi(t, \tau)$ as a function of its first argument. It is not difficult to pose a differential equation whose solution displays the behavior of $\Phi(t, \tau)$ with respect to the second argument.

4.5 Property   Given Z_o and t_o, the linear $n \times n$ matrix differential equation

$$\frac{d}{dt} Z(t) = -A^T(t)Z(t) , \quad Z(t_o) = Z_o \tag{12}$$

has the unique, continuously-differentiable solution

$$Z(t) = \Phi_A^T(t_o, t)Z_o \tag{13}$$

Verification of this property is left as an exercise, with the note that Exercise 3.2 provides the key to differentiating $Z(t)$. The associated $n \times 1$ linear state equation

$$\dot{z}(t) = -A^T(t)z(t) , \quad z(t_o) = z_o$$

is called the *adjoint state equation* for the linear state equation (11). Obviously the unique solution of the adjoint state equation is

$$z(t) = \Phi_{-A^T}(t, t_o)z_o = \Phi_A^T(t_o, t)z_o$$

4.6 Example   For

$$A(t) = \begin{bmatrix} 1 & \cos t \\ 0 & 0 \end{bmatrix} \tag{14}$$

Property 4.2 does not apply. Writing out the first four terms of the Peano-Baker series gives

$$\Phi_A(t,\ 0) = \begin{bmatrix} 1 & 0 \\ 0 & 1 \end{bmatrix} + \begin{bmatrix} t & \sin t \\ 0 & 0 \end{bmatrix} + \begin{bmatrix} t^2/2 & 1-\cos t \\ 0 & 0 \end{bmatrix}$$
$$+ \begin{bmatrix} t^3/3! & t-\sin t \\ 0 & 0 \end{bmatrix} + \cdots$$

where $\tau = 0$ has been assumed for simplicity. It is dangerous to guess the sum of this series, particularly the 1,2-entry, but Property 4.4 provides the relation

$$\frac{d}{dt}\Phi_A(t,\ 0) = A(t)\Phi_A(t,\ 0)\ ,\quad \Phi_A(0,\ 0) = I$$

that aids intelligent conjecture. Indeed,

$$\Phi_A(t,\ 0) = \begin{bmatrix} e^t & \frac{1}{2}(e^t + \sin t - \cos t) \\ 0 & 1 \end{bmatrix} \tag{15}$$

This is not quite enough to provide $\Phi_A(t,\ \tau)$ as an explicit function of τ, and therefore Property 4.5 cannot be used to obtain for free the transition matrix for

$$-A^T(t) = \begin{bmatrix} -1 & 0 \\ -\cos t & 0 \end{bmatrix}$$

However writing out the first few terms of the relevant Peano-Baker series and guessing with the aid of Property 4.5 yields

$$\Phi_{-A^T}(t,\ 0) = \begin{bmatrix} e^{-t} & 0 \\ -\frac{1}{2}+\frac{1}{2}e^{-t}(\cos t - \sin t) & 1 \end{bmatrix}$$

□□□

Property 4.4 leads directly to a clever proof of the following *composition property*. (Attempting a brute-force proof using the Peano-Baker series is not recommended.)

4.7 Property   For any values of t, τ, and σ, the transition matrix for $A(t)$ satisfies

$$\Phi(t,\ \tau) = \Phi(t,\ \sigma)\ \Phi(\sigma,\ \tau) \tag{16}$$

Proof   Choosing arbitrary but fixed values of τ and σ, let $R(t) = \Phi(t,\ \sigma)\ \Phi(\sigma,\ \tau)$. Then for all t,

$$\frac{d}{dt}R(t) = A(t)\Phi(t,\ \sigma)\ \Phi(\sigma,\ \tau) = A(t)R(t)$$

and, of course,

$$\frac{d}{dt}\Phi(t,\ \tau) = A(t)\Phi(t,\ \tau)$$

Also the "initial conditions" at $t = \sigma$ are the same for $R(t)$ and $\Phi(t,\ \tau)$ since $R(\sigma) = \Phi(\sigma,\ \sigma)\ \Phi(\sigma,\ \tau) = \Phi(\sigma,\ \tau)$. Then by the uniqueness of solutions to linear

matrix differential equations, we have $R(t) = \Phi(t, \tau)$, for all t. Since this argument works for every value of τ and σ, the proof is complete.
□□□

The approach in this proof is a useful extension of the approach in the proof of Property 4.2. That is, to prove that two continuously-differentiable functions are identical show that they agree at one point, that they satisfy the same linear differential equation, and then invoke uniqueness of solutions.

Property 4.7 can be interpreted in terms of a composition rule for solutions of the corresponding linear state equation (11). For example in (16) let $\tau = t_o$, $\sigma = t_1 > t_o$, and $t = t_2 > t_1$. Then, as illustrated in Figure 4.8, the composition property implies that the solution of (11) at time t_2 can be represented as

$$x(t_2) = \Phi(t_2, t_o)x(t_o)$$

or as

$$x(t_2) = \Phi(t_2, t_1)x(t_1)$$

where

$$x(t_1) = \Phi(t_1, t_o)x(t_o)$$

This interpretation also applies when, for instance, $t_1 < t_o$ by following trajectories backward in time.

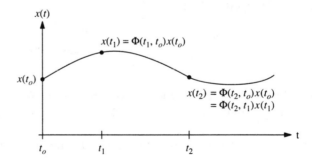

4.8 Figure   An application of the composition property.

The composition property can be applied to establish invertibility of transition matrices, but the next property and its proof are of surpassing elegance in this regard. (Recall the definition of the trace of a matrix in Chapter 1.)

4.9 Property   For any values of t and τ,

$$\det \Phi(t, \tau) = e^{\int_\tau^t \operatorname{tr}[A(\sigma)]\, d\sigma} \tag{17}$$

Proof   The key to the proof is to show that for any fixed τ the scalar function $det\ \Phi(t,\ \tau)$ satisfies the scalar differential equation

$$\frac{d}{dt}\ det\ \Phi(t,\ \tau) = tr\ [A\ (t)]\ det\ \Phi(t,\ \tau)\ ,\quad det\ \Phi(\tau,\ \tau) = 1 \tag{18}$$

Then (17) follows from Property 4.2, that is, from the solution of the scalar differential equation (18).

      To proceed with differentiation of $det\ \Phi(t,\ \tau)$, where τ is fixed, we use the chain rule with the following notation. Let $c_{ij}(t,\ \tau)$ be the cofactor of the entry $\phi_{ij}(t,\ \tau)$ of $\Phi(t,\ \tau)$, and denote the i,j-entry of the transpose of the cofactor matrix $C\ (t,\ \tau)$ by $c_{ij}^T(t,\ \tau)$. (That is, $c_{ij}^T = c_{ji}$.) Recognizing that the determinant is a differentiable function of matrix entries, in particular it is a sum of products of entries, the chain rule gives

$$\frac{d}{dt}\ det\ \Phi(t,\ \tau) = \sum_{i=1}^{n} \sum_{j=1}^{n} \left[\frac{\partial}{\partial \phi_{ij}}\ det\ \Phi(t,\ \tau) \right] \dot{\phi}_{ij}(t,\ \tau) \tag{19}$$

For any $j = 1, \ldots,\ n$, computation of the Laplace expansion of the determinant along the jth column gives

$$det\ \Phi(t,\ \tau) = \sum_{i=1}^{n} c_{ij}(t,\ \tau)\phi_{ij}(t,\ \tau)$$

so that

$$\frac{\partial}{\partial \phi_{ij}}\ det\ \Phi(t,\ \tau) = c_{ij}(t,\ \tau)$$

Therefore

$$\frac{d}{dt}\ det\ \Phi(t,\ \tau) = \sum_{i=1}^{n} \sum_{j=1}^{n} c_{ij}(t,\ \tau)\ \dot{\phi}_{ij}(t,\ \tau)$$

$$= \sum_{j=1}^{n} \sum_{i=1}^{n} c_{ji}^T(t,\ \tau)\ \dot{\phi}_{ij}(t,\ \tau)$$

The double summation on the right side can be rewritten to obtain

$$\frac{d}{dt}\ det\ \Phi(t,\ \tau) = tr\ \left[C^T(t,\ \tau)\ \frac{d}{dt}\ \Phi(t,\ \tau) \right]$$

$$= tr\ \left[C^T(t,\ \tau)\ A\ (t)\ \Phi(t,\ \tau) \right]$$

$$= tr\ \left[\Phi(t,\ \tau)\ C^T(t,\ \tau)\ A\ (t) \right] \tag{20}$$

(The last step uses the fact that the trace of a product of square matrices is independent of the ordering of the product.) Now the identity

$$I \det \Phi(t, \tau) = \Phi(t, \tau)C^T(t, \tau)$$

which is a consequence of the Laplace expansion of the determinant, gives

$$\frac{d}{dt} \det \Phi(t, \tau) = \text{tr} [A(t)] \det \Phi(t, \tau)$$

Since, trivially, $\det \Phi(\tau, \tau) = 1$, the proof is complete.

4.10 Property   The transition matrix is invertible for any values of t and τ, and

$$\Phi^{-1}(t, \tau) = \Phi(\tau, t) \tag{21}$$

Proof   Invertibility follows from Property 4.9, since $A(t)$ is continuous and thus the exponent is finite for any finite t and τ. The formula for the inverse follows from Property 4.7 by taking $t = \tau$ in (16).

4.11 Example   These last few properties provide the steps needed to compute the transition matrices in Example 4.6 as functions of two arguments. Beginning with

$$A(t) = \begin{bmatrix} 1 & \cos t \\ 0 & 0 \end{bmatrix}, \quad \Phi_A(t, 0) = \begin{bmatrix} e^t & 1/2(\sin t - \cos t + e^t) \\ 0 & 1 \end{bmatrix} \tag{22}$$

From Property 4.7,

$$\Phi_A(t, \tau) = \Phi_A(t, 0)\Phi_A(0, \tau)$$

and then Property 4.10 gives, after computing the inverse of $\Phi_A(t, 0)$,

$$\Phi_A(t, \tau) = \Phi_A(t, 0)\Phi_A^{-1}(\tau, 0)$$

$$= \begin{bmatrix} e^t & 1/2(e^t + \sin t - \cos t) \\ 0 & 1 \end{bmatrix} \begin{bmatrix} e^{-\tau} & -1/2(1 + e^{-\tau}\sin \tau - e^{-\tau}\cos \tau) \\ 0 & 1 \end{bmatrix}$$

$$= \begin{bmatrix} e^{t-\tau} & 1/2 e^{t-\tau}(\cos \tau - \sin \tau) + 1/2(\sin t - \cos t) \\ 0 & 1 \end{bmatrix} \tag{23}$$

Alternatively we can obtain $\Phi_A(0, \tau)$ from Example 4.6 as $[\Phi_{-A^T}(\tau, 0)]^T$. Similarly $\Phi_{-A^T}(t, \tau)$ can be computed directly from $\Phi_A(t, \tau)$ via Property 4.5.

Variable Changes

Often changes of state variables are of interest, and to stay within the class of linear state equations, only linear, time-dependent variable changes are considered. That is, for

$$\dot{x}(t) = A(t)x(t), \quad x(t_o) = x_o \tag{24}$$

suppose a new state vector is defined by

$$z(t) = P^{-1}(t)x(t)$$

where the $n \times n$ matrix $P(t)$ is invertible and continuously differentiable at each t.

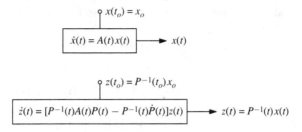

4.12 Figure   State variable change produces an equivalent linear state equation.

(Both assumptions are used explicitly in the following.) To find the state equation in terms of $z(t)$, write $x(t) = P(t)z(t)$ and differentiate to obtain

$$\dot{x}(t) = P(t)\dot{z}(t) + \dot{P}(t)z(t)$$

Also $A(t)x(t) = A(t)P(t)z(t)$, so substituting into the original state equation leads to

$$\dot{z}(t) = [\, P^{-1}(t)A(t)P(t) - P^{-1}(t)\dot{P}(t)\,]z(t) \,, \quad z(t_o) = P^{-1}(t_o)x_o \qquad (25)$$

This little calculation, and the juxtaposition of the linear state equations (24) and (25) in Figure 4.12, should motivate the relation between the respective transition matrices.

4.13 Property   Suppose $P(t)$ is a continuously-differentiable, $n \times n$ matrix function such that $P^{-1}(t)$ exists at each value of t. Then the transition matrix for

$$F(t) = P^{-1}(t)A(t)P(t) - P^{-1}(t)\dot{P}(t) \qquad (26)$$

is given by

$$\Phi_F(t, \tau) = P^{-1}(t)\Phi_A(t, \tau)P(\tau) \qquad (27)$$

Proof   First note that $F(t)$ in (26) is continuous, so the default assumptions are maintained. Then, for arbitrary but fixed τ, let

$$X(t) = P^{-1}(t)\ \Phi_A(t, \tau)P(\tau)$$

Clearly $X(\tau) = I$, and differentiating with the aid of Exercise 1.12 gives

$$\dot{X}(t) = -\ P^{-1}(t)\dot{P}(t)P^{-1}(t)\Phi_A(t, \tau)P(\tau) + P^{-1}(t)A(t)\Phi_A(t, \tau)P(\tau)$$

$$= \left[P^{-1}(t)A(t)P(t) - P^{-1}(t)\dot{P}(t) \right] P^{-1}(t)\Phi_A(t, \tau)P(\tau)$$

$$= F(t)X(t)$$

Since this is valid for any τ, by the characterization of transition matrices provided in Property 4.4 the proof is complete.

4.14 Example   A state variable change can be used to derive the solution of a linear state equation with nonzero input 'guessed' in Chapter 3. Given

$$\dot{x}(t) = A(t)x(t) + B(t)u(t) , \quad x(t_o) = x_o \tag{28}$$

Let

$$z(t) = P^{-1}(t)x(t) = \Phi^{-1}(t, t_o)x(t)$$

where it is clear that $P(t) = \Phi(t, t_o)$ satisfies all the hypotheses required for a state variable change. Substituting into (28) gives

$$A(t)\Phi(t, t_o)z(t) + \Phi(t, t_o)\dot{z}(t) = A(t)\Phi(t, t_o)z(t) + B(t)u(t) , \quad z(t_o) = x_o$$

or

$$\dot{z}(t) = \Phi^{-1}(t, t_o)B(t)u(t) , \quad z(t_o) = x_o \tag{29}$$

Both sides can be integrated from t_o to t to obtain

$$z(t) - x_o = \int_{t_o}^{t} \Phi^{-1}(\sigma, t_o)B(\sigma)u(\sigma)\, d\sigma$$

Replacing $z(t)$ by $P^{-1}(t)x(t)$ and rearranging using properties of the transition matrix gives

$$x(t) = \Phi(t, t_o)x_o + \int_{t_o}^{t} \Phi(t, \sigma)B(\sigma)u(\sigma)\, d\sigma$$

Of course if there is an output equation

$$y(t) = C(t)x(t) + D(t)u(t)$$

then we obtain immediately the complete solution formula for the output signal:

$$y(t) = C(t)\Phi(t, t_o)x_o + \int_{t_o}^{t} C(t)\Phi(t, \sigma)B(\sigma)u(\sigma)\, d\sigma + D(t)u(t) \tag{30}$$

This variable change approach can be viewed as an 'integrating factor' approach, as so often used in the scalar case. An expression equivalent to (28) is

$$\Phi^{-1}(t, t_o)\left[\dot{x}(t) - A(t)x(t) \right] = \Phi^{-1}(t, t_o)B(t)u(t) , \quad x(t_o) = x_o$$

and this simply is another form of (29).

EXERCISES

Exercise 4.1   If the $n \times n$ matrix function $X(t)$ is a solution of the matrix differential equation

$$\dot{X}(t) = A(t)X(t) , \quad X(t_o) = X_o$$

show that
(a) if X_o is invertible, then $X(t)$ is invertible for all t,
(b) if X_o is invertible, then for any t and τ the transition matrix for $A(t)$ is given by

$$\Phi(t, \tau) = X(t)X^{-1}(\tau)$$

Exercise 4.2   If $x(t)$ and $z(t)$ are the respective solutions of a linear state equation and its adjoint state equation, with initial conditions $x(t_o) = x_o$ and $z(t_o) = z_o$, derive a formula for $z^T(t)x(t)$.

Exercise 4.3   Compute the adjoint of the nth-order scalar differential equation

$$y^{(n)}(t) + a_{n-1}(t)y^{(n-1)}(t) + \cdots + a_0(t)y(t) = 0$$

by converting the adjoint of the corresponding linear state equation back into an nth-order scalar differential equation.

Exercise 4.4   For the time-invariant linear state equation

$$\dot{x}(t) = Ax(t) , \quad x(0) = x_o$$

show that given an x_o there exists a constant α such that

$$\det \left[x(t) \ Ax(t) \ \cdots \ A^{n-1}x(t) \right] = \alpha e^{\text{tr}[A]t}$$

Exercise 4.5   For the linear state equation

$$\dot{x}(t) = A(t)x(t) , \quad x(t_o) = x_o$$

show that

$$\|x(t)\| \le \|x_o\| e^{\int_{t_o}^{t} \|A(\sigma)\| d\sigma} , \quad t \ge t_o$$

Exercise 4.6   Show that

$$X(t) = e^{\int_0^t A(\sigma) d\sigma} F$$

is a solution of the $n \times n$ matrix equation

$$\dot{X}(t) = A(t)X(t)$$

if F is a constant matrix that satisfies

$$\left[A(t)[\int_0^t A(\sigma)\, d\sigma]^k - [\int_0^t A(\sigma)\, d\sigma]^k A(t) \right] F = 0 \,, \quad k = 1, 2, \cdots$$

(This can be useful if F has many zero entries.)

Exercise 4.7   For a continuous $n \times n$ matrix $A(t)$, prove that

$$A(t) \int_\tau^t A(\sigma)\, d\sigma = \int_\tau^t A(\sigma)\, d\sigma\, A(t)$$

for all t and τ if and only if

$$A(t)A(\tau) = A(\tau)A(t)$$

for all t and τ.

Exercise 4.8   Compute $\Phi(t, 0)$ for

$$A(t) = \begin{bmatrix} 0 & a(t) \\ -a(t) & 0 \end{bmatrix}$$

where $a(t)$ is a continuous scalar function. (*Hint*: Recognize the subsequences of even powers and odd powers.)

Exercise 4.9   Show that the time-varying linear state equation

$$\dot{x}(t) = A(t)x(t)$$

can be transformed to a time-invariant linear state equation by a state variable change if and only if the transition matrix for $A(t)$ can be written in the form

$$\Phi(t, 0) = T(t)e^{Rt}$$

where R is an $n \times n$ constant matrix, and $T(t)$ is $n \times n$ and invertible at each t.

Exercise 4.10   Suppose $A(t)$ is $n \times n$ and continuously differentiable. Prove that the transition matrix for $A(t)$ can be written as

$$\Phi(t, 0) = e^{A_1 t}\, e^{A_2 t}$$

where A_1 and A_2 are constant $n \times n$ matrices, if and only if

$$\dot{A}(t) = A_1 A(t) - A(t)A_1 \,, \quad A(0) = A_1 + A_2$$

Exercise 4.11   Suppose A_1 and A_2 are constant $n \times n$ matrices, and that $A(t)$ satisfies

$$\dot{A}(t) = A_1 A(t) - A(t)A_1 \,, \quad A(0) = A_1 + A_2$$

Show that the linear state equation $\dot{x}(t) = A(t)x(t)$ can be transformed to $\dot{z}(t) = A_2 z(t)$ by a state variable change.

Exercise 4.12   Show that if $A(t)$ is partitioned as

$$A(t) = \begin{bmatrix} A_{11}(t) & A_{12}(t) \\ 0 & A_{22}(t) \end{bmatrix}$$

where $A_{11}(t)$ and $A_{22}(t)$ are square, then

$$\Phi(t, \tau) = \begin{bmatrix} \Phi_{11}(t, \tau) & \Phi_{12}(t, \tau) \\ 0 & \Phi_{22}(t, \tau) \end{bmatrix}$$

where

$$\frac{\partial}{\partial t} \Phi_{jj}(t, \tau) = A_{jj}(t)\Phi_{jj}(t, \tau), \quad j = 1, 2$$

Can you find an expression for $\Phi_{12}(t, \tau)$ in terms of $\Phi_{11}(t, \tau)$ and $\Phi_{22}(t, \tau)$? (*Hint*: Use Exercise 3.9.)

Exercise 4.13 Using Exercise 4.12, prove that

$$F(t) = e^{At} \int_0^t e^{-A\sigma} B \, d\sigma$$

is given by $\Phi_{12}(t, 0)$, the upper-right partition of the transition matrix for

$$\begin{bmatrix} A & B \\ 0 & 0 \end{bmatrix}$$

Exercise 4.14 Compute $\Phi(t, 0)$ for

$$A(t) = \begin{bmatrix} -1 & e^{2t} \\ 0 & -1 \end{bmatrix}$$

What are the pointwise-in-time eigenvalues of $A(t)$? For every initial state x_o, are solutions of

$$\dot{x}(t) = A(t)x(t), \quad x(0) = x_o$$

bounded for $t \geq 0$?

Exercise 4.15 Show that the linear state equations

$$\dot{x}(t) = \begin{bmatrix} 0 & 1 \\ 2 - t^2 & 2t \end{bmatrix} x(t)$$

and

$$\dot{z}(t) = \begin{bmatrix} t & 1 \\ 1 & t \end{bmatrix} z(t)$$

are related by a change of state variables.

Exercise 4.16 For A and B constant, $n \times n$ matrices, show that the transition matrix for the linear state equation

$$\dot{x}(t) = e^{-At}Be^{At} x(t)$$

is

$$\Phi(t, t_o) = e^{-At} e^{(A+B)(t - t_o)} e^{At_o}$$

Exercise 4.17   Show that the transition matrix for $A_1(t) + A_2(t)$ can be written as

$$\Phi_{A_1+A_2}(t, \tau) = \Phi_{A_1}(t, 0)\Phi_{A_3}(t, \tau)\Phi_{A_1}(0, \tau)$$

where

$$A_3(t) = \Phi_{A_1}(0, t)A_2(t)\Phi_{A_1}(t, 0)$$

Exercise 4.18   Given a continuous $n \times n$ matrix $A(t)$ and a constant $n \times n$ matrix F, show how to define a state variable change that transforms the linear state equation

$$\dot{x}(t) = A(t)x(t)$$

into

$$\dot{z}(t) = Fz(t)$$

Exercise 4.19   For the linear state equation

$$\dot{x}(t) = A(t)x(t) + B(t)u(t), \quad x(t_o) = x_o$$

$$y(t) = C(t)x(t) + D(t)u(t)$$

suppose state variables are changed according to $z(t) = P^{-1}(t)x(t)$. If $z(t_o) = P^{-1}(t_o)x_o$, show directly from the complete solution formula that for any $u(t)$ the response $y(t)$ of the two state equations is identical.

Exercise 4.20   If there exists a constant α such that $\|A(t)\| \le \alpha$ for all t, prove that the transition matrix for $A(t)$ can be written as

$$\Phi(t+\sigma, \sigma) = e^{\bar{A}_t(\sigma)t} + R(t, \sigma), \quad t, \sigma > 0$$

where $\bar{A}_t(\sigma)$ is an "average"

$$\bar{A}_t(\sigma) = \frac{1}{t} \int_\sigma^{\sigma+t} A(\tau) \, d\tau$$

and $R(t, \sigma)$ satisfies

$$\|R(t, \sigma)\| \le \alpha^2 t^2 e^{\alpha t}, \quad t, \sigma > 0$$

NOTES

Note 4.1   The exponential nature of the transition matrix when $A(t)$ commutes with its integral, Property 4.2, is discussed in greater generality and detail in Chapter 7 of

D.L. Lukes, *Differential Equations: Classical to Controlled,* Academic Press, New York, 1982

Changes of state variable yielding a new state equation that satisfies the commutativity condition are considered in

J. Zhu, C.D. Johnson, "New results in the reduction of linear time-varying dynamical systems," *SIAM Journal on Control and Optimization,* Vol. 27, No. 3, pp. 476–494, 1989

Note 4.2 A power series representation for the transition matrix is derived in

W.B. Blair, ''Series solution to the general linear time varying system,'' *IEEE Transactions on Automatic Control,* Vol. 16, No. 2, pp. 210–211, 1971

Note 4.3 Higher-order $n \times n$ matrix differential equations also can be considered. See, for example,

T.M. Apostol, ''Explicit formulas for solutions of the second-order matrix differential equation $Y''(t) = AY(t)$,'' *American Mathematical Monthly,* Vol. 82, No. 2, pp. 159–162, 1975

Note 4.4 The notion of an adjoint state equation can be connected to the concept of the adjoint of a linear map on an inner product space. Exercise 4.2 indicates this connection, on viewing $z^T x$ as an inner product on R^n. For further discussion of the linear-system aspects of adjoints, see Section 9.3 of

T. Kailath, *Linear Systems,* Prentice Hall, Englewood Cliffs, New Jersey, 1980

5

TWO IMPORTANT CASES

Two classes of transition matrices are addressed in further detail in this chapter. The first is the case of constant $A(t)$, and the second is where $A(t)$ is a periodic matrix function of time. Special properties of the corresponding transition matrices are developed, and implications are drawn for the response characteristics of the associated linear state equations.

Constant Case

When $A(t) = A$, a constant $n \times n$ matrix, the transition matrix is the matrix exponential

$$\Phi(t, \tau) = e^{A(t-\tau)} = \sum_{k=0}^{\infty} \frac{1}{k!} A^k (t-\tau)^k \tag{1}$$

We first list properties of matrix exponentials that specialize from general transition matrix properties in Chapter 4, and then introduce some that do not. Since only the difference of arguments $(t - \tau)$ appears in (1), one variable can be discarded with no loss of generality. Therefore in the matrix exponential case we work with

$$\Phi(t, 0) = e^{At} \tag{2}$$

As noted in Chapter 4 this matrix exponential is an analytic function of t on any finite time interval.

The following properties are easy specializations of the properties in Chapter 4. In particular the first provides a characterization of the matrix exponential on choosing the initial condition $X_o = I$.

5.1 Property   The $n \times n$ matrix differential equation

$$\dot{X}(t) = AX(t), \quad X(0) = X_o \tag{3}$$

has the unique solution

$$X(t) = e^{At}X_o$$

5.2 Property   The $n \times n$ matrix differential equation

$$\dot{Z}(t) = -A^T Z(t) , \quad Z(0) = Z_o$$

has the unique solution

$$Z(t) = e^{-A^T t}Z_o$$

5.3 Property   For any t_a and t_b,

$$e^{A(t_a + t_b)} = e^{At_a}e^{At_b}$$

5.4 Property   For any t,

$$\det e^{At} = e^{\text{tr}[A]t}$$

5.5 Property   The matrix exponential is invertible for any t (regardless of A), and

$$\left[e^{At} \right]^{-1} = e^{-At}$$

5.6 Property   If P is an invertible, constant $n \times n$ matrix, then

$$e^{P^{-1}APt} = P^{-1}e^{At}P$$

Several additional properties of matrix exponentials do not devolve from general properties of transition matrices, but depend on specific features of the power series defining the matrix exponential. A few of the most important are developed in detail, with others left to the exercises.

5.7 Property   If A and F are $n \times n$ matrices, then

$$e^{At}e^{Ft} = e^{(A + F)t} , \quad \text{for all } t \tag{4}$$

if and only if $AF = FA$.

Proof   Assuming $AF = FA$, first note that

$$e^{(A + F)t}\bigg|_{t = 0} = e^{At}e^{Ft}\bigg|_{t = 0} = I$$

Since F commutes also with positive powers of A, and thus commutes with the terms in the power series for e^{At},

$$\frac{d}{dt} e^{At} e^{Ft} = A e^{At} e^{Ft} + e^{At} F e^{Ft}$$

$$= (A + F) e^{At} e^{Ft}$$

Clearly $e^{(A + F)t}$ satisfies the same linear matrix differential equation, and by uniqueness of solutions we have (4).

Conversely if (4) holds, then differentiating both sides twice gives

$$A^2 e^{At} e^{Ft} + A e^{At} F e^{Ft} + A e^{At} F e^{Ft} + e^{At} F^2 e^{Ft} = (A + F)^2 e^{(A + F)t}$$

Evaluating at $t = 0$ yields

$$A^2 + 2AF + F^2 = (A + F)^2$$

$$= A^2 + AF + FA + F^2$$

Subtracting $A^2 + AF + F^2$ from both sides shows that A and F commute.

5.8 Property   There exist analytic scalar functions $\alpha_0(t), \ldots, \alpha_{n-1}(t)$ such that

$$e^{At} = \sum_{k=0}^{n-1} \alpha_k(t) A^k \tag{5}$$

Proof   Using Property 5.1, the matrix differential equation characterizing the matrix exponential, we can establish (5) by computing analytic functions $\alpha_0(t), \ldots, \alpha_{n-1}(t)$ such that

$$\sum_{k=0}^{n-1} \dot{\alpha}_k(t) A^k = \sum_{k=0}^{n-1} \alpha_k(t) A^{k+1}, \quad \sum_{k=0}^{n-1} \alpha_k(0) A^k = I \tag{6}$$

The Cayley Hamilton theorem implies

$$A^n = -a_0 I - a_1 A - \cdots - a_{n-1} A^{n-1}$$

where $a_0, \ldots, a_{n-1}$ are the coefficients in the characteristic polynomial of A. Then (6) can be written solely in terms of $I, A, \ldots, A^{n-1}$ as

$$\sum_{k=0}^{n-1} \dot{\alpha}_k(t) A^k = \sum_{k=0}^{n-2} \alpha_k(t) A^{k+1} - \sum_{k=0}^{n-1} a_k \alpha_{n-1}(t) A^k$$

$$= -a_0 \alpha_{n-1}(t) I + \sum_{k=1}^{n-1} [\alpha_{k-1}(t) - a_k \alpha_{n-1}(t)] A^k, \quad \sum_{k=0}^{n-1} \alpha_k(0) A^k = I \tag{7}$$

The astute observation to be made is that (7) can be solved by considering the coefficient equation for each power of A separately. Equating coefficients of like powers of A yields the time-invariant linear state equation

$$
\begin{bmatrix} \dot\alpha_0(t) \\ \dot\alpha_1(t) \\ \vdots \\ \dot\alpha_{n-1}(t) \end{bmatrix} = \begin{bmatrix} 0 & 0 & \cdots & 0 & -a_0 \\ 1 & 0 & \cdots & 0 & -a_1 \\ \vdots & \vdots & \vdots & \vdots & \vdots \\ 0 & 0 & \cdots & 1 & -a_{n-1} \end{bmatrix} \begin{bmatrix} \alpha_0(t) \\ \alpha_1(t) \\ \vdots \\ \alpha_{n-1}(t) \end{bmatrix} , \quad \begin{bmatrix} \alpha_0(0) \\ \alpha_1(0) \\ \vdots \\ \alpha_{n-1}(0) \end{bmatrix} = \begin{bmatrix} 1 \\ 0 \\ \vdots \\ 0 \end{bmatrix}
$$

Thus existence of an analytic solution to this linear state equation shows existence of analytic functions $\alpha_0(t), \ldots, \alpha_{n-1}(t)$ that satisfy (7), and hence (6).
□□□

The Laplace transform can be used to develop a more-or-less explicit form for the matrix exponential that provides more insight than the power series definition. We need only deal with Laplace transforms that are rational functions of s, that is, ratios of poly-nomials in s. Recall the terminology that a rational function is *proper* if the degree of the numerator polynomial is no greater than the degree of the denominator polynomial, and *strictly proper* if the numerator polynomial degree is strictly less than the denomi-nator polynomial degree.

Taking the Laplace transform of both sides of the $n \times n$ matrix differential equa-tion

$$
\dot X(t) = AX(t) , \quad X(0) = I
$$

gives, after rearrangement,

$$
\mathsf{X}(s) = (sI - A)^{-1}
$$

Thus, by uniqueness properties of Laplace transforms, and uniqueness of solutions of linear matrix differential equations, the Laplace transform of e^{At} is $(sI - A)^{-1}$. This is an $n \times n$ matrix of strictly-proper rational functions of s, as is clear from counting polynomial-entry degrees in the formula

$$
(sI - A)^{-1} = \frac{\text{adj } (sI - A)}{\det (sI - A)} \tag{8}
$$

Specifically $\det (sI - A)$ is a degree-n polynomial in s, while each entry of $\text{adj } (sI - A)$ is a polynomial of degree at most $n-1$. Now suppose

$$
\det (sI - A) = (s - \lambda_1)^{\sigma_1} \cdots (s - \lambda_m)^{\sigma_m}
$$

where $\lambda_1, \ldots, \lambda_m$ are the distinct eigenvalues of A, with corresponding multiplicities $\sigma_1, \ldots, \sigma_m \geq 1$. Then the partial fraction expansion of each entry in $(sI - A)^{-1}$ gives

$$
(sI - A)^{-1} = \sum_{k=1}^{m} \sum_{j=1}^{\sigma_k} W_{kj} \frac{1}{(s - \lambda_k)^j}
$$

where each W_{kj} is an $n \times n$ matrix of partial fraction expansion coefficients. That is, each entry of W_{kj} is the coefficient of $1/(s - \lambda_k)^j$ in the expansion of the corresponding entry in the matrix $(sI - A)^{-1}$. (The matrix W_{kj} is complex if the corresponding eigen-value λ_k is complex.) In fact, using a formula for partial fraction expansion coefficients, W_{kj} can be written as

$$W_{kj} = \frac{1}{(\sigma_k - j)!} \frac{d^{\sigma_k - j}}{ds^{\sigma_k - j}} \left[(s - \lambda_k)^{\sigma_k} (sI - A)^{-1} \right] \Bigg|_{s = \lambda_k} \tag{9}$$

Taking the easy inverse Laplace transform gives an explicit form for the matrix exponential:

$$e^{At} = \sum_{k=1}^{m} \sum_{j=1}^{\sigma_k} W_{kj} \frac{t^{j-1}}{(j-1)!} e^{\lambda_k t} \tag{10}$$

Of course if some eigenvalues are complex, conjugate terms on the right side of (10) can be combined to give a real representation.

5.9 Example   For the *harmonic oscillator*, where

$$A = \begin{bmatrix} 0 & 1 \\ -1 & 0 \end{bmatrix}$$

a simple calculation gives

$$(sI - A)^{-1} = \begin{bmatrix} s & -1 \\ 1 & s \end{bmatrix}^{-1} = \frac{1}{s^2 + 1} \begin{bmatrix} s & 1 \\ -1 & s \end{bmatrix}$$

In this case partial fraction expansion is not needed, and a table of Laplace transforms can be used, if memory fails, to obtain

$$e^{At} = \begin{bmatrix} \cos t & \sin t \\ -\sin t & \cos t \end{bmatrix}$$

□□□

The *Jordan form* for a matrix is not used in any essential way in this book. But it may be familiar, and in conjunction with Property 5.6 it leads to another explicit form for the matrix exponential in terms of eigenvalues. We outline the development as an example of manipulations related to matrix exponentials. The Jordan form also is use-ful in constructing examples and counterexamples for various conjectures since it is only a state variable change away from a general A in a time-invariant linear state equa-tion. This utility is somewhat diminished by the fact that in the complex-eigenvalue case the variable change is complex, and thus coefficient matrices in the new state equa-tion typically are complex. A remedy for such unpleasantness is the 'real Jordan form' mentioned in Note 5.3.

5.10 Example   For a real $n \times n$ matrix A there exists an invertible $n \times n$ matrix P, not necessarily real, such that $J = P^{-1}AP$ has the following structure. The matrix J is block diagonal, with the kth diagonal block in the form

$$
J_k = \begin{bmatrix} \lambda & 1 & \cdots & & 0 \\ 0 & \lambda & \cdots & & 0 \\ \vdots & \vdots & & \vdots & \vdots \\ 0 & 0 & \cdots & & 1 \\ 0 & 0 & \cdots & & \lambda \end{bmatrix}
$$

where λ is an eigenvalue of A. There is at least one block for each eigenvalue of A, but the patterns of diagonal blocks that can arise for eigenvalues with high multiplicities are not of interest here. We need only know that the n eigenvalues of A are displayed on the diagonal of J. Of course, as reviewed in Chapter 1, if A has distinct eigenvalues, then P can be constructed from eigenvectors of A and J is diagonal. In general J (and P) are complex when A has complex eigenvalues. In any case Property 5.6 gives

$$
e^{At} = Pe^{Jt}P^{-1} \tag{11}
$$

and the structure of the right side is not difficult to describe.

Using the power series definition, we can show that the exponential of the block diagonal matrix J also is block diagonal, with the blocks given by $e^{J_k t}$. Furthermore writing $J_k = \lambda I + N_k$, where N_k has all zero entries except for 1's above the diagonal, and noting that λI commutes with N_k, Property 5.7 gives

$$
e^{J_k t} = e^{\lambda I t} e^{N_k t} = e^{\lambda t} e^{N_k t} \tag{12}
$$

Finally, since N_k is *nilpotent*, calculation of the finite power series for $e^{N_k t}$ shows that $e^{N_k t}$ is upper triangular, with nonzero entries given by

$$
\left[e^{N_k t} \right]_{ij} = \frac{t^{j-i}}{(j-i)!} , \quad i \le j \tag{13}
$$

Thus (11), (12), and (13) prescribe a general form for e^{At} in terms of the eigenvalues of A. (Again notice how simple the distinct eigenvalue case is.)

As a specific illustration the Jordan-form matrix

$$
J = \begin{bmatrix} 0 & 1 & 0 & 0 & 0 \\ 0 & 0 & 1 & 0 & 0 \\ 0 & 0 & 0 & 0 & 0 \\ 0 & 0 & 0 & 1 & 0 \\ 0 & 0 & 0 & 0 & 1 \end{bmatrix}
$$

has one 3×3 block corresponding to a multiplicity-3 eigenvalue at zero, and two scalar blocks corresponding to a multiplicity-2 unity eigenvalue. Thus (12) and (13) give

$$e^{At} = \begin{bmatrix} 1 & t & t^2/2 & 0 & 0 \\ 0 & 1 & t & 0 & 0 \\ 0 & 0 & 1 & 0 & 0 \\ 0 & 0 & 0 & e^t & 0 \\ 0 & 0 & 0 & 0 & e^t \end{bmatrix}$$

□□□

Special features of the transition matrix when $A(t)$ is constant naturally imply special properties of the response of a time-invariant linear state equation

$$\dot{x}(t) = Ax(t) + Bu(t), \quad x(t_o) = x_o$$
$$y(t) = Cx(t) + Du(t) \tag{14}$$

The complete solution formula in Chapter 3 becomes

$$y(t) = Ce^{A(t-t_o)}x_o + \int_{t_o}^{t} Ce^{A(t-\sigma)}Bu(\sigma)\,d\sigma + Du(t), \quad t \ge t_o$$

This exhibits the zero-state and zero-input response components for time-invariant linear state equations, and in particular shows that the integral term in the zero-state response is a convolution. If $t_o = 0$ the complete solution is

$$y(t) = Ce^{At}x_o + \int_{0}^{t} Ce^{A(t-\sigma)}Bu(\sigma)\,d\sigma + Du(t), \quad t \ge 0$$

A change of integration variable from σ to $\tau = t-\sigma$ in the convolution integral gives

$$y(t) = Ce^{At}x_o + \int_{0}^{t} Ce^{A\tau}Bu(t-\tau)\,d\tau + Du(t), \quad t \ge 0 \tag{15}$$

Replacing every t in (15) by $t - t_o$ shows that if the initial time is $t_o \ne 0$, then the complete response to the initial state $x(t_o) = x_o$ and input $u_o(t) = u(t-t_o)$ is $y_o(t) = y(t-t_o)$. In words, time shifting the input and initial time implies a corresponding time shift in the output signal. This proves that without loss of generality we can assume $t_o = 0$ for a time-invariant linear state equation.

Assuming a scalar input for simplicity, consider the zero-state response to a unit impulse $u(t) = \delta(t)$. (Recall that it is important for consistency reasons to interpret the initial time as $t = 0^-$ whenever an impulsive input signal is considered.) This unit impulse response is

$$y(t) = Ce^{At}B + D\delta(t)$$

For an ordinary input signal it follows from (15) that the zero-state response is given by a convolution of the input signal with the unit-impulse response. In other words in the single-input case, the unit-impulse response determines the zero-state response to any continuous input signal. It is not hard to show that in the multi-input case m impulse responses are required.

The Laplace transform often is used to represent the response of the linear time-invariant state equation (14). Using the convolution property of the transform, and the Laplace transform of the matrix exponential, (15) gives

$$Y(s) = C(sI - A)^{-1}x_o + \left[C(sI - A)^{-1}B + D \right] U(s) \qquad (16)$$

This formula also can be obtained by writing the state equation (14) in terms of Laplace transforms, and solving for $Y(s)$. (Again, the initial time should be interpreted as $t_o = 0^-$ for this calculation if impulsive inputs are permitted.)

It is easy to see, from (16) and (8), that if $U(s)$ is a proper rational function, then $Y(s)$ also is a proper rational function. Finally recall that the relation between $Y(s)$ and $U(s)$ under the assumption of zero initial state is called the *transfer function*. Namely the transfer function of a time-invariant linear state equation is the $p \times m$ matrix of rational functions

$$G(s) = C(sI - A)^{-1}B + D$$

Because of the presence of D, the entries of $G(s)$ in general are proper rational functions, but not strictly proper.

Periodic Case

The second special case we consider involves a restricted but important class of matrix functions of time. A continuous $n \times n$ matrix function $A(t)$ is called *T-periodic* if there exists a positive constant T such that

$$A(t + T) = A(t) \qquad (17)$$

for all t. (It is standard practice to assume that the *period T* is the least value for which (17) holds.) The basic result for this special case involves a particular representation for the transition matrix. This *Floquet decomposition* then can be used to investigate solution properties of T-periodic linear state equations.

5.11 Property   The transition matrix for a T-periodic $A(t)$ can be written in the form

$$\Phi(t, \tau) = P(t) e^{R(t-\tau)} P^{-1}(\tau) \qquad (18)$$

where R is a constant (possibly complex) $n \times n$ matrix, and $P(t)$ is a continuous, T-periodic, $n \times n$ matrix function that is invertible at each t.

Proof   Define the $n \times n$ matrix R by setting

$$e^{RT} = \Phi(T, 0) \qquad (19)$$

(This nontrivial step involves computing the *natural logarithm* of the invertible matrix $\Phi(T, 0)$, and a complex R can result. See Exercise 5.15 for further development, and Note 5.3 for citations.) Also define $P(t)$ by setting

$$P(t) = \Phi(t, 0)\, e^{-Rt} \tag{20}$$

Obviously $P(t)$ is invertible at each t, and it is easy to show that these definitions give the claimed decomposition. Indeed

$$\Phi(t, 0) = P(t)e^{Rt}$$

implies

$$\Phi(0, t) = \Phi^{-1}(t, 0) = e^{-Rt}P^{-1}(t)$$

so that, as claimed,

$$\Phi(t, \tau) = \Phi(t, 0)\Phi(0, \tau) = P(t)e^{R(t-\tau)}P^{-1}(\tau) \tag{21}$$

It remains to show that the $P(t)$ defined by (20) is T-periodic. From (20),

$$P(t + T) = \Phi(t + T, 0)e^{-R(t+T)}$$
$$= \Phi(t + T, T)\Phi(T, 0)e^{-RT}e^{-Rt}$$

and since $\Phi(T, 0)e^{-RT} = I$,

$$P(t + T) = \Phi(t + T, T)e^{-Rt} \tag{22}$$

Now we note that $\Phi(t + T, T)$ satisfies the matrix differential equation

$$\frac{d}{dt}\,\Phi(t + T, T) = \frac{d}{d(t+T)}\,\Phi(t + T, T) = A(t+T)\Phi(t + T, T)$$

$$= A(t)\Phi(t + T, T), \quad \Phi(t + T, T)\Big|_{t=0} = I$$

Therefore, by uniqueness of solutions, $\Phi(t + T, T) = \Phi(t, 0)$. Then (22) can be written as

$$P(t + T) = \Phi(t, 0)e^{-Rt} = P(t)$$

to conclude the proof.
□□□

Because of the unmotivated definitions of R and $P(t)$, the proof of Property 5.11 resembles theft more than honest work. However there is one case where the constant matrix R in (18) has a simple interpretation, and is easy to compute. From Property 4.2 we conclude that if the T-periodic $A(t)$ commutes with its integral, then R is the average value of $A(t)$ over one period.

5.12 Example   At the end of Example 4.6, in a different notation, the transition matrix for

$$A(t) = \begin{bmatrix} -1 & 0 \\ -\cos t & 0 \end{bmatrix}$$

is given as

$$\Phi(t,\ 0) = \begin{bmatrix} e^{-t} & 0 \\ -\frac{1}{2}+\frac{1}{2}e^{-t}(\cos t - \sin t) & 1 \end{bmatrix} \tag{23}$$

This result can be deconstructed to illustrate Property 5.11. Clearly $T = 2\pi$, and

$$\Phi(2\pi,\ 0) = \begin{bmatrix} e^{-2\pi} & 0 \\ -\frac{1}{2}+\frac{1}{2}e^{-2\pi} & 1 \end{bmatrix}$$

It is not difficult to verify that

$$R = \frac{1}{T}\ \ln \Phi(2\pi,\ 0) = \begin{bmatrix} -1 & 0 \\ -\frac{1}{2} & 0 \end{bmatrix}$$

by computing e^{Rt}, and evaluating the result at $t = 2\pi$. Then

$$e^{-Rt} = \begin{bmatrix} e^{t} & 0 \\ -\frac{1}{2}+\frac{1}{2}e^{t} & 1 \end{bmatrix}$$

and, from (20) and (23),

$$P(t) = \begin{bmatrix} 1 & 0 \\ -\frac{1}{2}+\frac{1}{2}(\cos t - \sin t) & 1 \end{bmatrix}$$

Thus the Floquet decomposition for $\Phi(t,\ 0)$ is

$$\Phi(t,\ 0) = \begin{bmatrix} 1 & 0 \\ -\frac{1}{2}+\frac{1}{2}(\cos t - \sin t) & 1 \end{bmatrix} \begin{bmatrix} e^{-t} & 0 \\ -\frac{1}{2}+\frac{1}{2}e^{-t} & 1 \end{bmatrix} \begin{bmatrix} 1 & 0 \\ 0 & 1 \end{bmatrix} \tag{24}$$

□□□

The representation in Property 5.11 for the transition matrix implies that if R is known and $P(t)$ is known tor $t \in [t_o,\ t_o + T)$, then $\Phi(t,\ t_o)$ can be computed for arbitrary values of t. Also the growth properties of $\Phi(t,\ t_o)$, and thus of solutions of the linear state equation

$$\dot{x}(t) = A(t)x(t)\ ,\quad x(t_o) = x_o \tag{25}$$

with T-periodic $A(t)$, depend on the eigenvalues of the constant matrix $e^{RT} = \Phi(T,\ 0)$. To see this, note that for any positive integer k repeated application of the composition property (Property 4.7) leads to

$$x(t + kT) = \Phi(t + kT,\ t_o)x_o$$

$$= \Phi(t+kT,\ t+(k-1)T)\ \Phi(t+(k-1)T,\ t+(k-2)T)$$

$$\cdots\ \Phi(t + T,\ t)\ \Phi(t,\ t_o)x_o$$

$$= P(t+kT)e^{RT}P^{-1}(t+(k-1)T)\ P(t+(k-1)T)e^{RT}P^{-1}(t+(k-2)T)$$

$$\cdots\ P(t+T)e^{RT}P^{-1}(t)x(t)$$

$$= P(t + kT)[e^{RT}]^{k}P^{-1}(t)x(t)$$

If, for example, the eigenvalues of e^{RT} all have magnitude strictly less than unity, then $[e^{RT}]^k \to 0$ as $k \to \infty$, as a Jordan-form argument shows. (Write the Jordan form of e^{RT} as the sum of a diagonal matrix and a nilpotent matrix, as in Example 5.10. Then, using commutativity, apply the binomial expansion to the kth-power of this sum to see that each entry of the result is zero, or approaches zero as $k \to \infty$.) Thus for any t, $x(t + kT) \to 0$ as $k \to \infty$. That is, $x(t) \to 0$ as $t \to \infty$ for every x_o. Similarly when at least one eigenvalue has magnitude greater than unity there are initial states for which $x(t)$ grows without bound as $t \to \infty$.

If e^{RT} has at least one unity eigenvalue, the existence of nonzero T-periodic solutions to (25) for appropriate initial states is established in the following development. We prove the converse also. Note that this is one setting where the solution for $t < t_o$ as well as for $t \geq t_o$ is considered, as dictated by the definition of periodicity: $x(t + T) = x(t)$ for all t.

5.13 Theorem   Suppose $A(t)$ is T-periodic. Given any t_o there exists a nonzero initial state x_o such that the solution of

$$\dot{x}(t) = A(t)x(t) , \quad x(t_o) = x_o \tag{26}$$

is T-periodic if and only if at least one eigenvalue of e^{RT} is unity.

Proof   Suppose that at least one eigenvalue of e^{RT} is unity, and let z_o be a corresponding eigenvector. Then z_o is real and nonzero, and it is easy to verify that for any t_o

$$z(t) = e^{R(t - t_o)}z_o \tag{27}$$

is T-periodic. (Simply compute $z(t + T)$ from (27).) Invoking the Floquet description for $\Phi(t, t_o)$ and letting $x_o = P(t_o)z_o$ yields the (nonzero) solution of (26):

$$x(t) = \Phi(t, t_o)x_o = P(t)e^{R(t - t_o)}P^{-1}(t_o)x_o$$

$$= P(t)z(t)$$

This solution clearly is T-periodic, since both $P(t)$ and $z(t)$ are T-periodic.

Now suppose that given t_o the nonzero initial state x_o is such that the corresponding solution $x(t)$ is T-periodic. Then, using the Floquet description,

$$x(t) = P(t)e^{R(t - t_o)}P^{-1}(t_o)x_o$$

and

$$x(t + T) = P(t + T)e^{R(t + T - t_o)}P^{-1}(t_o)x_o$$

$$= P(t)e^{R(t + T - t_o)}P^{-1}(t_o)x_o$$

Since $x(t) = x(t + T)$ for all t, these representations imply

$$e^{RT}P^{-1}(t_o)x_o = P^{-1}(t_o)x_o \tag{28}$$

But $P^{-1}(t_o)x_o \neq 0$, so (28) exhibits $P^{-1}(t_o)x_o$ as an eigenvector of e^{RT} corresponding to a unity eigenvalue.
□□□

Theorem 5.13 can be restated in terms of the matrix R rather than e^{RT}, since e^{RT} has a unity eigenvalue if and only if R has an eigenvalue that is an integer multiple of the purely imaginary number $2\pi i/T$. To prove this, if $(k\,2\pi i/T)$ is an eigenvalue of R with eigenvector z, then $(RT)^j z = R^j z T^j = (k\,2\pi i)^j z$. Thus, from the power series for the matrix exponential, $e^{RT}z = e^{k2\pi i}z = z$, and this shows that e^{RT} has a unity eigenvalue. The converse argument involves transformation of e^{RT} to Jordan form.

Now consider the case of a linear state equation where both $A(t)$ and $B(t)$ are T-periodic, and where the inputs of interest also are T-periodic. For simplicity such a state equation is written as

$$\dot{x}(t) = A(t)x(t) + f(t), \quad x(t_o) = x_o \tag{29}$$

We assume that both $A(t)$ and $f(t)$ are T-periodic, and $A(t)$ is continuous, as usual. However to accommodate a technical argument in the proof of Theorem 5.15 we permit $f(t)$ to be piecewise continuous.

5.14 Lemma   A solution $x(t)$ of the T-periodic state equation (29) is T-periodic if and only if $x(t_o + T) = x_o$.

Proof   Of course if $x(t)$ is T-periodic, then $x(t_o + T) = x(t_o)$. Conversely suppose x_o is such that the corresponding solution of (29) satisfies $x(t_o + T) = x_o$. Letting $z(t) = x(t + T) - x(t)$, it follows that $z(t_o) = 0$, and

$$\dot{z}(t) = \Big[A(t + T)x(t + T) + f(t + T) \Big] - \Big[A(t)x(t) + \dot{f}(t) \Big]$$
$$= A(t)z(t)$$

But uniqueness of solutions implies $z(t) = 0$ for all t, that is, $x(t)$ is T-periodic.
□□□

Using this lemma the next result provides conditions for the existence of a T-periodic solution for *any* T-periodic $f(t)$. (A refinement dealing with a *fixed* T-periodic $f(t)$ is suggested in Exercise 5.18.)

5.15 Theorem   Suppose $A(t)$ is T-periodic. Then for any t_o and any T-periodic $f(t)$ there exists x_o such that the solution of

$$\dot{x}(t) = A(t)x(t) + f(t), \quad x(t_o) = x_o \tag{30}$$

is T-periodic if and only if for any t_o there exists no $z_o \neq 0$ for which

$$\dot{z}(t) = A(t)z(t), \quad z(t_o) = z_o \tag{31}$$

has a T-periodic solution.

Proof   For any x_o, t_o, and T-periodic $f(t)$, the solution of (30) is

$$x(t) = \Phi(t,\ t_o)x_o + \int_{t_o}^{t} \Phi(t,\ \sigma)f(\sigma)\ d\sigma$$

By Lemma 5.14, $x(t)$ is T-periodic if and only if

$$\left[I - \Phi(t_o + T,\ t_o)\right]x_o = \int_{t_o}^{t_o + T} \Phi(t_o + T,\ \sigma)f(\sigma)\ d\sigma \tag{32}$$

Therefore, by Theorem 5.13, it must be shown that this algebraic equation has a solution for x_o given any t_o and any T-periodic $f(t)$ if and only if e^{RT} has no unity eigenvalues.

First suppose $e^{RT} = \Phi(T,\ 0)$ has no unity eigenvalues, that is,

$$\det\left[I - \Phi(T,\ 0)\right] \neq 0 \tag{33}$$

By invertibility of transition matrices, (33) is equivalent to the condition

$$0 \neq \det\left\{\Phi(t_o + T,\ T)\left[I - \Phi(T,\ 0)\right]\Phi(0,\ t_o)\right\}$$

$$= \det\left\{\Phi(t_o + T,\ T)\Phi(0,\ t_o) - \Phi(t_o + T,\ t_o)\right\}$$

Since $\Phi(t_o + T,\ T) = \Phi(t_o,\ 0)$, as shown in the proof of Property 5.11, we conclude that (33) is equivalent to invertibility of $[I - \Phi(t_o + T,\ t_o)]$ for any t_o. Thus (32) has a solution x_o for any t_o and any T-periodic $f(t)$.

Now suppose that (32) has a solution given any t_o and any T-periodic $f(t)$. Given t_o, corresponding to any $n \times 1$ vector f_o define a particular T-periodic, piecewise-continuous $f(t)$ by setting

$$f(t) = \Phi(t,\ t_o + T)f_o\ ,\quad t \in [t_o,\ t_o + T) \tag{34}$$

and extending this definition to all t by repeating. For such a piecewise-continuous, T-periodic $f(t)$,

$$\int_{t_o}^{t_o + T} \Phi(t_o + T,\ \sigma)f(\sigma)\ d\sigma = \int_{t_o}^{t_o + T} f_o\ d\sigma = Tf_o$$

and (32) becomes

$$\left[I - \Phi(t_o + T,\ t_o)\right]x_o = Tf_o \tag{35}$$

Given any $f(t)$ of the type constructed above, that is given any f_o, (35) has a solution for x_o by assumption. Therefore

$$\det\left[I - \Phi(t_o + T,\ t_o)\right] \neq 0$$

and, again, this is equivalent to (33), and thus to the statement that no eigenvalue of e^{RT} is unity.
□□□

Application of this general result to a situation that might be familiar is enlightening. The sufficiency portion of Theorem 5.15 immediately applies to the case where $f(t) = B(t)u(t)$, though necessity requires the notion of *controllability* discussed in Chapter 9 (to avoid, as a trivial instance, zero $B(t)$). Of course a time-invariant linear state equation is T-periodic for any value of $T > 0$.

5.16 Corollary   For the time-invariant linear state equation

$$\dot{x}(t) = Ax(t) + Bu(t) , \quad x(0) = x_o \tag{36}$$

suppose A has no eigenvalue with zero real part. Then given any T-periodic input $u(t)$, there exists an x_o such that the corresponding solution is T-periodic.

In particular it is worthwhile to contemplate this corollary in the single-input case where A has negative-real-part eigenvalues, and the input signal is $u(t) = sin(\omega t)$. By Corollary 5.16 there exists an initial state such that the complete response $x(t)$ is periodic with $T = 2\pi/\omega$. And it is clear from the Laplace transform representation of the solution that for any initial state the response $x(t)$ approaches periodicity as $t \to \infty$. Perhaps surprisingly, if A has (some, or all) eigenvalues with positive real part, but none with zero real part, then there still exists a periodic solution for some initial state. Evidently the unbounded terms in the zero-input response component are canceled by unbounded terms in the zero-state response.

EXERCISES

Exercise 5.1   For a constant $n \times n$ matrix A, show that the transition matrix for the transpose of A is the transpose of the transition matrix for A. Is this true for nonconstant $A(t)$? Is it true for the case where $A(t)$ commutes with its integral?

Exercise 5.2   Compute e^{At} for

$$A = \begin{bmatrix} -1 & 0 & 0 \\ 0 & -1 & 0 \\ 1 & 0 & -1 \end{bmatrix}$$

Exercise 5.3   Compute e^{At} for

$$A = \begin{bmatrix} 0 & 1 \\ 1 & 0 \end{bmatrix}$$

by two different methods.

Exercise 5.4   Compute $\Phi(t, 0)$ for

$$A(t) = \begin{bmatrix} t & 1 \\ 1 & t \end{bmatrix}$$

(*Hint*: One efficient way is to use the result of Exercise 5.3.)

Exercise 5.5   If A is a constant, invertible, $n \times n$ matrix, show that

$$\int_0^t e^{A\sigma} \, d\sigma = A^{-1} \left[e^{At} - I \right]$$

What additional condition on A will yield

$$A^{-1} = \int_\infty^0 e^{At} \, dt$$

Exercise 5.6   Suppose the $n \times n$ matrix $A(t)$ can be written in the form

$$A(t) = \sum_{j=1}^{r} f_j(t)A_j$$

where $f_1(t), \ldots, f_r(t)$ are continuous, scalar functions, and $A_1, \ldots, A_r$ are constant $n \times n$ matrices that satisfy

$$A_i A_j = A_j A_i \,, \quad i,j = 1, \ldots, r$$

Prove that the transition matrix for $A(t)$ can be written as

$$\Phi(t, t_o) = e^{A_1 \int_{t_o}^t f_1(\sigma)d\sigma} \cdots e^{A_r \int_{t_o}^t f_r(\sigma)d\sigma}$$

Use this result to compute $\Phi(t, 0)$ for

$$A(t) = \begin{bmatrix} \cos \omega t & \sin \omega t \\ -\sin \omega t & \cos \omega t \end{bmatrix}$$

Exercise 5.7   For the time-invariant, n-dimensional, single-input nonlinear state equation

$$\dot{x}(t) = Ax(t) + Dx(t)u(t) + bu(t) \,, \quad x(0) = 0$$

show that under appropriate additional hypotheses a solution is

$$x(t) = \int_0^t e^{A(t-\sigma)} e^{D\int_\sigma^t u(\tau)d\tau} \, bu(\sigma) \, d\sigma$$

Exercise 5.8   If A and F are $n \times n$ constant matrices, show that

$$e^{(A+F)t} - e^{At} = \int_0^t e^{A(t-\sigma)} F e^{(A+F)\sigma} \, d\sigma$$

Exercise 5.9   If A and F are $n \times n$ constant matrices, show that

$$e^A e^F - e^{A+F} = \int_0^1 e^{A\sigma}[e^{(A+F)(1-\sigma)}F - Fe^{(A+F)(1-\sigma)}]e^{F\sigma} \, d\sigma$$

Exercise 5.10 Compute $\Phi(t, 0)$ for the T-periodic state equation with

$$A(t) = \begin{bmatrix} -2+\cos 2t & 0 \\ 0 & -3+\cos 2t \end{bmatrix}$$

Compute $P(t)$ and R for the Floquet decomposition of the transition matrix.

Exercise 5.11 Suppose A is $n \times n$, and

$$\det(sI - A) = s^n + a_{n-1}s^{n-1} + \cdots + a_0$$

Verify the formula

$$\text{adj}(sI - A) = (s^{n-1}+a_{n-1}s^{n-2}+ \cdots +a_1)I + \cdots + (s+a_{n-1})A^{n-2} + A^{n-1}$$

and use it to show that there exist strictly-proper rational functions of s such that

$$(sI - A)^{-1} = \hat{\alpha}_0(s)I + \hat{\alpha}_1(s)A + \cdots + \hat{\alpha}_{n-1}(s)A^{n-1}$$

Exercise 5.12 Consider the linear state equation

$$\dot{x}(t) = Ax(t) + f(t), \quad x(t_o) = x_o$$

where all eigenvalues of A have negative real parts, and $f(t)$ is continuous and T-periodic. Show that

$$x(t) = \int_{-\infty}^{t} e^{A(t-\sigma)}f(\sigma) \, d\sigma$$

is a T-periodic solution, corresponding to

$$x_o = \int_{-\infty}^{t_o} e^{A(t_o-\sigma)}f(\sigma) \, d\sigma$$

Show that a solution corresponding to a different x_o, converges to this periodic solution as $t \to \infty$.

Exercise 5.13 Show that a linear state equation with T-periodic $A(t)$ can be transformed to a time-invariant linear state equation by a T-periodic variable change.

Exercise 5.14 Suppose that $A(t)$ is T-periodic, and t_o is fixed. Show that the transition matrix for $A(t)$ can be written in the form

$$\Phi(t, t_o) = Q(t, t_o)e^{S(t-t_o)}$$

where S is a (possibly complex) constant matrix (depending on t_o), and $Q(t, t_o)$ is continuous and invertible at each t, and satisfies

$$Q(t+T, t_o) = Q(t, t_o), \quad Q(t_o, t_o) = I$$

Exercise 5.15 Suppose M is an $n \times n$ invertible matrix with distinct eigenvalues. Show that there exists a possibly-complex, $n \times n$ matrix R such that

$$e^R = M$$

Exercise 5.16 Prove that a T-periodic linear state equation

$$\dot{x}(t) = A(t)x(t)$$

has unbounded solutions if

$$\int_0^T \text{tr} \, [A(\sigma)] \, d\sigma > 0$$

Exercise 5.17 Suppose $A(t)$ is $n \times n$, real, continuous, and T-periodic. Show that the transition matrix for $A(t)$ can be written as

$$\Phi(t, 0) = Q(t)e^{St}$$

where S is a constant, real, $n \times n$ matrix, and $Q(t)$ is $n \times n$, real, continuous, and $2T$-periodic. (*Hint*: It is a mathematical fact that if M is real and invertible, then there is a real S such that $e^S = M^2$.)

Exercise 5.18 For a T-periodic state equation with a fixed T-periodic input, establish the following refinement of Theorem 5.15. There exists an x_o such that the solution of

$$\dot{x}(t) = A(t)x(t) + f(t) , \quad x(t_o) = x_o$$

is T-periodic if and only if $f(t)$ is such that

$$\int_{t_o}^{t_o + T} z^T(t)f(t) \, dt = 0$$

for all T-periodic solutions $z(t)$ of the adjoint state equation

$$\dot{z}(t) = -A^T(t)z(t) , \quad z(t_o) = z_o$$

Exercise 5.19 Determine values of ω for which there exists an x_o such that the resulting solution of

$$\dot{x}(t) = \begin{bmatrix} 0 & 1 \\ -1 & 0 \end{bmatrix} x(t) + \begin{bmatrix} 0 \\ \sin \omega t \end{bmatrix} , \quad x(0) = x_o$$

is periodic. (*Hint*: Use the result of Exercise 5.18.)

NOTES

Note 5.1 In Property 5.7 necessity of the commutativity condition on A and F fails if equality of exponentials is postulated at a single value of t. Specifically there are non-commuting matrices A and F such that $e^A \, e^F = e^{A+F}$. For further details see

D.S. Bernstein, "Commuting matrix exponentials," Problem 88-1, *SIAM Review*, Vol. 31, No. 1, p. 125, 1989

and the solution and references that follow the problem statement.

Note 5.2   Further information about the functions $\alpha_k(t)$ in Property 5.8, including differential equations they individually satisfy, and linear independence properties, is provided in

M. Vidyasagar, "A characterization of e^{At} and a constructive proof of the controllability condition," *IEEE Transactions on Automatic Control,* Vol. 16, No. 4, pp. 370–371, 1971

Note 5.3   The Jordan form is treated in almost every book on matrices. The real version of the Jordan form (when A has complex eigenvalues) is less ubiquitous. See Section 3.4 of

R.A. Horn, C.A. Johnson, *Matrix Analysis,* Cambridge University Press, Cambridge, England, 1985

The natural logarithm of a matrix in the general case is a more complex issue than in the special case considered in Exercise 5.15. A Jordan-form argument is given in Section 3.4 of

R.K. Miller, A.N. Michel, *Ordinary Differential Equations,* Academic Press, New York, 1982

A more advanced treatment, along with a proof of the fact quoted in Exercise 5.17, can be found in Section 8.1 of

D.L. Lukes, *Differential Equations: Classical to Controlled,* Academic Press, New York, 1982

Note 5.4   Differential equations with periodic coefficients have a long history in mathematical physics, and associated phenomena such as parametric pumping are of current technological interest. A readable treatment is

J.A. Richards, *Analysis of Periodically Time-Varying Systems,* Springer-Verlag, New York, 1983

This book includes historical remarks, and introduces standard terminology ignored in our discussion. For example in Property 5.11 the eigenvalues of R are called *characteristic exponents,* and the eigenvalues of e^{RT} are called *characteristic multipliers.* Also Richards treats at length the classical *Hill equation,*

$$\ddot{y}(t) + \lceil \alpha + \beta a(t) \rfloor y(t) = 0$$

where $a(t)$ is T-periodic.

Note 5.5   Periodicity properties of solutions of linear state equations when $A(t)$ and $f(t)$ have time-symmetry properties (even, or odd) in addition to being periodic are discussed in

R.J. Mulholland, "Time symmetry and periodic solutions of the state equations," *IEEE Transactions on Automatic Control,* Vol. 16, No. 4, pp. 367–368, 1971

Note 5.6   Extension of the Laplace transform representation to time-varying linear systems has long been an appealing notion. Early work by L.A. Zadeh is reviewed in Section 8.17 of

W. Kaplan, *Operational Methods for Linear Systems,* Addison-Wesley, Reading, Massachusetts, 1962

See also Chapters 9 and 10 of

H. D'Angelo, *Linear Time-Varying Systems,* Allyn and Bacon, Boston, 1970

and, for more recent developments,

E.W. Kamen, "Poles and zeros of linear time varying systems," *Linear Algebra and Its Applications,* Vol. 98, pp. 263–289, 1988

Note 5.7   We have not exhausted known properties of transition matrices — a believable claim we support with two examples. Suppose

$$A(t) = \sum_{k=1}^{q} a_k(t) A_k$$

where $A_1, \ldots, A_q$ are constant $n \times n$ matrices, $a_1(t), \ldots, a_q(t)$ are scalar functions, and of course $q \leq n^2$. Then there exist scalar functions $f_1(t), \ldots, f_q(t)$ such that

$$\Phi(t, 0) = e^{A_1 f_1(t)} \cdots e^{A_q f_q(t)}$$

at least for t in a small neighborhood of $t = 0$. A discussion of this property, with references to the original mathematics literature, is in

R.J. Mulholland, "Exponential representation for linear systems," *IEEE Transactions on Automatic Control,* Vol. 16, No. 1, pp. 97–98, 1971

The second example is a formula that might be familiar from the scalar case:

$$e^A = \lim_{n \to \infty} \left[I + \frac{A}{n} \right]^n$$

Note 5.8   Numerical computation of the matrix exponential e^{At} can be approached in many ways, each with attendant weaknesses. A survey of about 20 methods is in

C. Moler, C. Van Loan, "Nineteen dubious ways to compute the exponential of a matrix," *SIAM Review,* Vol. 20, No. 4, pp. 801–836, 1978

6

INTERNAL STABILITY

Internal stability deals with boundedness properties and asymptotic behavior (as $t \to \infty$) of solutions of the zero-input linear state equation

$$\dot{x}(t) = A(t)x(t) , \quad x(t_o) = x_o \tag{1}$$

While bounds on solutions might be of interest for fixed t_o and x_o, or for various initial states at a fixed t_o, we focus on boundedness properties that hold regardless of the choice of t_o or x_o. In a similar fashion the concept we adopt relative to asymptotically-zero solutions is independent of the choice of initial time. The reason is that these 'uniform in t_o' concepts are most appropriate in relation to input-output stability properties of linear state equations developed in Chapter 12.

It is natural to begin by characterizing stability properties of the linear state equation (1) in terms of bounds on the transition matrix $\Phi(t, \tau)$ for $A(t)$. However this does not provide a generally useful stability test for specific examples because of the difficulty of computing $\Phi(t, \tau)$. More effective stability criteria are addressed in Chapter 7 and Chapter 8.

Uniform Stability

The first notion involves boundedness of solutions of (1). Because solutions are linear in the initial state, it is convenient to express the bound as a linear function of the norm of the initial state.

6.1 Definition   The linear state equation (1) is called *uniformly stable* if there exists a finite positive constant γ such that for any t_o and x_o the corresponding solution satisfies

$$\|x(t)\| \leq \gamma \|x_o\| , \quad t \geq t_o \tag{2}$$

Evaluation of (2) at $t = t_o$ shows that the constant γ must satisfy $\gamma \geq 1$. The adjective *uniform* in the definition refers precisely to the fact that γ must not depend on the choice of initial time, as illustrated in Figure 6.2. A 'nonuniform' stability concept can be defined by permitting γ to depend on the initial time, but this is not considered here except to show that there is a difference via a standard example.

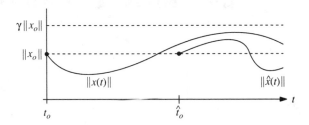

6.2 Figure   Uniform stability implies the γ-bound is independent of t_o.

6.3 Example   The scalar linear state equation

$$\dot{x}(t) = [4t\sin t - 2t]x(t) , \quad x(t_o) = x_o$$

has the readily-verifiable solution

$$x(t) = \exp \left[4\sin t - 4t\cos t - t^2 - 4\sin t_o + 4t_o \cos t_o + t_o^2 \right] x_o \qquad (3)$$

It is easy to show that for *fixed* t_o there is a γ such that (3) is bounded by $\gamma|x_o|$ for all $t \geq t_o$, since the $(-t^2)$ term takes over the exponent as t increases. However the state equation is not uniformly stable. With fixed initial state x_o consider a sequence of initial times $t_o = 2k\pi$, where $k = 0, 1, \cdots$, and the values of the respective solutions at times π units later:

$$x(2k\pi + \pi) = \exp[(4k + 1)\pi(4 - \pi)] \, x_o$$

Clearly there is no bound on the exponential factor that is independent of k. In other words, a candidate γ must be ever larger as k, and the corresponding initial time, gets larger.
□□□

We emphasize again that Definition 6.1 is stated in a form specific to linear state equations. Equivalence to a more general definition of uniform stability that is used also in the nonlinear case is the subject of Exercise 6.1.

The basic characterization of uniform stability is readily discernible from Definition 6.1, though the proof requires a bit of finesse.

6.4 Theorem   The linear state equation (1) is uniformly stable if and only if there exists a finite positive constant γ such that

$$\| \Phi(t, \tau) \| \leq \gamma \qquad (4)$$

for all t, τ such that $t \geq \tau$.

Proof   First suppose such a γ exists. Then for any t_o and x_o, the solution of (1) satisfies

$$\|x(t)\| = \|\Phi(t,\ t_o)x_o\| \le \|\Phi(t,\ t_o)\|\,\|x_o\| \le \gamma\|x_o\|\ ,\quad t \ge t_o$$

and uniform stability is established.

For the reverse implication suppose that the state equation (1) is uniformly stable. Then there is a finite γ such that, for any t_o and x_o, solutions satisfy

$$\|x(t)\| \le \gamma\|x_o\|\ ,\quad t \ge t_o$$

Given any t_o and $t_a \ge t_o$, let x_a be such that

$$\|x_a\| = 1\ ,\quad \|\Phi(t_a,\ t_o)x_a\| = \|\Phi(t_a,\ t_o)\|$$

(Such an x_a exists by definition of the induced norm.) Then the initial state $x(t_o) = x_a$ yields a solution of (1) that at time t_a satisfies

$$\|x(t_a)\| = \|\Phi(t_a,\ t_o)x_a\| = \|\Phi(t_a,\ t_o)\|\,\|x_a\| \le \gamma\|x_a\| \tag{5}$$

Since $\|x_a\| = 1$, this shows that $\|\Phi(t_a,\ t_o)\| \le \gamma$. Because such an x_a can be selected for any t_o and $t_a \ge t_o$, the proof is complete.

Uniform Exponential Stability

Next we consider a stability property for (1) that addresses both the boundedness of solutions, and the asymptotic behavior of solutions. It *implies* uniform stability, and imposes an additional requirement that all solutions approach zero exponentially as $t \to \infty$.

6.5 Definition   The linear state equation (1) is called *uniformly exponentially stable* if there exist finite positive constants γ, λ such that for any t_o and x_o the corresponding solution satisfies

$$\|x(t)\| \le \gamma e^{-\lambda(t - t_o)}\|x_o\|\ ,\quad t \ge t_o \tag{6}$$

Again γ is no less than unity, and the adjective *uniform* refers to the fact that γ and λ are independent of t_o. This is illustrated in Figure 6.6. The property of uniform exponential stability can be expressed in terms of an exponential bound on the transition matrix. The proof is similar to that of Theorem 6.4, and so is left as Exercise 6.9.

6.7 Theorem   The linear state equation (1) is uniformly exponentially stable if and only if there exist finite positive constants γ and λ such that

$$\|\Phi(t,\ \tau)\| \le \gamma e^{-\lambda(t - \tau)} \tag{7}$$

for all t, τ such that $t \ge \tau$.

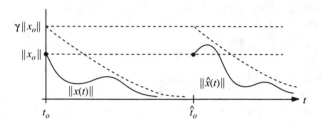

6.6 Figure   A decaying-exponential bound independent of t_o.

Uniform stability and uniform exponential stability are the only internal stability concepts used in the sequel. Uniform exponential stability is the most important of the two, and another theoretical characterization of uniform exponential stability for the bounded coefficient case will prove useful.

6.8 Theorem   Suppose there exists a finite positive constant α such that $\|A(t)\| \le \alpha$ for all t. Then the linear state equation (1) is uniformly exponentially stable if and only if there exists a finite positive constant β such that

$$\int_\tau^t \|\Phi(t, \sigma)\| \, d\sigma \le \beta \tag{8}$$

for all t, τ such that $t \ge \tau$.

Proof   If the state equation is uniformly exponentially stable, then by Theorem 6.7 there exist finite γ, $\lambda > 0$ such that

$$\|\Phi(t, \sigma)\| \le \gamma e^{-\lambda(t-\sigma)}$$

for all t, σ such that $t \ge \sigma$. Then

$$\int_\tau^t \|\Phi(t, \sigma)\| \, d\sigma \le \int_\tau^t \gamma e^{-\lambda(t-\sigma)} \, d\sigma$$

$$= \gamma(1 - e^{-\lambda(t-\tau)})/\lambda$$

$$\le \gamma/\lambda$$

for all t, τ such that $t \ge \tau$. Thus (8) is established with $\beta = \gamma/\lambda$.

Conversely suppose (8) holds. A fundamental theorem of calculus permits the representation

$$\Phi(t, \tau) = I - \int_\tau^t \frac{\partial}{\partial \sigma} \Phi(t, \sigma) \, d\sigma$$

$$= I + \int_\tau^t \Phi(t, \sigma)A(\sigma) \, d\sigma$$

and thus

$$\| \Phi(t, \tau) \| \le 1 + \alpha \int_{\tau}^{t} \| \Phi(t, \sigma) \| \, d\sigma$$

$$\le 1 + \alpha \beta \tag{9}$$

for all t, τ such that $t \ge \tau$. In completing this proof the composition property of the transition matrix is crucial. So long as $t \ge \tau$ we can write, cleverly,

$$\| \Phi(t, \tau) \| (t - \tau) = \int_{\tau}^{t} \| \Phi(t, \tau) \| \, d\sigma$$

$$\le \int_{\tau}^{t} \| \Phi(t, \sigma) \| \, \| \Phi(\sigma, \tau) \| \, d\sigma$$

$$\le \beta (1 + \alpha \beta)$$

Therefore letting $T = 2\beta (1 + \alpha \beta)$ and $t = \tau + T$ gives

$$\| \Phi(\tau + T, \tau) \| \le \tfrac{1}{2} \tag{10}$$

for all τ. Applying (9) and (10), the following inequalities on time intervals of the form $[\tau + kT, \tau + (k+1)T)$, where τ is arbitrary, are transparent:

$$\| \Phi(t, \tau) \| \le 1 + \alpha \beta, \quad t \in [\tau, \tau + T)$$

$$\| \Phi(t, \tau) \| = \| \Phi(t, \tau + T) \Phi(\tau + T, \tau) \| \le \| \Phi(t, \tau + T) \| \, \| \Phi(\tau + T, \tau) \|$$

$$\le \frac{1 + \alpha \beta}{2}, \quad t \in [\tau + T, \tau + 2T)$$

$$\| \Phi(t, \tau) \| = \| \Phi(t, \tau + 2T) \Phi(\tau + 2T, \tau + T) \Phi(\tau + T, \tau) \|$$

$$\le \| \Phi(t, \tau + 2T) \| \, \| \Phi(\tau + 2T, \tau + T) \| \, \| \Phi(\tau + T, \tau) \|$$

$$\le \frac{1 + \alpha \beta}{2^2}, \quad t \in [\tau + 2T, \tau + 3T)$$

Continuing in this fashion shows that, for any value of τ,

$$\| \Phi(t, \tau) \| \le \frac{1 + \alpha \beta}{2^k}, \quad t \in [\tau + kT, \tau + (k+1)T) \tag{11}$$

Finally choose $\lambda = (-1/T) \ln(1/2)$ and $\gamma = 2(1 + \alpha \beta)$. Figure 6.9 presents a plot of the corresponding decaying exponential and the bound (11), from which it is clear that

$$\| \Phi(t, \tau) \| \le \gamma e^{-\lambda(t - \tau)}$$

for all t, τ such that $t \ge \tau$. Uniform exponential stability thus is a consequence of Theorem 6.7.

□□□

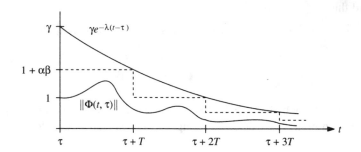

6.9 Figure   Bounds constructed in the proof of Theorem 6.8.

For time-invariant linear state equations, where $A(t) = A$ and $\Phi(t, \sigma) = e^{A(t - \sigma)}$, an integration-variable change in (8) shows that uniform exponential stability is equivalent to finiteness of

$$\int_0^\infty \| e^{At} \| \, dt \tag{12}$$

The adjective 'uniform' is superfluous in the time-invariant case, and we will drop it in clear contexts. Though exponential stability usually is called asymptotic stability in the context of time-invariant linear state equations, we retain the term exponential stability.

An explicit representation for e^{At} discussed in Chapter 5 combined with the finiteness condition on (12) yields a better-known characterization of exponential stability.

6.10 Theorem   A linear state equation (1) with constant $A(t) = A$ is exponentially stable if and only if all eigenvalues of A have negative real parts.

Proof   Suppose the eigenvalue condition holds. Then writing e^{At} in the explicit form in Chapter 5, where $\lambda_1, \ldots, \lambda_m$ are the distinct eigenvalues of A, gives

$$\int_0^\infty \| e^{At} \| \, dt = \int_0^\infty \| \sum_{k=1}^m \sum_{j=1}^{\sigma_k} W_{kj} \frac{t^{j-1}}{(j-1)!} e^{\lambda_k t} \| \, dt$$

$$\leq \sum_{k=1}^m \sum_{j=1}^{\sigma_k} \| W_{kj} \| \int_\tau^t \frac{t^{j-1}}{(j-1)!} | e^{\lambda_k t} | \, dt \tag{13}$$

Since $| e^{\lambda_k t} | = e^{\mathrm{Re}[\lambda_k] t}$, an exercise in integration by parts shows that the right side is finite, and exponential stability follows.

If the negative-real-part eigenvalue condition on A fails, then appropriate selection of an eigenvector of A as an initial state can be used to show that the linear state equation is not exponentially stable. Suppose first that a real eigenvalue λ is nonnegative, and let p be an associated eigenvector. Then the power series representation for the matrix exponential easily shows that

$$e^{At} p = e^{\lambda t} p$$

For the initial state $x_o = p$, it is clear that the corresponding solution of (1), $x(t) = e^{\lambda t} p$, does not go to zero as $t \to \infty$. Thus the state equation is not exponentially stable.

Now suppose that $\lambda = \sigma + i\omega$ is a complex eigenvalue of A with $\sigma \geq 0$. Again let p be an eigenvector associated with λ, and note that the conjugates $\bar{\lambda}, \bar{p}$ give another eigenvalue-eigenvector pair for A. Then

$$e^{At}(p + \bar{p}) = e^{\lambda t} p + e^{\bar{\lambda} t} \bar{p} = 2\mathrm{Re}[e^{\lambda t} p \,]$$

$$= 2\mathrm{Re}\left[e^{\sigma t} (\cos \omega t + i \sin \omega t) (\mathrm{Re}\,[p] + i\,\mathrm{Im}\,[p]) \right]$$

$$= 2\mathrm{Re}\,[p]\, e^{\sigma t} \cos \omega t - 2\mathrm{Im}\,[p]\, e^{\sigma t} \sin \omega t$$

and the (real) initial state $x_o = p + \bar{p}$ yields a solution that does not approach zero as $t \to \infty$.

□□□

This proof, with a bit of elaboration, shows also that $\lim_{t \to \infty} e^{At} = 0$ is a necessary and sufficient condition for uniform exponential stability in the time-invariant case. The corresponding statement is not true for time-varying linear state equations.

6.11 Example   Consider a scalar linear state equation (1) with

$$A(t) = \frac{-2t}{t^2 + 1} \tag{14}$$

Quick computation gives

$$\Phi(t,\, t_o) = \frac{t_o^2 + 1}{t^2 + 1}$$

and it is obvious that $\Phi(t,\, t_o) \to 0$ as $t \to \infty$, for any t_o. However the state equation is not uniformly exponentially stable, for suppose there exist positive λ and γ such that

$$\|\Phi(t,\, \tau)\| = \frac{\tau^2 + 1}{t^2 + 1} \leq \gamma e^{-\lambda(t - \tau)}$$

for all t, τ such that $t \geq \tau$. Taking $\tau = 0$, this inequality implies

$$1 \leq (t^2 + 1)\gamma e^{-\lambda t} , \quad t \geq 0$$

but L'Hospital's rule easily proves that the right side goes to zero as $t \to \infty$. This contradiction shows that the condition for uniform exponential stability cannot be satisfied.

Uniform Asymptotic Stability

Example 6.11 raises the interesting puzzle of what is needed in addition to $\lim_{t \to \infty} \Phi(t,\, t_o) = 0$ for uniform exponential stability in the time-varying case. The answer turns out to be a uniformity condition, and perhaps the best way to explore this issue is to start afresh with another stability definition.

6.12 Definition   The linear state equation (1) is called *uniformly asymptotically stable* if it is uniformly stable, and if given any positive constant δ there exists a positive T such that for any t_o and x_o the corresponding solution satisfies

$$\|x(t)\| \le \delta\|x_o\| \, , \quad t \ge t_o + T \tag{15}$$

Note that the elapsed time T until the solution satisfies the bound (15) must be independent of the initial time. (It is easy to verify that the state equation in Example 6.11 does not have this feature.) Some of the same tools used in proving Theorem 6.8 can be used to show that this 'elapsed-time uniformity' is the key to uniform exponential stability.

6.13 Theorem   The linear state equation (1) is uniformly asymptotically stable if and only if it is uniformly exponentially stable.

Proof   Suppose that the state equation is uniformly exponentially stable, that is, there exist finite, positive γ and λ such that $\|\Phi(t, \tau)\| \le \gamma e^{-\lambda(t-\tau)}$ whenever $t \ge \tau$. Then the state equation clearly is uniformly stable. To show it is uniformly asymptotically stable, for a given $\delta > 0$ pick T such that $e^{-\lambda T} \le \delta/\gamma$. Then for any t_o and x_o, and $t \ge t_o + T$,

$$\|x(t)\| = \|\Phi(t, t_o)x_o\| \le \|\Phi(t, t_o)\| \, \|x_o\|$$
$$\le \gamma e^{-\lambda(t-t_o)}\|x_o\| \le \gamma e^{-\lambda T}\|x_o\| \le \delta\|x_o\| \, , \quad t \ge t_o + T$$

This demonstrates uniform asymptotic stability.

Conversely suppose the state equation is uniformly asymptotically stable. Uniform stability is implied by definition, so there exists a positive γ such that

$$\|\Phi(t, \tau)\| \le \gamma \tag{16}$$

for all t, τ such that $t \ge \tau$. Select $\delta = \frac{1}{2}$, and by Definition 6.12 let T be such that (15) is satisfied. Then given a t_o, let x_a be such that $\|x_a\| = 1$, and

$$\|\Phi(t_o + T, t_o)x_a\| = \|\Phi(t_o + T, t_o)\|$$

With the initial state $x(t_o) = x_a$, the solution of (1) satisfies

$$\|x(t_o + T)\| = \|\Phi(t_o + T, t_o)x_a\| = \|\Phi(t_o + T, t_o)\| \, \|x_a\|$$
$$\le \frac{1}{2}\|x_a\|$$

from which

$$\|\Phi(t_o + T, t_o)\| \le \frac{1}{2} \tag{17}$$

Of course such an x_a exists for any given t_o, so the argument compels (17) for any t_o. Now uniform exponential stability is implied by (16) and (17), exactly as in the proof of Theorem 6.8.

Lyapunov Transformations

The stability concepts under discussion are properties of a particular linear state equation that presumably represents a system of interest in terms of physically-meaningful variables. A basic question involves preservation of stability properties under a state variable change. Since time-varying variable changes are permitted, simple scalar examples can be generated to show that, for example, uniform stability can be created or destroyed by variable change. To circumvent this difficulty we must limit attention to a particular class of state variable changes.

6.14 Definition   An $n \times n$ matrix $P(t)$ that is continuously differentiable and invertible at each t is called a *Lyapunov transformation* if there exist finite positive constants ρ and η such that for all t,

$$\|P(t)\| \le \rho , \quad |\det P(t)| \ge \eta \tag{18}$$

A condition equivalent to (18) is existence of a finite positive constant ρ such that for all t,

$$\|P(t)\| \le \rho , \quad \|P^{-1}(t)\| \le \rho$$

Exercise 1.7 shows that the lower bound on $|\det P(t)|$ implies an upper bound on $\|P^{-1}(t)\|$, and Exercise 1.15 provides the converse.

Reflecting on the effect of a state variable change on the transition matrix, a detailed proof that Lyapunov transformations preserve stability properties is perhaps belaboring the evident.

6.15 Theorem   Suppose the $n \times n$ matrix $P(t)$ is a Lyapunov transformation. Then the linear state equation (1) is uniformly stable (respectively, uniformly exponentially stable) if and only if the state equation

$$\dot{z}(t) = \left[P^{-1}(t)A(t)P(t) - P^{-1}(t)\dot{P}(t) \right] z(t) \tag{19}$$

is uniformly stable (respectively, uniformly exponentially stable).

Proof   The linear state equations (1) and (19) are related by the variable change $z(t) = P^{-1}(t)x(t)$, as shown in Chapter 4, and we note that the properties required of a Lyapunov transformation subsume those required of a variable change. Thus the relation between the two transition matrices is

$$\Phi_z(t, \tau) = P^{-1}(t)\Phi_x(t, \tau)P(\tau)$$

Now suppose (1) is uniformly stable. Then there exists γ such that $\|\Phi_x(t, \tau)\| \le \gamma$ for all t, τ such that $t \ge \tau$, and, from (18) and Exercise 1.7,

$$\|\Phi_z(t, \tau)\| = \|P^{-1}(t)\Phi_x(t, \tau)P(\tau)\|$$

$$\le \|P^{-1}(t)\| \, \|\Phi_x(t, \tau)\| \, \|P(\tau)\|$$

$$\le \frac{\gamma\rho^n}{\eta} \tag{20}$$

for all t, τ such that $t \geq \tau$. This shows that (19) is uniformly stable. An obviously similar argument applied to

$$\Phi_x(t, \tau) = P(t)\Phi_z(t, \tau)P^{-1}(\tau)$$

shows that if (19) is uniformly stable, then (1) is uniformly stable. The corresponding demonstrations for uniform exponential stability are similar.
□□□

The Floquet decomposition for T-periodic state equations, Property 5.11, provides a general illustration. Since $P(t)$ is the product of a transition matrix and a matrix exponential, it is continuously differentiable with respect to t. Since $P(t)$ is invertible, by continuity arguments there exist ρ, $\eta > 0$ such that (18) holds for all t in any interval of length T. By periodicity these bounds then hold for all t, and it follows that $P(t)$ is a Lyapunov transformation. It is easy to verify that $z(t) = P^{-1}(t)x(t)$ yields the time-invariant linear state equation

$$\dot{z}(t) = Rz(t)$$

By this connection the stability properties of the original T-periodic state equation are equivalent to the stability properties of a time-invariant linear state equation.

6.16 Example   Revisiting Example 5.12, the stability properties of

$$\dot{x}(t) = \begin{bmatrix} -1 & 0 \\ -\cos t & 0 \end{bmatrix} x(t) \tag{21}$$

are equivalent to the stability properties of

$$\dot{z}(t) = \begin{bmatrix} -1 & 0 \\ -1/2 & 0 \end{bmatrix} z(t)$$

From the computation

$$e^{Rt} = \begin{bmatrix} e^{-t} & 0 \\ -1/2 + 1/2 e^{-t} & 1 \end{bmatrix} \tag{22}$$

in Example 5.12, or from the solution of Exercise 6.7, it follows that (21) is uniformly stable, but not uniformly exponentially stable.

EXERCISES

Exercise 6.1   Show that uniform stability of the linear state equation

$$\dot{x}(t) = A(t)x(t), \quad x(t_o) = x_o$$

is equivalent to the following property. Given any positive constant ε there exists a positive constant δ such that, regardless of t_o, if $\|x_o\| \leq \delta$, then the corresponding solution satisfies $\|x(t)\| \leq \varepsilon$ for all $t \geq t_o$.

Exercise 6.2   Suppose there exists a finite constant α such that $\|A(t)\| \leq \alpha$ for all t. Prove that given a finite $\delta > 0$ there exists a finite $\gamma > 0$ such that $\|\Phi(t,\ \tau)\| \leq \gamma$ for all t, τ such that $|t - \tau| \leq \delta$.

Exercise 6.3   If $A(t) = -A^T(t)$, show that the linear state equation

$$\dot{x}(t) = A(t)x(t)$$

is uniformly stable. Show also that $P(t) = \Phi(t, 0)$ is a Lyapunov transformation.

Exercise 6.4   Show that the linear state equation $\dot{x}(t) = A(t)x(t)$ is uniformly exponentially stable if and only if the state equation $\dot{z}(t) = A^T(-t)z(t)$ is uniformly exponentially stable. Show by example that this equivalence does not hold for $\dot{z}(t) = A^T(t)z(t)$. (*Hint*: See Exercise 3.3 and Exercise 5.1.)

Exercise 6.5   Suppose $\Phi_1(t,\ \tau)$ is the transition matrix for $\frac{1}{2}[A(t) - A^T(t)]$, and let $P(t) = \Phi_1(t, 0)$. For the state equation $\dot{x}(t) = A(t)x(t)$, suppose the variable change $z(t) = P^{-1}(t)x(t)$ is used to obtain $\dot{z}(t) = F(t)z(t)$. Compute a simple expression for $F(t)$, and show that $F(t)$ is symmetric. Combine this with the Exercise 6.3 to show that for stability purposes only state equations with a symmetric coefficient matrix need be considered.

Exercise 6.6   Is the linear state equation

$$\dot{x}(t) = \begin{bmatrix} 1 & e^{-t} \\ -e^{-t} & 1 \end{bmatrix} x(t)$$

uniformly stable?

Exercise 6.7   For a time invariant linear state equation

$$\dot{x}(t) = Ax(t)$$

derive necessary and sufficient conditions for uniform stability in terms of the eigenvalues of A.

Exercise 6.8   Suppose the linear state equation $\dot{x}(t) - A(t)x(t)$ is uniformly stable. Then given x_o and t_o, show that the solution of

$$\dot{x}(t) = A(t)x(t) + f(t),\quad x(t_o) = x_o$$

is bounded if there exists a finite constant η such that

$$\int_{t_o}^{\infty} \|f(\sigma)\| d\sigma \leq \eta$$

Give a simple example to show that if $f(t)$ is a constant, then unbounded solutions can occur.

Exercise 6.9   Prove that the state equation $\dot{x}(t) = A(t)x(t)$ is uniformly exponentially stable if and only if there exist finite positive constants γ, λ such that

$$\|\Phi(t,\ \tau)\| \leq \gamma e^{-\lambda(t-\tau)}$$

for all t and τ satisfying $t \geq \tau$.

Exercise 6.10   Show that the linear state equation $\dot{x}(t) = A(t)x(t)$ with T-periodic $A(t)$ is uniformly exponentially stable if and only if $\lim_{t \to \infty} \Phi(t, t_o) = 0$ for each t_o.

Exercise 6.11   Suppose there exist finite constant α such that $\|A(t)\| \leq \alpha$ for all t, and finite β such that

$$\int_{\tau}^{t} \|\Phi(t, \sigma)\| \, d\sigma \leq \beta$$

for all t, τ with $t \geq \tau$. Show there exists a finite constant γ such that

$$\int_{\tau}^{t} \|\Phi(t, \sigma)\|^2 \, d\sigma \leq \gamma$$

for all t, τ such that $t \geq \tau$.

Exercise 6.12   Suppose there exists a finite constant α such that $\|A(t)\| \leq \alpha$ for all t. Prove that the linear state equation

$$\dot{x}(t) = A(t)x(t)$$

is uniformly exponentially stable if and only if there exists a finite constant β such that

$$\int_{\tau}^{t} \|\Phi(\sigma, \tau)\| \, d\sigma \leq \beta$$

for all t, τ such that $t \geq \tau$.

Exercise 6.13   Show that there exists a Lyapunov transformation $P(t)$ such that the linear state equation $\dot{x}(t) = A(t)x(t)$ is transformed to $\dot{z}(t) = 0$ by the state variable change $z(t) = P^{-1}(t)x(t)$ if and only if there exists a finite constant γ such that

$$\|\Phi(t, \tau)\| \leq \gamma$$

for all t and τ.

Exercise 6.14   Consider the linear state equation

$$\dot{x}(t) = a(t)Ax(t)$$

where $a(t)$ is a continuous, T-periodic, scalar function, and A is a constant $n \times n$ matrix. Derive a necessary and sufficient condition for uniform stability. (*Hint*: Use Exercise 6.7.)

NOTES

Note 6.1   There is a huge literature on stability theory for ordinary differential equations. The terminology is not completely standard, and careful attention to definitions is important when consulting alternative sources. For example we have defined uniform stability in a form specific to the linear case. Stability definitions in the more general context of nonlinear state equations are cast in terms of stability of an equilibrium state. Since zero always is an equilibrium state for a zero-input linear state equation, this aspect has been suppressed. Further, stability definitions for nonlinear state equations are local in nature: bounds and asymptotic properties of solutions for initial states sufficiently close to an equilibrium. In the linear case this restriction is superfluous. Books that provide a broader look at the subjects we cover include

R. Bellman, *Stability Theory of Differential Equations,* McGraw-Hill, New York, 1953

W.A. Coppel, *Stability and Asymptotic Behavior of Differential Equations,* Heath, Boston, 1965

J.L. Willems, *Stability Theory of Dynamical Systems,* John Wiley, New York, 1970

C.J. Harris, J.F. Miles, *Stability of Linear Systems,* Academic Press, New York, 1980

Note 6.2   Typically the definition of Lyapunov transformation includes a bound $\|\dot{P}(t)\| \leq \gamma$ for all t. This added condition preserves boundedness of $A(t)$ under state variable change, but is not needed for preservation of stability properties. Thus the condition has been dropped from Definition 6.14.

7

LYAPUNOV STABILITY CRITERIA

The origin of Lyapunov's so-called *direct method* for stability assessment is the notion that total energy of an unforced, dissipative mechanical system decreases as the state of the system evolves in time. Therefore the state vector approaches a constant value corresponding to zero energy as time increases. Phrased more generally, stability properties involve the growth properties of solutions of the state equation, and these properties can be measured by a suitable (energy-like) scalar function of the state vector. The problem is to find a suitable scalar function.

Introduction

To illustrate the basic idea we consider conditions that imply all solutions of the linear state equation

$$\dot{x}(t) = A(t)x(t) , \quad x(t_o) = x_o \tag{1}$$

are such that $\| x(t) \|^2$ monotonically decreases as $t \to \infty$. For any solution $x(t)$ of (1), the derivative of the scalar function

$$\| x(t) \|^2 = x^T(t)x(t) \tag{2}$$

with respect to t can be written as

$$\frac{d}{dt} \| x(t) \|^2 = \dot{x}^T(t)x(t) + x^T(t)\dot{x}(t)$$

$$= x^T(t)[A^T(t) + A(t)]x(t) \tag{3}$$

In this computation $\dot{x}(t)$ is replaced by $A(t)x(t)$ precisely because $x(t)$ is a solution of (1). Suppose that the quadratic form on the right side of (3) is negative definite, that is, suppose the matrix $A^T(t) + A(t)$ is negative definite at each t. Then, as shown in

98

Figure 7.1, $\|x(t)\|^2$ decreases as t increases. Further we can show that if this negative definiteness does not asymptotically vanish, that is, if there is a constant $\nu > 0$ such that $A^T(t) + A(t) \leq -\nu I$ for all t, then $\|x(t)\|^2$ goes to zero as $t \rightarrow \infty$. Notice that the transition matrix for $A(t)$ is not needed in this calculation, and growth properties of the scalar function (2) depend on sign-definiteness properties of the quadratic form in (3). Admittedly this calculation results in a restrictive sufficient condition — negative definiteness of $A^T(t) + A(t)$ — for a type of asymptotic stability. However more general scalar functions than (2) can be considered.

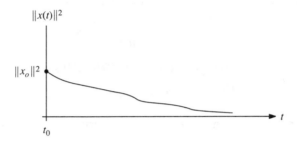

7.1 Figure   If $A^T(t) + A(t) < 0$ at each t, the solution norm decreases for $t \geq t_o$.

Formalization of the above discussion involves somewhat intricate definitions of time-dependent quadratic forms that are useful as scalar functions of the state vector of (1) for stability purposes. Such quadratic forms are called *quadratic Lyapunov functions*. They can be written as $x^T Q(t)x$, where $Q(t)$ is assumed to be symmetric and continuously-differentiable for all t. If $x(t)$ is any solution of (1) for $t \geq t_o$, then we are interested in the behavior of the real quantity $x^T(t)Q(t)x(t)$ for $t \geq t_o$. This behavior can be assessed by computing the time derivative using the product rule, and replacing $\dot{x}(t)$ by $A(t)x(t)$ to obtain

$$\frac{d}{dt}\left[x^T(t)Q(t)x(t) \right] = x^T(t)[A^T(t)Q(t) + Q(t)A(t) + \dot{Q}(t)]x(t) \qquad (4)$$

To analyze stability properties, various bounds are required on quadratic Lyapunov functions, and on the quadratic forms (4) that arise as their derivatives along solutions of (1). These bounds can be expressed in alternative ways. For example the condition that there exists a positive constant η such that

$$Q(t) \geq \eta I$$

for all t is equivalent by definition to existence of a positive η such that

$$x^T Q(t)x \geq \eta \|x\|^2$$

for all t and all $n \times 1$ vectors x. Yet another way to write this is to require that there exists a symmetric, positive-definite constant matrix M such that

$$x^T Q(t)x \geq x^T M x$$

for all t and all $n \times 1$ vectors x. The choice is largely a matter of taste, and the most economical form is adopted here.

Uniform Stability

We begin with a sufficient condition for uniform stability. The presentation style throughout is to list requirements on $Q(t)$ so that the corresponding quadratic form can be used to prove the desired stability property.

7.2 Theorem   The linear state equation (1) is uniformly stable if there exists an $n \times n$ matrix $Q(t)$ that for all t is symmetric, continuously differentiable, and such that

$$\eta I \le Q(t) \le \rho I \tag{5}$$

$$A^T(t)Q(t) + Q(t)A(t) + \dot{Q}(t) \le 0 \tag{6}$$

where η and ρ are finite positive constants.

Proof   Given any t_o and x_o, the corresponding solution $x(t)$ of (1) is such that, from (4) and (6),

$$x^T(t)Q(t)x(t) - x_o^TQ(t_o)x_o = \int_{t_o}^{t} \frac{d}{d\sigma} x^T(\sigma)Q(\sigma)x(\sigma)\, d\sigma$$

$$\le 0, \quad t \ge t_o$$

Using the inequalities in (5) we obtain

$$x^T(t)Q(t)x(t) \le x_o^TQ(t_o)x_o \le \rho\|x_o\|^2, \quad t \ge t_o$$

and then

$$\eta\|x(t)\|^2 \le \rho\|x_o\|^2, \quad t \ge t_o$$

Therefore

$$\|x(t)\| \le \sqrt{\rho/\eta}\,\|x_o\|, \quad t \ge t_o \tag{7}$$

Since (7) holds for any x_o and t_o, the state equation (1) is uniformly stable by definition. □□□

Typically it is profitable to use a quadratic Lyapunov function to obtain stability conditions for a family of linear state equations, rather than a particular instance.

7.3 Example   Consider the linear state equation

$$\dot{x}(t) = \begin{bmatrix} 0 & 1 \\ -1 & -a(t) \end{bmatrix} x(t) \tag{8}$$

where $a(t)$ is a continuous function defined for all t. Choose $Q(t) = I$, so that $x^T(t)Q(t)x(t) = x^T(t)x(t) = \|x(t)\|^2$, as suggested at the beginning of this chapter. Then (5) is satisfied by $\eta = \rho = 1$, and

$$A^T(t)Q(t) + Q(t)A(t) + \dot{Q}(t) = A^T(t) + A(t)$$

$$= \begin{bmatrix} 0 & 0 \\ 0 & -2a(t) \end{bmatrix}$$

If $a(t) \geq 0$ for all t, then the hypotheses in Theorem 7.2 are satisfied. Therefore we have proved (8) is uniformly stable if $a(t)$ is continuous and nonnegative for all t. Perhaps it should be emphasized that a more sophisticated choice of $Q(t)$ could yield uniform stability under weaker conditions on $a(t)$.

Uniform Exponential Stability

For uniform exponential stability Theorem 7.2 does not suffice — the choice $Q(t) = I$ proves that (8) with zero $a(t)$ is uniformly stable, but Example 5.9 shows this case is not exponentially stable. The strengthening of conditions in the following result appears slight at first glance, but this is deceptive. For example the strengthened conditions fail to hold in Example 7.3 with $Q(t) = I$ for any choice of $a(t)$.

7.4 Theorem   The linear state equation (1) is uniformly exponentially stable if there exists an $n \times n$ matrix $Q(t)$ that for all t is symmetric, continuously differentiable, and such that

$$\eta I \leq Q(t) \leq \rho I \tag{9}$$

$$A^T(t)Q(t) + Q(t)A(t) + \dot{Q}(t) \leq -\nu I \tag{10}$$

where η, ρ and ν are finite positive constants.

Proof   For any t_o, x_o, and corresponding solution $x(t)$ of the state equation, the inequality (10) gives

$$\frac{d}{dt}[x^T(t)Q(t)x(t)] \leq -\nu \|x(t)\|^2 , \quad t \geq t_o$$

Also from (9),

$$x^T(t)Q(t)x(t) \leq \rho \|x(t)\|^2 , \quad t \geq t_o$$

so that

$$-\|x(t)\|^2 \leq -\frac{1}{\rho} x^T(t)Q(t)x(t) , \quad t \geq t_o$$

Therefore

$$\frac{d}{dt}[x^T(t)Q(t)x(t)] \leq \frac{-\nu}{\rho} x^T(t)Q(t)x(t) , \quad t \geq t_o \tag{11}$$

and this implies, after multiplication by the appropriate exponential integrating factor, and integrating from t_o to t,

$$x^T(t)Q(t)x(t) \le e^{-\frac{\nu}{\rho}(t-t_o)} x_o^T Q(t_o)x_o , \quad t \ge t_o$$

Summoning (9) again,

$$\|x(t)\|^2 \le \frac{1}{\eta} x^T(t)Q(t)x(t)$$

$$\le \frac{1}{\eta} e^{-\frac{\nu}{\rho}(t-t_o)} x_o^T Q(t_o)x_o , \quad t \ge t_o$$

which in turn gives

$$\|x(t)\|^2 \le \frac{\rho}{\eta} e^{-\frac{\nu}{\rho}(t-t_o)} \|x_o\|^2 , \quad t \ge t_o \qquad (12)$$

Noting that (12) holds for any x_o and t_o, and taking the positive square root of both sides, uniform exponential stability is established.
□□□

For $n = 2$ and constant $Q(t) = Q$, Theorem 7.4 admits a simple pictorial representation. The condition (9) implies that Q is positive definite, and therefore the level curves of the real-valued function $x^T Q x$ are ellipses in the (x_1, x_2)-plane. The condition (10) implies that for any solution $x(t)$ of the state equation, the value of $x^T(t)Q x(t)$ is decreasing as t increases. Thus a plot of the solution $x(t)$ on the (x_1, x_2)-plane crosses smaller-value level curves as t increases, as shown in Figure 7.5. Under the

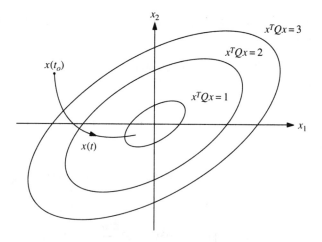

7.5 Figure   A solution $x(t)$ in relation to level curves for $x^T Q x$.

same assumptions, a similar pictorial interpretation can be given for Theorem 7.2. Note that if $Q(t)$ is not constant, the level curves vary with t and the picture is much less informative.

Just in case it appears that stability of linear state equations is reasonably intuitive, consider again the state equation (8) in Example 7.3 with a view to establishing uniform exponential stability. A first guess is that the state equation is uniformly exponentially stable if $a(t)$ is continuous and positive for all t, though suspicions might arise if $a(t) \to 0$ as $t \to \infty$. These suspicions would be well founded, but what is more surprising is that there are other obstructions to uniform exponential stability.

7.6 Example   A particular linear state equation of the form considered in Example 7.3 is

$$\dot{x}(t) = \begin{bmatrix} 0 & 1 \\ -1 & -(2 + e^t) \end{bmatrix} x(t) \tag{13}$$

Here $a(t) \geq 2$ for all t, and we have uniform stability, but the state equation is not uniformly exponentially stable. To see this, verify that a solution is

$$x(t) = \begin{bmatrix} 1 + e^{-t} \\ -e^{-t} \end{bmatrix}$$

Clearly this solution does not approach zero as $t \to \infty$.
□□□

The stability criteria provided by the preceding theorems are sufficient conditions that depend on skill in selecting an appropriate $Q(t)$. It is comforting to show that there indeed exists a suitable $Q(t)$ for a large class of uniformly exponentially stable linear state equations. The dark side is that it can be roughly as hard to compute $Q(t)$ as it is to compute the transition matrix for $A(t)$.

7.7 Theorem   Suppose that the linear state equation (1) is uniformly exponentially stable, and there exists a finite constant α such that $\|A(t)\| \leq \alpha$ for all t. Then

$$Q(t) = \int_t^\infty \Phi^T(\sigma, t)\Phi(\sigma, t)\, d\sigma \tag{14}$$

satisfies all the hypotheses of Theorem 7.4.

Proof   First we show that the integral converges for each t, so that $Q(t)$ is well defined. Since the state equation is uniformly exponentially stable, there exist positive γ and λ such that

$$\|\Phi(t, t_o)\| \leq \gamma e^{-\lambda(t - t_o)}$$

for all t, t_o such that $t \geq t_o$. Thus

$$\left\| \int_t^\infty \Phi^T(\sigma, t)\Phi(\sigma, t)\, d\sigma \right\| \leq \int_t^\infty \|\Phi^T(\sigma, t)\| \cdot \|\Phi(\sigma, t)\|\, d\sigma$$

$$\leq \int_t^\infty \gamma^2 e^{-2\lambda(\sigma-t)}\, d\sigma$$

$$= \gamma^2/(2\lambda)$$

for all t. This calculation also defines ρ in (9). Since $Q(t)$ clearly is symmetric and continuously differentiable at each t, it remains only to show that there exist η, $\nu > 0$ as needed in (9) and (10). To obtain ν, differentiation of (14) gives

$$\dot{Q}(t) = -I + \int_t^\infty \left[-A^T(t)\Phi^T(\sigma, t)\Phi(\sigma, t) - \Phi^T(\sigma, t)\Phi(\sigma, t)A(t) \right] d\sigma$$

$$= -I - A^T(t)Q(t) - Q(t)A(t) \tag{15}$$

That is

$$A^T(t)Q(t) + Q(t)A(t) + \dot{Q}(t) = -I$$

and clearly a valid choice for ν in (10) is $\nu = 1$. Finally it must be shown that there exists a positive η such that $Q(t) \geq \eta I$ for all t, and for this we set up an adroit maneuver. A differentiation followed by application of Exercise 1.4 gives, for any x and t,

$$\frac{d}{d\sigma}[\, x^T\Phi^T(\sigma, t)\Phi(\sigma, t)x\,] = x^T\Phi^T(\sigma, t)\,[\, A^T(\sigma) + A(\sigma)\,]\,\Phi(\sigma, t)x$$

$$\geq -\|A^T(\sigma) + A(\sigma)\|\, x^T\Phi^T(\sigma, t)\Phi(\sigma, t)x$$

$$\geq -2\alpha\, x^T\Phi^T(\sigma, t)\Phi(\sigma, t)x$$

Using the fact that $\Phi(\sigma, t)$ approaches zero exponentially as $\sigma \to \infty$, we integrate both sides to obtain

$$\int_t^\infty \frac{d}{d\sigma}\,[\, x^T\Phi^T(\sigma, t)\Phi(\sigma, t)x\,]\, d\sigma \geq -2\alpha \int_t^\infty x^T\Phi^T(\sigma, t)\Phi(\sigma, t)x\, d\sigma$$

$$= -2\alpha\, x^TQ(t)x \tag{16}$$

Evaluating the integral gives

$$-x^Tx \geq -2\alpha\, x^TQ(t)x$$

or

$$Q(t) \geq \frac{1}{2\alpha} I \tag{17}$$

for all t. Thus with the choice $\eta = 1/(2\alpha)$ all hypotheses of Theorem 7.4 are satisfied.
□□□

Exercise 7.14 shows that in fact there is a large family of matrices $Q(t)$ that can be used to prove uniform exponential stability under the hypotheses of Theorem 7.4.

Instability

Quadratic Lyapunov functions also can be used to develop instability criteria of various types. One example is the following result that, except for one value of t, does not involve a sign-definiteness assumption on $Q(t)$.

7.8 Theorem   Suppose there exists an $n \times n$ matrix $Q(t)$ that for all t is symmetric, continuously differentiable, and such that

$$\|Q(t)\| \leq \rho \tag{18}$$

$$A^T(t)Q(t) + Q(t)A(t) + \dot{Q}(t) \leq -\nu I \tag{19}$$

where ρ and ν are finite positive constants. Also suppose there exists a t_a such that $Q(t_a)$ is not positive semidefinite. Then the linear state equation (1) is not uniformly stable.

Proof   Suppose $x(t)$ is the solution of (1) with $t_o = t_a$ and $x_o = x_a$ such that $x_a^T Q(t_a) x_a < 0$. Then, from (19),

$$x^T(t)Q(t)x(t) - x_o^T Q(t_o)x_o = \int_{t_o}^{t} \frac{d}{d\sigma} x^T(\sigma)Q(\sigma)x(\sigma) \, d\sigma$$

$$\leq -\nu \int_{t_o}^{t} x^T(\sigma)x(\sigma) \, d\sigma \leq 0, \quad t \geq t_o$$

One consequence of this inequality, (18), and the choice of x_o and t_o, is

$$-\rho \|x(t)\|^2 \leq x^T(t)Q(t)x(t) \leq x_o^T Q(t_o)x_o < 0, \quad t \geq t_o \tag{20}$$

and a further consequence is that

$$\nu \int_{t_o}^{t} x^T(\sigma)x(\sigma) \, d\sigma \leq x_o^T Q(t_o)x_o - x^T(t)Q(t)x(t)$$

$$\leq |x^T(t)Q(t)x(t)| + |x_o^T Q(t_o)x_o|$$

$$\leq 2|x^T(t)Q(t)x(t)|, \quad t \geq t_o \tag{21}$$

Using (18), (21) gives

$$\int_{t_o}^{t} x^T(\sigma)x(\sigma)\,d\sigma \le \frac{2\rho}{\nu}\|x(t)\|^2 , \quad t \ge t_o \tag{22}$$

The state equation can be shown to be not uniformly stable by proving that $x(t)$ is unbounded. This we do by a contradiction argument. Suppose that there exists a finite γ such that $\|x(t)\| \le \gamma$, for all $t \ge t_o$. Then (22) gives

$$\int_{t_o}^{t} x^T(\sigma)x(\sigma)\,d\sigma \le \frac{2\rho\gamma^2}{\nu} , \quad t \ge t_o$$

and the integrand, which is a continuously-differentiable scalar function, must go to zero as $t \to \infty$. Therefore $x(t)$ must also go to zero, and this implies that (20) is violated for sufficiently large t. The contradiction proves that $x(t)$ cannot be a bounded solution.

7.9 Example   Consider a linear state equation with

$$A(t) = \begin{bmatrix} 0 & 1 \\ -a_1(t) & -a_2(t) \end{bmatrix}$$

The choice

$$Q(t) = \begin{bmatrix} a_1(t) & 0 \\ 0 & 1 \end{bmatrix} \tag{23}$$

gives

$$Q(t)A(t) + A^T(t)Q(t) + \dot{Q}(t) = \begin{bmatrix} \dot{a}_1(t) & 0 \\ 0 & -2a_2(t) \end{bmatrix}$$

Suppose that $a_1(t)$ is continuously differentiable, and there exists a finite constant ρ such that $|a_1(t)| \le \rho$ for all t. Further suppose there exists t_a such that $a_1(t_a) < 0$, and a positive constant ν such that, for all t,

$$\dot{a}_1(t) \le -\nu , \quad a_2(t) \ge \frac{\nu}{2}$$

Then it is easy to check that all assumptions of Theorem 7.8 are satisfied, so that under these conditions on $a_1(t)$ and $a_2(t)$ the state equation is not uniformly stable. The unkind might view this result as disappointing, since the obvious special case of constant A is not captured by the conditions on $a_1(t)$ and $a_2(t)$.

Time Invariant Case

In the time-invariant case quadratic Lyapunov functions with constant Q can be used to connect Theorem 7.4 with the usual eigenvalue condition for exponential stability. If Q is symmetric and positive definite, then (9) is satisfied automatically. However, rather than specifying such a Q and checking to see if a positive v exists such that (10) is satisfied, the approach can be reversed. Choose a positive definite matrix M, for example $M = vI$, where $v > 0$. If there exists a symmetric, positive-definite Q such that

$$QA + A^T Q = -M \qquad (24)$$

then all the hypotheses of Theorem 7.4 are satisfied. Therefore the associated linear state equation

$$\dot{x}(t) = Ax(t), \quad x(0) = x_o$$

is exponentially stable, and from Theorem 6.10 we conclude that all eigenvalues of A have negative real parts. Conversely the eigenvalues of A enter the existence question for solutions of the *Lyapunov equation* (24).

7.10 Theorem If the $n \times n$ matrix A has negative-real-part eigenvalues, then for each symmetric $n \times n$ matrix M there exists a unique solution of (24) given by

$$Q = \int_0^\infty e^{A^T t} M e^{At} \, dt \qquad (25)$$

Furthermore if M is positive definite, then Q is positive definite.

Proof If all eigenvalues of A have negative real parts, it is obvious that the integral in (25) converges, so Q is well defined. To show that Q is a solution of (24), we calculate

$$A^T Q + QA = \int_0^\infty A^T e^{A^T t} M e^{At} \, dt + \int_0^\infty e^{A^T t} M e^{At} A \, dt$$

$$= \int_0^\infty \frac{d}{dt} [e^{A^T t} M e^{At}] \, dt$$

$$= e^{A^T t} M e^{At} \Big|_0^\infty = -M$$

To prove this solution is unique, suppose Q_a also is a solution. Then

$$[Q_a - Q]A + A^T[Q_a - Q] = 0 \qquad (26)$$

But this implies

$$e^{A^T t}[Q_a - Q]Ae^{At} + e^{A^T t}A^T[Q_a - Q]e^{At} = 0, \quad t \geq 0$$

from which

$$\frac{d}{dt}[e^{A^T t}(Q_a - Q)e^{At}] = 0, \quad t \geq 0$$

Integrating both sides from 0 to ∞ gives

$$0 = e^{A^T t}(Q_a - Q)e^{At} \Big|_0^\infty = -(Q_a - Q)$$

That is, $Q_a = Q$.

Now suppose that M is positive definite. Clearly Q is symmetric. To show it is positive definite simply note that for a nonzero $n \times 1$ vector x,

$$x^T Q x = \int_0^\infty x^T e^{A^T t} M e^{At} x \, dt > 0$$

since the integrand is a positive scalar function. (In detail, $e^{At}x \ne 0$ for $t \ge 0$, so positive definiteness of M shows that the integrand is positive for all $t \ge 0$.)
□□□

When A has negative-real-part eigenvalues the Lyapunov equation (24) has a unique, positive-definite solution under weaker conditions on M. This is addressed in later chapters. (Also (24) has solutions under weaker hypotheses on A, though these results are not pursued.)

EXERCISES

Exercise 7.1   For a linear state equation such that $A(t) = -A^T(t)$, find a $Q(t)$ that demonstrates uniform stability. Can you find a $Q(t)$ that demonstrates exponential stability?

Exercise 7.2   State and prove a Lyapunov instability theorem that guarantees every nonzero initial state yields an unbounded solution.

Exercise 7.3   Using

$$Q(t) = Q = \begin{bmatrix} 1 & 1/2 \\ 1/2 & 1/2 \end{bmatrix}$$

find the weakest conditions on $a(t)$ such that

$$\dot{x}(t) = \begin{bmatrix} 0 & 1 \\ -a(t) & -2 \end{bmatrix} x(t)$$

can be shown to be uniformly stable.

Exercise 7.4   For a linear state equation with

$$A(t) = \begin{bmatrix} 0 & 1 \\ -a(t) & -2 \end{bmatrix}$$

consider the choice

$$Q(t) = \begin{bmatrix} a(t) & 0 \\ 0 & 1 \end{bmatrix}$$

Find the least restrictive conditions on $a(t)$ so that uniform exponential stability can be concluded. Does there exist an $a(t)$ satisfying the conditions?

Exercise 7.5   For a linear state equation with

$$A(t) = \begin{bmatrix} 0 & 1 \\ -a_1(t) & -a_2(t) \end{bmatrix}$$

use the choice

$$Q(t) = \begin{bmatrix} 1 & 0 \\ 0 & \dfrac{1}{a_1(t)} \end{bmatrix}$$

to determine conditions on $a_1(t)$ and $a_2(t)$ such that the state equation is uniformly stable.

Exercise 7.6   For a linear state equation with

$$A(t) = \begin{bmatrix} 0 & 1 \\ -a_1(t) & -a_2(t) \end{bmatrix}$$

use the choice

$$Q(t) = \begin{bmatrix} a_1(t) & 0 \\ 0 & 1 \end{bmatrix}$$

to determine conditions on $a_1(t)$ and $a_2(t)$ such that the state equation is uniformly stable. Do there exist coefficients $a_1(t)$ and $a_2(t)$ such that this $Q(t)$ demonstrates uniform exponential stability?

Exercise 7.7   For a linear state equation with

$$A(t) = \begin{bmatrix} 0 & 1 \\ -a_1(t) & -a_2(t) \end{bmatrix}$$

use the choice

$$Q(t) = \begin{bmatrix} a_1(t) + a_2(t) + \dfrac{a_1(t)}{a_2(t)} & 1 \\ 1 & 1 + \dfrac{1}{a_2(t)} \end{bmatrix}$$

to derive sufficient conditions for uniform exponential stability.

Exercise 7.8   For a linear state equation with

$$A(t) = \begin{bmatrix} 0 & 1 \\ -1 & -a(t) \end{bmatrix}$$

use the choice

$$Q(t) = \begin{bmatrix} a(t) + \dfrac{2}{a(t)} & 1 \\ 1 & \dfrac{2}{a(t)} \end{bmatrix}$$

to determine conditions on $a(t)$ such that the state equation is uniformly stable.

Exercise 7.9   Show that all eigenvalues of the matrix A have real parts less than $-\mu$, $\mu > 0$, if and only if for every symmetric, positive-definite M there exists a unique, symmetric, positive-definite Q such that

$$A^T Q + QA + 2\mu Q = -M$$

Exercise 7.10   Suppose that for given constant $n \times n$ matrices A and M there exists a constant, $n \times n$ matrix Q that satisfies

$$A^T Q + QA = -M$$

Show that for all $t \geq 0$,

$$Q = e^{A^T t} Q e^{At} + \int_0^t e^{A^T \sigma} M e^{A\sigma} \, d\sigma$$

Exercise 7.11   For a given constant, $n \times n$ matrix A, suppose M and Q are symmetric, positive definite, $n \times n$ matrices such that

$$QA + A^T Q = -M$$

Using the (in general complex) eigenvectors of A in a clever way, show that all eigenvalues of A have negative real parts.

Exercise 7.12   Develop a sufficient condition for existence of a unique solution, and an explicit solution formula, for the linear equation

$$FQ + QA = -M$$

where F, A, and M are constant $n \times n$ matrices.

Exercise 7.13   Suppose the $n \times n$ matrix A has negative-real-part eigenvalues, and M is an $n \times n$, symmetric, positive-definite matrix. Prove that if Q satisfies

$$QA + A^T Q = -M$$

then

$$\max_{0 \leq t < \infty} \|e^{At}\| \leq \sqrt{\|Q\| \|Q^{-1}\|}$$

Hint: At any $t \geq 0$ use a particular $n \times 1$ vector x and the Rayleigh-Ritz inequality for

$$\int_t^\infty x^T e^{A^T \sigma} M e^{A\sigma} x \, d\sigma$$

Exercise 7.14   Suppose that all eigenvalues of A have real parts less than $-\mu < 0$. Show that for any ε satisfying $0 < \varepsilon < \mu$,

$$\|e^{At}\| \leq \sqrt{2\|Q\|(\|A\| + \mu - \varepsilon)} \; e^{-(\mu - \varepsilon)t} \, , \quad t \geq 0$$

where Q is the unique solution of

$$A^T Q + QA + 2(\mu - \varepsilon)Q = -I$$

Hint: Use Theorem 7.10 to conclude

$$Q = \int_0^\infty e^{[A^T+(\mu-\epsilon)I]t} e^{[A+(\mu-\epsilon)I]t} \, dt$$

Then show that for any $n \times 1$ vector x and any $t \geq 0$,

$$\int_t^\infty \frac{d}{d\sigma} \left[x^T e^{[A^T+(\mu-\epsilon)I]\sigma} e^{[A+(\mu-\epsilon)I]\sigma} x \right] d\sigma \geq -2 \left[\, \|A\| + \mu - \epsilon \, \right] x^T Q x$$

Exercise 7.15   State and prove a generalized version of Theorem 7.7 using

$$Q(t) = \int_t^\infty \Phi^T(\sigma, t) P(\sigma) \Phi(\sigma, t) \, d\sigma$$

under appropriate assumptions on the $n \times n$ matrix $P(\sigma)$.

Exercise 7.16   Given the linear state equation $\dot{x}(t) = A(t)x(t)$, suppose there exists a real function $v(t, x)$ that is continuous with respect to t and x, and that satisfies the following conditions.
(a) There exist continuous, strictly increasing real functions $\alpha(\cdot)$ and $\beta(\cdot)$ such that $\alpha(0) = \beta(0) = 0$, and

$$\alpha(\|x\|) \leq v(t, x) \leq \beta(\|x\|)$$

for all t and all x.
(b) If $x(t)$ is any solution of the state equation, then the time function $v(t, x(t))$ is nonincreasing. Prove that the state equation is uniformly stable. (This shows that attention need not be restricted to quadratic Lyapunov functions, and that smoothness assumptions can be weakened.) *(Hint:* Use the characterization of uniform stability in Exercise 6.1.)

Exercise 7.17   If the state equation $\dot{x}(t) = A(t)x(t)$ is uniformly stable, prove that there exists a function $v(t, x)$ that has the properties listed in Exercise 7.15. *(Hint:* Writing the solution of the state equation with $x(t_o) = x_o$ as $x(t; x_o, t_o)$, let

$$v(t, x) = \sup_{\sigma \geq 0} \|x(t+\sigma; x, t)\|$$

where *supremum* denotes the least upper bound.)

NOTES

Note 7.1   The Lyapunov method is a powerful tool in the setting of nonlinear state equations as well. Scalar energy-like functions of the state more general than quadratic functions are used, and this requires general definitions of concepts such as positive definiteness. A standard reference is

W. Hahn, *Stability of Motion,* Springer-Verlag, New York, 1967

The subject also is treated in many introductory texts in nonlinear systems. For example,

H.K. Khalil, *Nonlinear Systems,* Macmillan, New York, 1992

Note 7.2   The conditions

$$0 < \eta I \leq Q(t) \leq \rho I$$

$$A^T(t)Q(t) + Q(t)A(t) + \dot{Q}(t) \leq -\nu I < 0$$

for uniform exponential stability can be weakened in various ways. These more general criteria involve concepts such as controllability and observability that are discussed in Chapter 9. Early results can be found in

B.D.O. Anderson, J.B. Moore, "New results in linear system stability," *SIAM Journal on Control,* Vol. 7, No. 3, pp. 398–414, 1969

B.D.O. Anderson, "Exponential stability of linear equations arising in adaptive identification," *IEEE Transactions on Automatic Control,* Vol. 22, No. 1, pp. 83–88, 1977

Also see

R. Ravi, A.M. Pascoal, P.P. Khargonekar, "Normalized coprime factorizations and the graph metric for linear time-varying systems," *Systems & Control Letters,* 1992

For the time-invariant case the more general results relate to the existence of positive-definite solutions of the Lyapunov equation (24) when M is positive semidefinite. Again this involves the concepts of controllability and observability. See Exercise 9.9 for a sample result.

8

ADDITIONAL STABILITY CRITERIA

In addition to the Lyapunov stability criteria in Chapter 7, other types of stability conditions often are useful. Typically these are sufficient conditions that are proved by application of the Lyapunov stability theorems, or the Gronwall-Bellman inequality (Exercise 3.4), though sometimes either technique can be used, and sometimes both are used in the same proof.

Eigenvalue Conditions

At first it might be thought that the the pointwise-in-time eigenvalues of $A(t)$ could be used to characterize internal stability properties of a linear state equation

$$\dot{x}(t) = A(t)x(t) , \quad x(t_o) = x_o \tag{1}$$

but this is not generally true. One example is provided by Exercise 4.14, and in case the unboundedness of $A(t)$ in that example is suspected as the difficulty, we give a well-known example with bounded $A(t)$.

8.1 Example For the linear state equation (1) with

$$A(t) = \begin{bmatrix} -1 + \alpha \cos^2 t & 1 - \alpha \sin t \cos t \\ -1 - \alpha \sin t \cos t & -1 + \alpha \sin^2 t \end{bmatrix} \tag{2}$$

where α is a positive constant, the pointwise eigenvalues are constants, given by

$$\lambda(t) = \lambda = \frac{\alpha - 2 \pm \sqrt{\alpha^2 - 4}}{2}$$

It is not difficult to verify that

$$\Phi(t, 0) = \begin{bmatrix} e^{(\alpha-1)t} \cos t & e^{-t} \sin t \\ -e^{(\alpha-1)t} \sin t & e^{-t} \cos t \end{bmatrix}$$

113

Thus while the pointwise eigenvalues of $A(t)$ have negative real parts if $0 < \alpha < 2$, the state equation has unbounded solutions if $\alpha > 1$.
□□□

Despite such examples the eigenvalue idea is not completely daft. At the end of this chapter we show, via a rather complicated Lyapunov argument, that for slowly-time-varying linear state equations uniform exponential stability is implied by negative-real-part eigenvalues of $A(t)$. Before that a number of simpler eigenvalue conditions and perturbation results are discussed, the first of which is a straightforward application of the Rayleigh-Ritz inequality reviewed in Chapter 1.

8.2 Theorem   For the linear state equation (1), denote the largest and smallest pointwise eigenvalues of $A(t) + A^T(t)$ by $\lambda_{max}(t)$ and $\lambda_{min}(t)$. Then for any x_o and t_o the solution of (1) satisfies

$$\|x_o\| e^{\frac{1}{2}\int_{t_o}^{t} \lambda_{min}(\sigma)\, d\sigma} \le \|x(t)\| \le \|x_o\| e^{\frac{1}{2}\int_{t_o}^{t} \lambda_{max}(\sigma)\, d\sigma} \quad , \quad t \ge t_o \tag{3}$$

Proof   First note that since the eigenvalues of a matrix are continuous functions of the entries of the matrix, and the entries of $A(t) + A^T(t)$ are continuous functions of t, the pointwise eigenvalues $\lambda_{min}(t)$ and $\lambda_{max}(t)$ are continuous functions of t. Thus the integrals in (3) are well defined. Suppose $x(t)$ is a solution of the state equation corresponding to a given t_o and nonzero x_o. Using

$$\frac{d}{dt} \|x(t)\|^2 = \frac{d}{dt} x^T(t) x(t) = x^T(t)[A^T(t) + A(t)]x(t)$$

the Rayleigh-Ritz inequality gives

$$\|x(t)\|^2 \lambda_{min}(t) \le \frac{d}{dt} \|x(t)\|^2 \le \|x(t)\|^2 \lambda_{max}(t) \ , \quad t \ge t_o$$

Dividing through by $\|x(t)\|^2$, which is positive at each t, and integrating from t_o to any $t \ge t_o$, yields

$$\int_{t_o}^{t} \lambda_{min}(\sigma)\, d\sigma \le \ln\|x(t)\|^2 - \ln\|x_o\|^2 \le \int_{t_o}^{t} \lambda_{max}(\sigma)\, d\sigma \ , \quad t \ge t_o$$

Exponentiation followed by taking the nonnegative square root gives (3).
□□□

Theorem 8.2 leads to easy proofs of some simple stability criteria based on the eigenvalues of $A(t) + A^T(t)$.

8.3 Corollary   The linear state equation (1) is uniformly stable if there exists a finite constant γ such that the largest pointwise eigenvalue of $A(t) + A^T(t)$ satisfies

$$\int_{\tau}^{t} \lambda_{max}(\sigma)\, d\sigma \le \gamma \tag{4}$$

for all t, τ such that $t \ge \tau$.

8.4 Corollary   The linear state equation (1) is uniformly exponentially stable if there exist a finite constant γ and finite positive constant λ such that the largest pointwise eigenvalue of $A(t) + A^T(t)$ satisfies

$$\int_{\tau}^{t} \lambda_{\max}(\sigma)\, d\sigma \leq -\lambda(t - \tau) + \gamma \tag{5}$$

for all t, τ such that $t \geq \tau$.

These criteria are quite conservative in the sense that many uniformly stable, or uniformly exponentially stable, linear state equations do not satisfy the respective conditions (4) and (5).

Perturbation Results

Another approach is to consider state equations that are close, in some sense, to a linear state equation that has a particular stability property. While explicit, tight bounds sometimes are of interest, the focus here is on simple calculations that establish the desired property.

8.5 Theorem   Suppose the linear state equation (1) is uniformly stable. Then the linear state equation

$$\dot{z}(t) = [A(t) + F(t)]z(t) \tag{6}$$

is uniformly stable if there exists a finite constant β such that for all τ

$$\int_{\tau}^{\infty} \|F(\sigma)\|\, d\sigma \leq \beta \tag{7}$$

Proof   For any t_o and z_o the solution of (6) satisfies

$$z(t) = \Phi_A(t, t_o)z_o + \int_{t_o}^{t} \Phi_A(t, \sigma)F(\sigma)z(\sigma)\, d\sigma$$

where, of course, $\Phi_A(t, \tau)$ denotes the transition matrix for $A(t)$. By uniform stability of (1) there exists a constant γ such that $\|\Phi_A(t, \tau)\| \leq \gamma$ for all t, τ such that $t \geq \tau$. Therefore, taking norms,

$$\|z(t)\| \leq \gamma\|z_o\| + \int_{t_o}^{t} \gamma\|F(\sigma)\|\,\|z(\sigma)\|\, d\sigma, \quad t \geq t_o$$

Applying the Gronwall inequality (Lemma 3.2) gives

$$\|z(t)\| \leq \gamma\|z_o\|\, e^{\int_{t_o}^{t} \gamma\|F(\sigma)\|\, d\sigma}, \quad t \geq t_o$$

Then the bound (7) yields

$$\| z(t) \| \le \gamma\, e^{\gamma\beta}\, \| z_o \| , \quad t \ge t_o$$

and since this same bound can be obtained for any value of t_o, uniform stability of (6) is established.

8.6 Theorem   Suppose the linear state equation (1) is uniformly exponentially stable and there exists a finite constant α such that $\| A(t) \| \le \alpha$ for all t. Then there exists a positive constant β such that the linear state equation

$$\dot{z}(t) = [A(t) + F(t)]z(t) \tag{8}$$

is uniformly exponentially stable if $\| F(t) \| \le \beta$ for all t.

Proof   Since (1) is uniformly exponentially stable and $A(t)$ is bounded, by Theorem 7.7

$$Q(t) = \int_t^\infty \Phi_A^T(\sigma, t)\Phi_A(\sigma, t)\, d\sigma \tag{9}$$

is such that all the hypotheses of Theorem 7.4 are satisfied for (1). Next we show that $Q(t)$ also satisfies all the hypotheses of Theorem 7.4 for the perturbed linear state equation (8). A quick check of the required properties reveals that it only remains to show existence of a positive constant ν such that, for all t,

$$[A(t) + F(t)]^T Q(t) + Q(t)[A(t) + F(t)] + \dot{Q}(t) \le -\nu I$$

By calculation of $\dot{Q}(t)$ from (9), this requirement can be rewritten as

$$F^T(t)Q(t) + Q(t)F(t) \le (1 - \nu)I \tag{10}$$

for all t. Denoting the bound on $\| Q(t) \|$ by ρ and choosing $\beta = 1/4\rho$ gives

$$\| F^T(t)Q(t) + Q(t)F(t) \| \le 2\| F(t) \|\, \| Q(t) \| \le 1/2$$

for all t, and thus (10) is satisfied with $\nu = {}^1\!/_2$.
□□□

The different types of perturbations that preserve the different stability properties in Theorems 8.5 and 8.6 are significant. For example the scalar state equation with $A(t) = 0$ is uniformly stable, but a perturbation with $F(t) = \beta$, for any positive β, no matter how small, clearly yields unbounded solutions. See also Exercise 8.6 and Note 8.4.

Slowly Varying Systems

Now a basic result involving an eigenvalue condition for uniform exponential stability of linear state equations with slowly-varying $A(t)$ is presented. The proof offered here makes use of the Kronecker product of matrices, which is defined as follows. If A

is an $n_A \times m_A$ matrix with entries a_{ij}, and B is an $n_B \times m_B$ matrix, then the *Kronecker product* $A \otimes B$ is given by

$$A \otimes B = \begin{bmatrix} a_{11}B & \cdots & a_{1m_A}B \\ \vdots & \vdots & \vdots \\ a_{n_A 1}B & \cdots & a_{n_A m_A}B \end{bmatrix} \tag{11}$$

Obviously $A \otimes B$ is an $n_A n_B \times m_A m_B$ matrix, and any two matrices are conformable with respect to this product. Less clear is the fact that the Kronecker product has many interesting properties. However the only property we need involves expressions of the form $I \otimes A + A \otimes I$, where both A and the identity are $n \times n$ matrices. It is not difficult to show that the n^2 eigenvalues of $I \otimes A + A \otimes I$ are simply the n^2 sums $\lambda_i + \lambda_j$, $i, j = 1, \ldots, n$, where $\lambda_1, \ldots, \lambda_n$ are the eigenvalues of A. Indeed this is transparent in the case of diagonal A.

8.7 Theorem Suppose for the linear state equation (1) with $A(t)$ continuously differentiable there exist finite positive constants α, μ such that, for all t, $\|A(t)\| \leq \alpha$ and every pointwise eigenvalue of $A(t)$ satisfies $\text{Re}[\lambda(t)] \leq -\mu$. Then there exists a positive constant β such that if the time-derivative of $A(t)$ satisfies $\|\dot{A}(t)\| \leq \beta$ for all t, the state equation is uniformly exponentially stable.

Proof For each t let $Q(t)$ be the solution of

$$A^T(t)Q(t) + Q(t)A(t) = -I \tag{12}$$

Existence, uniqueness, and positive definiteness of $Q(t)$ for each t is guaranteed by Theorem 7.10, and furthermore

$$Q(t) = \int_0^\infty e^{A^T(t)\sigma} e^{A(t)\sigma} \, d\sigma \tag{13}$$

The strategy of the proof is to show that $Q(t)$ can be used to satisfy the requirements of Theorem 7.4, so uniform exponential stability of (1) follows.

First the Kronecker product is used to show boundedness of $Q(t)$. Let e_i denote the ith column of I, and $q_i(t)$ denote the ith column of $Q(t)$. Defining the $n^2 \times 1$ vectors

$$e = \begin{bmatrix} e_1 \\ \vdots \\ e_n \end{bmatrix}, \quad q(t) = \begin{bmatrix} q_1(t) \\ \vdots \\ q_n(t) \end{bmatrix}$$

the $n \times n$ matrix equation (12) can be written as the $n^2 \times 1$ vector equation

$$[A^T(t) \otimes I + I \otimes A^T(t)]q(t) = -e \tag{14}$$

The following argument shows that $q(t)$ is bounded, and thus that there exists a finite ρ such that $Q(t) \leq \rho I$ for all t. If $\lambda_1(t), \ldots, \lambda_n(t)$ are the pointwise eigenvalues of $A(t)$, then the n^2 pointwise eigenvalues of $[A^T(t) \otimes I + I \otimes A^T(t)]$ are

$$\lambda_{i,j}(t) = \lambda_i(t) + \lambda_j(t), \quad i, j = 1, \ldots, n$$

Therefore $\mathrm{Re}[\lambda_{i,j}(t)] \leq -2\mu$, for all t, from which

$$| \det [A^T(t) \otimes I + I \otimes A^T(t)] | = | \prod_{i,j=1}^{n} \lambda_{i,j}(t) | \geq (2\mu)^{n^2}$$

for all t. Therefore $A^T(t) \otimes I + I \otimes A^T(t)$ is invertible at each t. Since $A(t)$ is bounded, $A^T(t) \otimes I + I \otimes A^T(t)$ is bounded, and hence the inverse

$$[A^T(t) \otimes I + I \otimes A^T(t)]^{-1}$$

is bounded for all t by Exercise 1.7. The right side of (14) is constant, and therefore we conclude that $q(t)$ is bounded.

Clearly $Q(t)$ is symmetric and continuously differentiable, and next we show that there exists a $\nu > 0$ such that

$$A^T(t)Q(t) + Q(t)A(t) + \dot{Q}(t) \leq -\nu I$$

for all t. Using (12) this requirement can be rewritten as

$$\dot{Q}(t) \leq (1 - \nu)I \tag{15}$$

Differentiation of (12) with respect to t yields

$$A^T(t)\dot{Q}(t) + \dot{Q}(t)A(t) = -\dot{A}^T(t)Q(t) - Q(t)\dot{A}(t)$$

At each t this Lyapunov equation has a unique solution

$$\dot{Q}(t) = \int_0^\infty e^{A^T(t)\sigma} [\dot{A}^T(t)Q(t) + Q(t)\dot{A}(t)] e^{A(t)\sigma} \, d\sigma$$

again since the eigenvalues of $A(t)$ have negative real parts at each t. To derive a bound on $\|\dot{Q}(t)\|$ we use the boundedness of $\|Q(t)\|$. For any $n \times 1$ vector x and any t,

$$| x^T e^{A^T(t)\sigma} [\dot{A}^T(t)Q(t) + Q(t)\dot{A}(t)] e^{A(t)\sigma}x |$$
$$\leq \|\dot{A}^T(t)Q(t) + Q(t)\dot{A}(t)\| \, x^T e^{A^T(t)\sigma} e^{A(t)\sigma}x$$

Thus

$$| x^T \dot{Q}(t)x | = | \int_0^\infty x^T e^{A^T(t)\sigma} [\dot{A}^T(t)Q(t) + Q(t)\dot{A}(t)] e^{A(t)\sigma}x \, d\sigma |$$

$$\leq \| \dot{A}^T(t)Q(t) + Q(t)\dot{A}(t) \| \, x^T Q(t)x$$

$$\leq 2\|\dot{A}(t)\| \, \|Q(t)\| \, x^T Q(t)x \tag{16}$$

Maximizing the right side over unity norm x, Exercise 1.5 gives, for all x such that $\|x\| = 1$,

$$| x^T \dot{Q}(t)x | \leq 2\|\dot{A}(t)\| \, \|Q(t)\|^2 \tag{17}$$

This yields, on maximization of the left side of (17) over unity norm x,

$$\|\dot{Q}(t)\| \leq 2\|\dot{A}(t)\| \, \|Q(t)\|^2$$

for all t. Using the bound on $\|Q(t)\|$, the bound β on $\|\dot{A}(t)\|$ can be chosen so that, for example, $\|\dot{Q}(t)\| \leq 1/2$. Then the choice $\nu = 1/2$ can be made for (15).

It only remains to show that there exists a positive η such that $Q(t) \geq \eta I$ for all t, and this involves a maneuver similar to one in the proof of Theorem 7.7. For any t and any $n \times 1$ vector x,

$$\frac{d}{d\sigma} [x^T e^{A^T(t)\sigma} e^{A(t)\sigma}x] = x^T e^{A^T(t)\sigma} [A^T(t) + A(t)] e^{A(t)\sigma}x$$

$$\geq -2\alpha x^T e^{A^T(t)\sigma} e^{A(t)\sigma}x \tag{18}$$

Therefore, since $e^{A(t)\sigma}$ goes to zero exponentially as $\sigma \to \infty$,

$$-x^T x = \int_0^\infty \frac{d}{d\sigma} [x^T e^{A^T(t)\sigma} e^{A(t)\sigma}x] \, d\sigma \geq -2\alpha x^T Q(t)x \tag{19}$$

That is

$$Q(t) \geq \frac{1}{2\alpha} I$$

for any t, and the proof is complete.

EXERCISES

Exercise 8.1   Derive a necessary and sufficient condition for uniform exponential stability of a scalar linear state equation.

Exercise 8.2   Show that the linear state equation $\dot{x}(t) = A(t)x(t)$ is not uniformly stable if for some t_o

$$\lim_{t \to \infty} \int_{t_o}^{t} \mathrm{tr}[A(\sigma)] \, d\sigma = \infty$$

Exercise 8.3   Theorem 8.2 implies that the linear time-invariant state equation

$$\dot{x}(t) = Ax(t)$$

is exponentially stable if all eigenvalues of $A + A^T$ are negative. Does the converse hold?

Exercise 8.4   Show that not all solutions $y(t)$ of the nth-order linear differential equation

$$y^{(n)}(t) + a_{n-1}(t)y^{(n-1)}(t) + \cdots + a_0(t)y(t) = 0$$

approach zero as $t \to \infty$ if for some t_o there is a positive constant α such that

$$\lim_{t \to \infty} \int_{t_o}^{t} a_{n-1}(\sigma) \, d\sigma \leq \alpha$$

Exercise 8.5   For the time-invariant linear state equation

$$\dot{x}(t) = [A + F]x(t)$$

suppose constants α and K are such that

$$\|e^{At}\| \leq Ke^{\alpha t}, \quad t \geq 0$$

Show that

$$\|e^{(A+F)t}\| \leq Ke^{(\alpha + K\|F\|)t}, \quad t \geq 0$$

Exercise 8.6   Suppose that the linear state equation

$$\dot{x}(t) = A(t)x(t)$$

is uniformly exponentially stable. Prove that if there exists a finite constant β such that

$$\int_{\tau}^{\infty} \|F(t)\| \, dt \leq \beta$$

for all τ, then the state equation

$$\dot{x}(t) = [A(t) + F(t)]x(t)$$

is uniformly exponentially stable.

Exercise 8.7   Suppose the linear state equation

$$\dot{x}(t) = [A + F(t)]x(t), \quad x(t_o) = x_o$$

is such that the constant matrix A has negative-real-part eigenvalues, and the continuous matrix function $F(t)$ satisfies

$$\lim_{t \to \infty} \|F(t)\| = 0$$

Prove that given any t_o and x_o the resulting solution satisfies

$$\lim_{t \to \infty} x(t) = 0$$

Exercise 8.8   For a differentiable $n \times n$ matrix function $A(t)$, suppose there exist positive constants α, β, μ such that, for all t, $\|A(t)\| \leq \alpha$, $\|\dot{A}(t)\| \leq \beta$, and the pointwise eigenvalues of $A(t)$ satisfy $\mathrm{Re}[\lambda(t)] \leq -\mu$. If $Q(t)$ is the unique positive definite solution of

$$A^T(t)Q(t) + Q(t)A(t) = -I$$

show that the linear state equation

$$\dot{x}(t) = [A(t) - \tfrac{1}{2}Q^{-1}(t)\dot{Q}(t)]x(t)$$

is uniformly exponentially stable.

Exercise 8.9   Extend Exercise 8.8 to a proof of Theorem 8.7 by using the Gronwall-Bellman inequality to prove that if β is sufficiently small, then uniform exponential stability of the linear state equation

$$\dot{z}(t) = A(t)z(t)$$

is implied by uniform exponential stability of the state equation

$$\dot{x}(t) = [A(t) - \tfrac{1}{2}Q^{-1}(t)\dot{Q}(t)]x(t)$$

Exercise 8.10   Suppose $A(t)$ satisfies the hypotheses of Theorem 8.7.

$$F(t) = A(t) + (\mu/2)I$$

$$Q(t) = \int_0^\infty e^{F^T(t)\sigma} e^{F(t)\sigma}\, d\sigma$$

and let ρ be such that $Q(t) \leq \rho I$, as in the proof of Theorem 8.7. Show that, for any value of t,

$$\|e^{A(t)\tau}\| \leq \rho\sqrt{2\alpha + \mu}\, e^{-(\mu/2)\tau}, \quad \tau \geq 0$$

(*Hint*: See the hint for Exercise 7.14.)

Exercise 8.11   Consider the single-input, n-dimensional, nonlinear state equation

$$\dot{x}(t) = A(u(t))x(t) + b(u(t)), \quad x(0) = x_o$$

where the entries of $A(\cdot)$ and $b(\cdot)$ are twice-continuously-differentiable functions of the input. Suppose that for each constant u_o satisfying $-\infty < u_{\min} \leq u_o \leq u_{\max} < \infty$ the eigenvalues of $A(u_o)$ have negative real parts. For a continuously-differentiable input signal $u(t)$ that satisfies $u_{\min} \leq u(t) \leq u_{\max}$ and $|\dot{u}(t)| \leq \delta$ for all $t \geq 0$, let

$$q(t) = -A^{-1}(u(t))b(u(t))$$

Show that if δ is sufficiently small and $\|x_o - q(0)\|$ is small, then $\|x(t) - q(t)\|$ remains small for all $t \geq 0$.

Exercise 8.12   Consider the nonlinear state equation

$$\dot{x}(t) = [A + F(t)]x(t) + g(t, x(t)), \quad x(t_o) = x_o$$

where A is a constant $n \times n$ matrix with negative-real-part eigenvalues, $F(t)$ is a continuous $n \times n$ matrix that satisfies $F(t) \leq \beta$ for all t, and $g(t, x)$ is a continuous function that satisfies

$\|g(t, x)\| \le \delta \|x\|$ for all t, x. Suppose $x(t)$ is a continuously differentiable solution defined for all $t \ge t_o$. Show that if β and δ are sufficiently small, then there exists finite positive constants γ, λ such that

$$\|x(t)\| \le \gamma e^{-\lambda(t-t_o)} \|x_o\|$$

for all $t \ge t_o$.

NOTES

Note 8.1   Example 8.1 is from

L. Markus, H. Yamabe, "Global stability criteria for differential systems," *Osaka Mathematical Journal,* Vol. 12, pp. 305–317, 1960

A survey of results on uniform exponential stability under the hypothesis that pointwise eigenvalues of the slowly-varying $A(t)$ have negative real parts is in

A. Ilchmann, D.H. Owens, D. Pratzel-Wolters, "Sufficient conditions for stability of linear time-varying systems," *Systems & Control Letters,* Vol. 9, pp. 157–163, 1987

An influential paper not cited in this reference is

C.A. Desoer, "Slowly varying system $\dot{x} = A(t)x$," *IEEE Transactions on Automatic Control,* Vol. 14, pp. 780–781, 1969

A sufficient condition for exponential decay of solutions in the case where $A(t)$ commutes with its integral is that the matrix function

$$\frac{1}{t} \int_{t_o}^{t} A(\sigma)\, d\sigma$$

be bounded and have negative-real-part eigenvalues for all $t \ge t_o$. This is proved in Section 7.7 of

D.L. Lukes, *Differential Equations: Classical to Controlled,* Academic Press, New York, 1982

Note 8.2   An example of a uniformly exponentially stable linear state equation where $A(t)$ has a pointwise eigenvalue with positive real part for all t, but is slowly varying, is provided in

R.A. Skoog, G.Y. Lau, "Instability of slowly varying systems," *IEEE Transactions on Automatic Control,* Vol. 17, No. 1, pp. 86–92, 1972

Recent work has produced stability results for slowly-varying linear state equations where eigenvalues can have positive real parts, so long as they have negative real parts 'on average.' See

V. Solo, "On the stability of slowly time-varying linear systems with applications to adaptive control," *SIAM Journal on Control and Optimization,* to appear

Note 8.3   Tighter bounds of the type given in Theorem 8.2 can be derived by using the *matrix measure*. This concept is developed and applied to the treatment of stability in

W.A. Coppel, *Stability and Asymptotic Behavior of Differential Equations,* D.C. Heath, Boston, 1965

Note 8.4 Finite-integral perturbations of the type in Theorem 8.5 can induce unbounded solutions when the unperturbed state equation has bounded solutions that approach zero asymptotically. For an example see Section 2.5 of

R. Bellman, *Stability Theory of Differential Equations,* McGraw-Hill, New York, 1953

In Section 1.14 state-variable changes to a time-variable diagonal form are considered. This approach is used to develop perturbation results for linear state equations of the form

$$\dot{x}(t) = [A + F(t)]x(t)$$

For additional results using a diagonal form for $A(t)$, consult

M.Y. Wu, "Stability of linear time-varying systems," *International Journal of System Sciences,* Vol. 15, pp. 137–150, 1984

Note 8.5 Extensive information on the Kronecker product is provided in

A. Graham, *Kronecker Products and Matrix Calculus: with Applications,* Halsted Press, New York, 1981

Note 8.6 Averaging techniques provide stability criteria for rapidly-varying periodic linear state equations. An entry into this literature is

R. Bellman, J. Bentsman, S.M. Meerkov, "Stability of fast periodic systems," *IEEE Transactions on Automatic Control,* Vol. 30, No. 3, pp. 289–291, 1985

9

CONTROLLABILITY AND OBSERVABILITY

The fundamental concepts of controllability and observability for an m-input, p-output, n-dimensional linear state equation

$$\dot{x}(t) = A(t)x(t) + B(t)u(t), \quad x(t_o) = x_o$$

$$y(t) = C(t)x(t) + D(t)u(t) \tag{1}$$

are introduced in this chapter. Controllability involves the influence of the input signal on the state vector, and does not involve the output equation. Observability deals with the influence of the state vector on the output, and does not involve the effect of a known input signal. In addition to their operational definitions in terms of driving the state with the input, and ascertaining the state from the output, these concepts play fundamental roles in the basic structure of linear state equations. The latter aspects are addressed in Chapter 10, and, using stronger notions of controllability and observability, in Chapter 11. For the time-invariant case further developments occur in Chapter 13 and Chapter 18.

Controllability

For a time-varying linear state equation the connection of the input signal to the state variables can change with time. Therefore the concept of controllability must be tied to a specific time interval, denoted $[t_o, t_f]$ with $t_f > t_o$.

9.1 Definition   The linear state equation (1) is called *controllable on* $[t_o, t_f]$ if given any x_o there exists a continuous input signal $u(t)$ defined on $[t_o, t_f]$ such that the corresponding solution of (1) satisfies $x(t_f) = 0$.

124

The continuity requirement on the input signal is consonant with our default technical setting, though typically much smoother input signals can be used to drive the state of a controllable linear state equation to zero.

As we develop criteria for controllability the observant will notice that contradiction proofs, or proofs of the contrapositive, often are used. Such proofs sometimes are criticized on the grounds that they are unenlightening. In any case the contradiction proofs are relatively simple, and they do explain why a claim *must* be true.

9.2 Theorem   The linear state equation (1) is controllable on $[t_o, t_f]$ if and only if the $n \times n$ matrix

$$W(t_o, t_f) = \int_{t_o}^{t_f} \Phi(t_o, t)B(t)B^T(t)\Phi^T(t_o, t) \, dt \tag{2}$$

is invertible.

Proof   Suppose $W(t_o, t_f)$ is invertible. Then given an $n \times 1$ vector x_o, choose

$$u(t) = -B^T(t)\Phi^T(t_o, t)W^{-1}(t_o, t_f)x_o \; , \quad t \in [t_o, t_f] \tag{3}$$

(This choice is completely unmotivated in the present context, though it is natural from a more-general viewpoint mentioned in Note 9.2.) The input signal (3) is continuous, and the corresponding solution of (1) with $x(t_o) = x_o$ can be written as

$$x(t_f) = \Phi(t_f, t_o)x_o + \int_{t_o}^{t_f} \Phi(t_f, \sigma)B(\sigma)u(\sigma) \, d\sigma$$

$$= \Phi(t_f, t_o)x_o - \int_{t_o}^{t_f} \Phi(t_f, \sigma)B(\sigma)B^T(\sigma)\Phi^T(t_o, \sigma)W^{-1}(t_o, t_f)x_o \, d\sigma$$

Using the composition property of the transition matrix gives

$$x(t_f) = \Phi(t_f, t_o)x_o - \Phi(t_f, t_o) \int_{t_o}^{t_f} \Phi(t_o, \sigma)B(\sigma)B^T(\sigma)\Phi^T(t_o, \sigma) \, d\sigma \, W^{-1}(t_o, t_f)x_o$$

$$= 0$$

Thus the state equation is controllable on $[t_o, t_f]$.

To show the reverse implication suppose that the linear state equation (1) is controllable on $[t_o, t_f]$ and that $W(t_o, t_f)$ is not invertible. On obtaining a contradiction we can conclude that $W(t_o, t_f)$ must be invertible. Since $W(t_o, t_f)$ is not invertible there exists a nonzero $n \times 1$ vector x_a such that

$$0 = x_a^T W(t_o, t_f)x_a = \int_{t_o}^{t_f} x_a^T \Phi(t_o, t)B(t)B^T(t)\Phi^T(t_o, t)x_a \, dt \tag{4}$$

The integrand in this expression is the nonnegative continuous function $\| x_a^T \Phi(t_o, t)B(t) \|^2$, and it follows that

$$x_a^T \Phi(t_o, t)B(t) = 0 \ , \quad t \in [t_o, t_f] \tag{5}$$

Because the state equation is controllable on $[t_o, t_f]$, choosing $x_o = x_a$ there exists a continuous input $u(t)$ such that

$$0 = \Phi(t_f, t_o)x_a + \int_{t_o}^{t_f} \Phi(t_f, \sigma)B(\sigma)u(\sigma) \, d\sigma$$

or

$$x_a = - \int_{t_o}^{t_f} \Phi(t_o, \sigma)B(\sigma)u(\sigma) \, d\sigma$$

Multiplying through by x_a^T and using (5) gives

$$x_a^T x_a = - \int_{t_o}^{t_f} x_a^T \Phi(t_o, \sigma)B(\sigma)u(\sigma) \, d\sigma = 0 \tag{6}$$

and this contradicts $x_a \neq 0$.
□□□

The *controllability Gramian* $W(t_o, t_f)$ has many properties, some of which are explored in Exercises. For every $t_f > t_o$ it is symmetric and positive semidefinite. Thus the linear state equation (1) is controllable on $[t_o, t_f]$ if and only if $W(t_o, t_f)$ is positive definite. If the state equation is not controllable on $[t_o, t_f]$, it might become so if t_f is increased. And controllability can be lost if t_f is lowered. Analogous observations can be made in regard to changing t_o.

Computing $W(t_o, t_f)$ from the definition (2) is not a happy prospect. Indeed $W(t_o, t_f)$ usually is computed by numerically solving matrix differential equations satisfied by $W(t, t_f)$ that are the subject of Exercise 9.4. However if we assume smoothness properties stronger than continuity for the coefficient matrices, the Gramian condition in Theorem 9.2 leads to a sufficient condition that is easier to check. Key to the proof is the fact that $W(t_o, t_f)$ fails to be invertible if and only if (5) holds for some $x_a \neq 0$. Since (5) corresponds to a type of linear dependence condition on the rows of $\Phi(t_o, t)B(t)$, controllability criteria have roots in concepts of linear independence of vector functions of time. However this viewpoint is not emphasized here.

9.3 Definition   Corresponding to the linear state equation (1), subject to existence and continuity of the indicated derivatives define a sequence of $n \times m$ matrices by

$$K_0(t) = B(t)$$

$$K_j(t) = -A(t)K_{j-1}(t) + \dot{K}_{j-1}(t) \ , \quad j = 1, 2, \cdots$$

An easy induction proof shows that

$$\frac{\partial^j}{\partial\sigma^j}\left[\Phi(t,\ \sigma)B(\sigma)\right]=\Phi(t,\ \sigma)K_j(\sigma)\ ,\quad j=0,\ 1,\ \cdots \tag{7}$$

Specifically the claim obviously holds for $j=0$. With J a nonnegative integer, suppose that

$$\frac{\partial^J}{\partial\sigma^J}\left[\Phi(t,\ \sigma)B(\sigma)\right]=\Phi(t,\ \sigma)K_J(\sigma)$$

Then, using this inductive hypothesis,

$$\frac{\partial^{J+1}}{\partial\sigma^{J+1}}\left[\Phi(t,\ \sigma)B(\sigma)\right]=\frac{\partial}{\partial\sigma}\left[\Phi(t,\ \sigma)K_J(\sigma)\right]$$

$$=-\Phi(t,\ \sigma)A(\sigma)K_J(\sigma)+\Phi(t,\ \sigma)\frac{d}{d\sigma}K_J(\sigma)$$

$$=\Phi(t,\ \sigma)K_{J+1}(\sigma)$$

Therefore the proof is complete.

Evaluation of (7) at $\sigma=t$ gives a simple interpretation of the matrices in Definition 9.3:

$$K_j(t)=\frac{\partial^j}{\partial\sigma^j}\left[\Phi(t,\ \sigma)B(\sigma)\right]\Bigg|_{\sigma=t}\ ,\quad j=0,\ 1,\ \cdots \tag{8}$$

9.4 Theorem   Suppose q is a positive integer such that, for $t\in[t_o,\ t_f]$, $B(t)$ is q-times continuously differentiable, and $A(t)$ is $(q-1)$-times continuously differentiable. Then the linear state equation (1) is controllable on $[t_o,\ t_f]$ if for some $t_c\in[t_o,\ t_f]$

$$\text{rank}\left[K_0(t_c)\ \ K_1(t_c)\ \ \cdots\ \ K_q(t_c)\right]=n \tag{9}$$

Proof   Suppose for some $t_c\in[t_o,\ t_f]$ the rank condition holds. To set up a contradiction argument suppose that the state equation is not controllable on $[t_o,\ t_f]$. Then $W(t_o,\ t_f)$ is not invertible and, as in the proof of Theorem 9.2, there exists a nonzero $n\times1$ vector x_a such that

$$x_a^T\Phi(t_o,\ t)B(t)=0\ ,\quad t\in[t_o,\ t_f] \tag{10}$$

Letting x_b be the nonzero vector $x_b=\Phi^T(t_o,\ t_c)x_a$, we have from (10) that

$$x_b^T\Phi(t_c,\ t)B(t)=0\ ,\quad t\in[t_o,\ t_f]$$

In particular this gives, at $t=t_c$, $x_b^TK_0(t_c)=0$. Next, differentiating (10) with respect to t gives

$$x_b^T\Phi(t_c,\ t)K_1(t)=0\ ,\quad t\in[t_o,\ t_f]$$

from which $x_b^T K_1(t_c) = 0$. Continuing this process gives, in general,

$$\frac{d^j}{dt^j} \left[x_b^T \Phi(t_c, t) B(t) \right] \bigg|_{t = t_c} = x_b^T K_j(t_c) = 0 , \quad j = 0, 1, \ldots, q$$

Therefore

$$x_b^T \left[K_0(t_c) \quad K_1(t_c) \quad \cdots \quad K_q(t_c) \right] = 0$$

and this contradicts the linear independence of the n rows implied by the rank condition in (9). Thus the state equation is controllable on $[t_o, t_f]$.

□□□

For a time-invariant linear state equation,

$$\dot{x}(t) = Ax(t) + Bu(t)$$

$$y(t) = Cx(t) + Du(t) \tag{11}$$

the most familiar test for controllability can be motivated from Theorem 9.4 by noting that

$$K_j(t) = (-1)^j A^j B , \quad j = 0, 1, \cdots$$

However to obtain a necessary as well as sufficient condition we base the proof on Theorem 9.2.

9.5 Theorem   The time-invariant linear state equation (11) is controllable on $[t_o, t_f]$ if and only if the $n \times nm$ *controllability matrix* satisfies

$$\text{rank} \left[B \quad AB \quad \cdots \quad A^{n-1}B \right] = n \tag{12}$$

Proof   We prove that the rank condition (12) fails if and only if the controllability Gramian

$$W(t_o, t_f) = \int_{t_o}^{t_f} e^{A(t_o-t)} BB^T e^{A^T(t_o-t)} \, dt$$

is not invertible. If the rank condition fails, then there exists a nonzero $n \times 1$ vector x_a such that

$$x_a^T A^k B = 0 , \quad k = 0, \ldots, n-1$$

This implies, using the matrix-exponential representation in Property 5.8,

$$x_a^T W(t_o, t_f) = \int_{t_o}^{t_f} \left[\sum_{k=0}^{n-1} \alpha_k(t_o-t) x_a^T A^k B \right] B^T e^{A^T(t_o-t)} \, dt$$

$$= 0 \tag{13}$$

and thus $W(t_o, t_f)$ is not invertible.

Conversely if the controllability Gramian is not invertible, then there exists a nonzero x_a such that

$$x_a^T W(t_o, t_f) x_a = 0$$

This implies, exactly as in the proof of Theorem 9.2,

$$x_a^T e^{A(t_o - t)} B = 0, \quad t \in [t_o, t_f]$$

At $t = t_o$ we obtain $x_a^T B = 0$, and differentiating k times and evaluating the result at $t = t_o$ gives

$$(-1)^k x_a^T A^k B = 0, \quad k = 0, \ldots, n-1 \tag{14}$$

Therefore

$$x_a^T \begin{bmatrix} B & AB & \cdots & A^{n-1}B \end{bmatrix} = 0$$

which proves that the rank condition (12) fails.

9.6 Example   Consider the linear state equation

$$\dot{x}(t) = \begin{bmatrix} a_1 & 0 \\ 0 & a_2 \end{bmatrix} x(t) + \begin{bmatrix} b_1(t) \\ b_2(t) \end{bmatrix} u(t), \quad t \in [t_o, t_f] \tag{15}$$

where the constants a_1 and a_2 are not equal. For constant values $b_1(t) = b_1$, $b_2(t) = b_2$, we can call on Theorem 9.5 to show that the state equation is controllable if and only if both b_1 and b_2 are nonzero. However for the nonzero, time-varying coefficients

$$b_1(t) = e^{a_1 t}, \quad b_2(t) = e^{a_2 t}$$

another straightforward calculation shows that

$$W(t_o, t_f) = \begin{bmatrix} e^{2a_1 t_o} & e^{(a_1 + a_2)t_o} \\ e^{(a_1 + a_2)t_o} & e^{2a_2 t_o} \end{bmatrix} (t_f - t_o)$$

Since $det\ W(t_o, t_f) = 0$ the time-varying linear state equation is not controllable on any interval $[t_o, t_f]$. Clearly pointwise-in-time interpretations of the controllability property can be misleading.
□□□

Since the rank condition (12) is independent of t_o and t_f, the controllability property for (11) is independent of the particular interval $[t_o, t_f]$. Thus for time-invariant linear state equations, the term *controllable* is used without reference to a time interval.

Observability

The second concept of interest for (1) involves the influence of the state vector on the output of the linear state equation. It is simplest to consider the case of zero input, and this does not entail loss of generality since the concept is unchanged in the presence of a known input signal. Specifically the zero-state response due to a known input signal can be computed, and subtracted from the complete response, leaving the zero-input response. Therefore we consider the unforced state equation

$$\dot{x}(t) = A(t)x(t), \quad x(t_o) = x_o$$

$$y(t) = C(t)x(t) \tag{16}$$

9.7 Definition   The linear state equation (16) is called *observable on* $[t_o, t_f]$ if any initial state x_o is uniquely determined by the corresponding response $y(t)$, for $t \in [t_o, t_f]$.

The basic characterization of observability is similar in form to the controllability case, though the proof is a bit simpler.

9.8 Theorem   The linear state equation (16) is observable on $[t_o, t_f]$ if and only if the $n \times n$ matrix

$$M(t_o, t_f) = \int_{t_0}^{t_f} \Phi^T(t, t_o)C^T(t)C(t)\Phi(t, t_0) \, dt \tag{17}$$

is invertible.

   Proof   Multiplying the solution expression

$$y(t) = C(t)\Phi(t, t_o)x_o$$

on both sides by $\Phi^T(t, t_o)C^T(t)$ and integrating yields

$$\int_{t_o}^{t_f} \Phi^T(t, t_0)C^T(t)y(t) \, dt = M(t_o, t_f)x_o \tag{18}$$

The left side is determined by $y(t)$, $t \in [t_o, t_f]$, and therefore (18) represents a linear algebraic equation for x_o. If $M(t_o, t_f)$ is invertible, then x_o is uniquely determined. On the other hand, if $M(t_o, t_f)$ is not invertible, then there exists a nonzero $n \times 1$ vector x_a such that $M(t_o, t_f)x_a = 0$. This implies $x_a^T M(t_o, t_f)x_a = 0$ and, just as in the proof of Theorem 9.2, it follows that

$$C(t)\Phi(t, t_o)x_a = 0, \quad t \in [t_o, t_f]$$

Thus $x(t_o) = x_o + x_a$ yields the same zero-input response for (16) on $[t_o, t_f]$ as $x(t_o) = x_o$, and the state equation fails to be observable on $[t_o, t_f]$.
□□□

The proof of Theorem 9.8 shows that for an observable linear state equation the initial state is uniquely determined by a linear algebraic equation, thus clarifying a vague aspect of Definition 9.7. Of course this equation is beset by the interrelated difficulties of computing the transition matrix and computing $M(t_o, t_f)$.

The *observability Gramian* $M(t_o, t_f)$, just as the controllability Gramian $W(t_o, t_f)$, has several interesting properties. It is symmetric and positive semidefinite, and positive definite if and only if the state equation is observable on $[t_o, t_f]$. Also $M(t_o, t_f)$ can be computed by numerically solving certain matrix differential equations. See the Exercises for profitable activities that avow the dual nature of controllability and observability.

We can develop more convenient criteria for observability just as in the controllability case.

9.9 Definition   Corresponding to the linear state equation (16), subject to existence of the indicated derivatives define a sequence of $p \times n$ matrices by

$$L_0(t) = C(t)$$
$$L_j(t) = L_{j-1}(t)A(t) + \dot{L}_{j-1}(t), \quad j = 1, 2, \cdots \tag{19}$$

It is easy to show by induction that

$$L_j(t) = \frac{\partial^j}{\partial t^j} \left[C(t)\Phi(t, \sigma) \right] \bigg|_{\sigma = t}, \quad j = 0, 1, \cdots \tag{20}$$

9.10 Theorem   Suppose q is a positive integer such that, for $t \in [t_o, t_f]$, $C(t)$ is q-times continuously differentiable, and $A(t)$ is $(q-1)$-times continuously differentiable. Then the linear state equation (16) is observable on $[t_o, t_f]$ if for some $t_a \in [t_o, t_f]$,

$$\text{rank} \begin{bmatrix} L_0(t_a) \\ \vdots \\ L_q(t_a) \end{bmatrix} = n \tag{21}$$

9.11 Theorem   If $A(t) = A$ and $C(t) = C$ in (16), then the time-invariant linear state equation is observable on $[t_o, t_f]$ if and only if the $np \times n$ *observability matrix* satisfies

$$\text{rank} \begin{bmatrix} C \\ CA \\ \vdots \\ CA^{n-1} \end{bmatrix} = n \tag{22}$$

As for controllability, the concept of observability for time-invariant linear state equations is independent of the particular (nonzero) time interval. Also comparing (15) and (22) we see that

$$\dot{x}(t) = Ax(t) + Bu(t)$$

is controllable if and only if

$$\dot{z}(t) = A^T z(t)$$
$$y(t) = B^T z(t)$$

is observable. This permits easy translation of algebraic consequences of controllability for time-invariant linear state equations into corresponding results for observability. (See for example Exercises 9.7 — 9.9.)

EXERCISES

Exercise 9.1   Prove Theorem 9.11.

Exercise 9.2   Consider a controllable, time-invariant linear state equation with two different $p \times 1$ outputs:

$$\dot{x}(t) = Ax(t) + Bu(t) , \quad x(0) = 0$$
$$y_a(t) = C_a x(t)$$
$$y_b(t) = C_b x(t)$$

Show that if the impulse response of the two outputs is identical, then $C_a = C_b$.

Exercise 9.3   Consider the linear state equation

$$\dot{x}(t) = \begin{bmatrix} 0 & 1 \\ 0 & 0 \end{bmatrix} x(t) + \begin{bmatrix} b_1(t) \\ 1 \end{bmatrix} u(t)$$

Is this state equation controllable on $[0, 1]$ for $b_1(t) = b_1$, an arbitrary constant? Is it controllable on $[0, 1]$ for every continuous function $b_1(t)$?

Exercise 9.4   Show that the controllability Gramian satisfies the matrix differential equation

$$\frac{d}{dt} W(t, t_f) = A(t)W(t, t_f) + W(t, t_f)A^T(t) - B(t)B^T(t) , \quad W(t_f, t_f) = 0$$

Also prove that the inverse of the controllability Gramian satisfies

$$\frac{d}{dt} W^{-1}(t, t_f) = -A^T(t)W^{-1}(t, t_f) - W^{-1}(t, t_f)A(t) + W^{-1}(t, t_f)B(t)B^T(t)W^{-1}(t, t_f)$$

for values of t such that the inverse exists, of course. Finally, show that

$$W(t_o, t_f) = W(t_o, t) + \Phi(t_o, t)W(t, t_f)\Phi^T(t_o, t)$$

Exercise 9.5 Establish properties of the observability Gramian $M(t_o, t_f)$ corresponding to the properties of $W(t_o, t_f)$ in Exercise 9.4.

Exercise 9.6 For the linear state equation

$$\dot{x}(t) = A(t)x(t) + B(t)u(t)$$

with associated controllability Gramian $W(t_o, t_f)$, show that the transition matrix for

$$\begin{bmatrix} A(t) & B(t)B^T(t) \\ 0 & -A^T(t) \end{bmatrix}$$

is given by

$$\begin{bmatrix} \Phi_A(t, \tau) & \Phi_A(t, \tau)W(\tau, t) \\ 0 & \Phi_A^T(\tau, t) \end{bmatrix}$$

Exercise 9.7 If β is a real constant, show that the time-invariant linear state equation

$$\dot{x}(t) = Ax(t) + Bu(t)$$

is controllable if and only if

$$\dot{z}(t) = (A - \beta I)z(t) + Bu(t)$$

is controllable.

Exercise 9.8 Suppose that the time-invariant linear state equation

$$\dot{x}(t) = Ax(t) + Bu(t)$$

is controllable, and that A has negative-real-part eigenvalues. Show that there exists a symmetric, positive-definite Q such that

$$AQ + QA^T = -BB^T$$

Exercise 9.9 Suppose the time-invariant linear state equation

$$\dot{x}(t) = Ax(t) + Bu(t)$$

is controllable, and that there exists a symmetric, positive-definite Q such that

$$AQ + QA^T = -BB^T$$

Show that all eigenvalues of A have negative real parts. (*Hint*: Use the (in general complex) left eigenvectors of A in a clever way.)

Exercise 9.10 The linear state equation

$$\dot{x}(t) = A(t)x(t) + B(t)u(t)$$

$$y(t) = C(t)x(t)$$

is called *output controllable on* $[t_o, t_f]$ if for any given x_o there exists a continuous input signal $u(t)$ defined for $t \in [t_o, t_f]$ such that the corresponding solution satisfies $y(t_f) = 0$. Assuming *rank* $C(t_f) = p$, show that a necessary and sufficient condition for output controllability on $[t_o, t_f]$

is invertibility of the $p \times p$ matrix

$$\int_{t_o}^{t_f} C(t_f)\Phi(t_f, t)B(t)B^T(t)\Phi^T(t_f, t)C^T(t_f)\, dt$$

Explain the role of the rank assumption on $C(t_f)$. For the special cases $m = 1$ and $p = 1$ express the condition in terms of the impulse response of the state equation.

Exercise 9.11   For a time-invariant linear state equation

$$\dot{x}(t) = Ax(t) + Bu(t)$$

$$y(t) = Cx(t)$$

with *rank C* $= p$, continue Exercise 9.10 by deriving a necessary and sufficient condition for output controllability similar to the condition in Theorem 9.5. If $m = p = 1$, characterize an output controllable state equation in terms of its impulse response, and its transfer function.

Exercise 9.12   It is interesting that continuity of $C(t)$ is crucial to the basic Gramian condition for observability. Show this by considering observability on $[0,1]$ for the scalar linear state equation with zero $A(t)$ and

$$C(t) = \begin{cases} 1, & t = 0 \\ 0, & t > 0 \end{cases}$$

Is continuity of $B(t)$ crucial in controllability?

Exercise 9.13   Suppose the single-input, single-output, n-dimensional, time-invariant linear state equation

$$\dot{x}(t) = Ax(t) + bu(t)$$

$$y(t) = cx(t)$$

is controllable and observable. Show that A and bc do not commute if $n \geq 2$.

Exercise 9.14   The linear state equation

$$\dot{x}(t) = A(t)x(t) + B(t)u(t), \quad x(t_o) = x_o$$

is called *reachable on* $[t_o, t_f]$ if for $x_o = 0$ and any given $n \times 1$ vector x_f there exists a (continuous) input signal $u(t)$ defined for $t \in [t_o, t_f]$ such that the corresponding solution satisfies $x(t_f) = x_f$. Show that the state equation is reachable on $[t_o, t_f]$ if and only if the $n \times n$ *reachability Gramian*

$$W_R(t_o, t_f) = \int_{t_o}^{t_f} \Phi(t_f, t)B(t)B^T(t)\Phi^T(t_f, t)\, dt$$

is invertible. Show also that the state equation is reachable on $[t_o, t_f]$ if and only if it is controllable on $[t_o, t_f]$.

Exercise 9.15   Based on Exercise 9.14, define a natural concept of *output reachability* for a time-varying linear state equation. Develop a basic Gramian criterion for output reachability in the style of Exercise 9.10.

Exercise 9.16   For the single-input, single-output state equation

$$\dot{x}(t) = A(t)x(t) + b(t)u(t)$$
$$y(t) = c(t)x(t)$$

suppose that

$$\overline{M}(t) = \begin{bmatrix} L_0(t) \\ L_1(t) \\ \vdots \\ L_{n-1}(t) \end{bmatrix}$$

is invertible for all t. Show that $y(t)$ satisfies a linear n^{th}-order differential equation of the form

$$y^{(n)}(t) - \sum_{j=0}^{n-1} \alpha_j(t) y^{(j)}(t) = \sum_{j=0}^{n} \beta_j(t) u^{(j)}(t)$$

where

$$[\ \alpha_0(t) \quad \cdots \quad \alpha_{n-1}(t)\ ] = L_n(t)\overline{M}^{-1}(t)$$

(A recursive formula for the β coefficients can be derived through a messy calculation.)

NOTES

Note 9.1   As indicated in Exercise 9.14, the term 'reachability' usually is associated with the ability to drive the state vector from zero to any desired state in finite time. In the setting of continuous-time linear state equations, this property is equivalent to the property of controllability, and the two terms sometimes are used interchangeably. However under certain types of uniformity conditions that are imposed in later chapters the equivalence is not preserved. Also for discrete-time linear state equations the corresponding concepts of controllability and reachability are not equivalent. Similar remarks apply to observability and the concept of 'reconstructibility,' defined roughly as follows. A linear state equation is *reconstructible on* $[t_o, t_f]$ if $x(t_f)$ can be determined from a knowledge of $y(t)$ for $t \in [t_o, t_f]$. This issue arises in the discussion of observers in Chapter 15.

Note 9.2   The concepts of controllability and observability introduced here can be refined to consider controllability of a particular state to the origin in finite time, or determination of a particular initial state from finite-time output observation. See for example the treatment in

R.W. Brockett, *Finite Dimensional Linear Systems,* John Wiley, New York, 1970

For time-invariant linear state equations, we pursue this refinement in Chapter 18 in the course of developing a geometric theory. A treatment of controllability and observability that emphasizes the role of linear independence of time functions is in

C.T. Chen, *Linear Systems Theory and Design,* Holt, Rinehart and Winston, New York, 1984

In many references a more sophisticated mathematical viewpoint is adopted for these topics. For controllability, the solution formula for a linear state equation shows that a state transfer from

$x(t_o) = x_o$ to $x(t_f) = 0$ is described by a linear map taking $m \times 1$ input signals into $n \times 1$ vectors. Setting up a suitable Hilbert space for the input space and equipping R^n with the usual inner product, basic linear operator theory involving adjoint operators and so on can be applied to the problem. Incidentally this formulation provides an interpretation of the mystery input signal in the proof of Theorem 9.2 as a minimum-energy input that accomplishes the transfer from x_o to zero.

Note 9.3   State transfers in a controllable time-invariant linear state equation can be accomplished with input signals that are polynomials in t of reasonable degree. Consult

A. Ailon, L. Baratchart, J. Grimm, G. Langholz, "On polynomial controllability with polynomial state for linear constant systems," *IEEE Transactions on Automatic Control,* Vol. 31, No. 2, pp. 155–156, 1986

D. Aeyels, "Controllability of linear time-invariant systems," *International Journal on Control,* Vol. 46, No. 6, pp. 2027–2034, 1987

For T-periodic linear state equations, controllability on any nonempty time interval is equivalent to controllability on $[0, nT]$, where n is the dimension of the state equation. This is established in

P. Brunovsky, "Controllability and linear closed-loop controls in linear periodic systems," *Journal of Differential Equations,* Vol. 6, pp. 296–313, 1969

Attempts to reduce this interval, and alternate definitions of controllability in the periodic case are discussed in

S. Bittanti, P. Colaneri, G. Guardabassi, "H-controllability and observability of linear periodic systems," *SIAM Journal on Control and Optimization,* Vol. 22, No. 6, pp. 889–893, 1984

H. Kano, T. Nishimura, "Controllability, stabilizability, and matrix Riccati equations for periodic systems," *IEEE Transactions on Automatic Control,* Vol. 30, No. 11, pp. 1129–1131, 1985

Note 9.4   For a linear state equation where $A(t)$ and $B(t)$ are analytic, Theorem 9.4 can be restated as a necessary and sufficient condition at any point $t_c \in [t_o, t_f]$. That is, an analytic linear state equation is controllable on the interval if and only if for some nonnegative integer j,

$$\text{rank} \left[K_0(t_c) \ K_1(t_c) \ \cdots \ K_j(t_c) \right] = n$$

The proof of necessity requires two technical facts related to analyticity, neither obvious. First, an analytic function that is not identically zero can be zero only at isolated points, and second, $\Phi(t, \tau)$ is analytic since $A(t)$ is analytic. In particular it is not true that a uniformly convergent series of analytic functions converges to an analytic function. Therefore the proof of analyticity of $\Phi(t, \tau)$ must be specific to properties of analytic differential equations. Consult Section 3.5 and Appendix C of

E.D. Sontag, *Mathematical Control Theory,* Springer-Verlag, New York, 1990

Note 9.5   Controllability is a point-to-point concept, in which the connecting trajectory is immaterial. The property of making the state follow a preassigned trajectory over a specified time interval is called *functional reproducibility* or *path controllability.* Consult

K. A. Grasse, "Sufficient conditions for the functional reproducibility of time-varying, input-output systems," *SIAM Journal on Control and Optimization,* Vol. 26, No. 1, pp. 230–249, 1988

See also the references on the closely-related notion of linear system inversion in Note 12.3.

Note 9.6   Additional aspects of controllabilty and observability, some of which arise in Chapter 11, are discussed in

L.M. Silverman, H.E. Meadows, "Controllability and observability in time-variable linear systems," *SIAM Journal on Control and Optimization,* Vol. 5, No. 1, pp. 64–73, 1967

In the time-invariant case important additional criteria for controllability and observability are introduced in Chapter 13.

10

REALIZABILITY

In this chapter we begin to address questions related to the input-output (zero-state) behavior of the standard linear state equation

$$\dot{x}(t) = A(t)x(t) + B(t)u(t), \quad x(t_o) = 0$$

$$y(t) = C(t)x(t) + D(t)u(t) \tag{1}$$

With zero initial state assumed, the output signal $y(t)$ corresponding to a given input signal $u(t)$, defined for $t \geq t_o$, is described by

$$y(t) = \int_{t_o}^{t} G(t, \sigma)u(\sigma)\, d\sigma + D(t)u(t), \quad t \geq t_o \tag{2}$$

where

$$G(t, \sigma) = C(t)\Phi(t, \sigma)B(\sigma)$$

Of course given the state equation (1), in principle $G(t, \sigma)$ can be computed so that the input-output behavior is known according to (2). Our interest here is in the reversal of this computation, and in particular we want to establish conditions on a specified $G(t, \sigma)$ that guarantee existence of a corresponding linear state equation. Aside from a certain theoretical symmetry, general motivation for our interest is provided by problems of implementing linear input/output behavior. Linear state equations can be constructed in hardware, as discussed in Chapter 1, or programmed in software for numerical solution.

Some terminology mentioned in Chapter 3 that goes with (2) bears repeating. The input-output behavior is *causal* since, for any $t_a \geq t_o$, the output value $y(t_a)$ does not depend on values of the input at times greater than t_a. Also the input-output behavior is *linear* since the response to a (constant-coefficient) linear combination of

138

input signals, $\alpha u_a(t) + \beta u_b(t)$, is $\alpha y_a(t) + \beta y_b(t)$, in the obvious notation. (In particular the response to zero input is $y(t) = 0$ for all t.) Thus we are interested in linear state equation representations for causal, linear input-output behavior.

Formulation

While the realizability question involves existence of a linear state equation (1) corresponding to a given $G(t, \sigma)$ and $D(t)$, it is obvious that $D(t)$ plays an unessential role. Therefore we assume henceforth that $D(t) = 0$, for all t, to simplify matters.

When there exists one linear state equation corresponding to a specified $G(t, \sigma)$, there exist many, since a change of state variables leaves $G(t, \sigma)$ unaffected. Also there exist linear state equations of different dimensions that yield a specified $G(t, \sigma)$. In particular new state variables that are disconnected from the input, the output, or both, can be added to a state equation without changing the corresponding input-output behavior.

10.1 Example   If the linear state equation (1) corresponds to a given input-output behavior, then a state equation of the form

$$\begin{bmatrix} \dot{x}(t) \\ \dot{z}(t) \end{bmatrix} = \begin{bmatrix} A(t) & 0 \\ 0 & F(t) \end{bmatrix} \begin{bmatrix} x(t) \\ z(t) \end{bmatrix} + \begin{bmatrix} B(t) \\ 0 \end{bmatrix} u(t)$$

$$y(t) = \begin{bmatrix} C(t) & 0 \end{bmatrix} \begin{bmatrix} x(t) \\ z(t) \end{bmatrix} + D(t)u(t) \tag{3}$$

yields the same input-output behavior. This is clear from Figure 10.2, or, since the transition matrix for (3) is block diagonal, from the easy calculation

$$\begin{bmatrix} C(t) & 0 \end{bmatrix} \begin{bmatrix} \Phi_A(t, \sigma) & 0 \\ 0 & \Phi_F(t, \sigma) \end{bmatrix} \begin{bmatrix} B(t) \\ 0 \end{bmatrix} = C(t)\Phi_A(t, \sigma)B(\sigma)$$

□□□

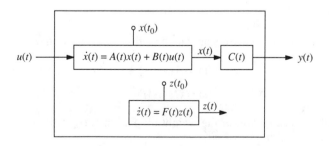

10.2 Figure   Structure of the linear state equation (3).

Example 10.1 shows that if a linear state equation of dimension n has the input-output behavior specified by $G(t, \sigma)$, then for any positive integer k, there are state equations of dimension $n + k$ that also have input-output behavior described by $G(t, \sigma)$. Thus our main theoretical interest is to consider least-dimension linear state equations corresponding to a specified $G(t, \sigma)$. But this is in accord with prosaic considerations: a least-dimension linear state equation is in some sense a simplest linear state equation yielding input-output behavior characterized by $G(t, \sigma)$.

There is a more vexing technical issue that should be addressed at the outset. Since the response computation in (2) involves values of $G(t, \sigma)$ only for $t \geq \sigma$, it seems most natural to assume that the input-output behavior is specified by $G(t, \sigma)$ only for arguments satisfying $t \geq \sigma$. With this restriction on arguments $G(t, \sigma)$ often is called an *impulse response,* for reasons that should be evident. However if $G(t, \sigma)$ arises from a linear state equation such as (1), then as a mathematical object $G(t, \sigma)$ is defined for all t, σ. And of course its values for $\sigma > t$ might not be completely determined by its values for $t \geq \sigma$. Delicate matters arise here. Some involve mathematical technicalities such as smoothness assumptions on $G(t, \sigma)$, and on the coefficient matrices in the state equations. Others involve subtleties in the mathematical representation of causality. A simple resolution is to insist that linear input-output behavior be specified by a $p \times m$ matrix function $G(t, \sigma)$ defined and, for compatibility with our default assumptions, continuous for all t, σ. Such a $G(t, \sigma)$ is called a *weighting pattern.*

A hint of the difficulties that arise in the realization problem when $G(t, \sigma)$ is specified only for $t \geq \sigma$ is provided by considering Exercise 10.4 in light of Theorem 10.6. For strong hypotheses that avert trouble with the impulse response, see the further consideration of the realization problem in Chapter 11. Finally notice that for a time-invariant linear state equation the distinction between the weighting pattern and impulse response is immaterial since values of $Ce^{A(t-\sigma)}B$ for $t \geq \sigma$ completely determine the values for $t < \sigma$. Namely for $\sigma > t$ the exponential $e^{A(t-\sigma)}$ is the inverse of $e^{A(\sigma-t)}$.

Realizability

Terminology that aids discussion of the realizability problem can be formalized as follows.

10.3 Definition   A linear state equation of dimension n

$$\dot{x}(t) = A(t)x(t) + B(t)u(t) , \quad x(t_o) = 0$$

$$y(t) = C(t)x(t) \tag{4}$$

is called a *realization* of the weighting pattern $G(t, \sigma)$ if

$$G(t, \sigma) = C(t)\Phi(t, \sigma)B(\sigma) \tag{5}$$

for all t and σ. If such a realization exists, then the weighting pattern is called *realizable*. If no realization of dimension less than n exists, then (4) is called a *minimal realization*.

10.4 Theorem   The weighting pattern $G(t, \sigma)$ is realizable if and only if there exist a $p \times n$ matrix function $H(t)$ and an $n \times m$ matrix function $F(t)$, both continuous for all t, such that

$$G(t, \sigma) = H(t)F(\sigma) \tag{6}$$

for all t and σ.

   Proof   Suppose there exist continuous matrix functions $F(t)$ and $H(t)$ such that (6) is satisfied. Then the linear state equation (with continuous coefficient matrices)

$$\dot{x}(t) = F(t)u(t)$$

$$y(t) = H(t)x(t) \tag{7}$$

is a realization of $G(t, \sigma)$ since the transition matrix for zero is the identity.
   Conversely suppose that $G(t, \sigma)$ is realizable. We can assume that the linear state equation (4) is one realization. Then using the composition property of the transition matrix we write

$$G(t, \sigma) = C(t)\Phi(t, \sigma)B(\sigma) = C(t)\Phi(t, 0)\Phi(0, \sigma)B(\sigma)$$

and by defining $H(t) = C(t)\Phi(t, 0)$ and $F(t) = \Phi(0, t)B(t)$ the proof is complete.
□□□
   While Theorem 10.4 provides the basic realizability criterion for weighting patterns, often it is not very useful because determining if $G(t, \sigma)$ can be factored in the requisite way can be difficult. In addition a simple example shows that the realization (7) can be displeasing compared to alternatives.

10.5 Example   For the weighting pattern

$$G(t, \sigma) = e^{-(t-\sigma)}$$

an obvious factorization gives the dimension-one realization corresponding to (7) as

$$\dot{x}(t) = e^{t}u(t)$$

$$y(t) = e^{-t}x(t)$$

While this linear state equation has an unbounded coefficient and clearly is not uniformly exponentially stable, neither of these ills is shared by the dimension-one realization

$$\dot{x}(t) = -x(t) + u(t)$$

$$y(t) = x(t) \tag{8}$$

Minimal Realization

We now consider the problem of characterizing minimal realizations of a realizable weighting pattern. It is convenient to make use of some simple observations mentioned in earlier chapters, but perhaps not emphasized. The first is that properties of controllability on $[t_o, t_f]$ and observability on $[t_o, t_f]$ are not effected by a change of state variables. Second, if (4) is an n-dimensional realization of a given weighting pattern, then the linear state equation obtained by changing variables according to $z(t) = P^{-1}(t)x(t)$ also is an n-dimensional realization of the same weighting pattern. In particular it is easy to verify that $P(t) = \Phi_A(t, t_o)$ satisfies

$$P^{-1}(t)A(t)P(t) - P^{-1}(t)\dot{P}(t) = 0$$

for all t, so the linear state equation in the new state $z(t)$ has the economical form

$$\dot{z}(t) = P^{-1}(t)B(t)u(t), \quad z(t_o) = 0$$
$$y(t) = C(t)P(t)z(t)$$

Therefore we often postulate realizations with $A(t)$ zero for simplicity, and without loss of generality.

It is not surprising, in view of Example 10.1, that controllability and observability play a role in characterizing minimality. However it might be a surprise that these concepts tell the whole story.

10.6 Theorem   Suppose the linear state equation (4) is a realization of the weighting pattern $G(t, \sigma)$. Then (4) is a minimal realization of $G(t, \sigma)$ if and only if for some t_o and $t_f > t_o$ it is both controllable and observable on $[t_o, t_f]$.

Proof   Sufficiency is proved via the contrapositive, by supposing that an n-dimensional realization (4) is not minimal. Without loss of generality it can be assumed that $A(t) = 0$ for all t. Then there is a lower-dimension realization of $G(t, \sigma)$, and again it can be assumed to have the form

$$\dot{z}(t) = F(t)u(t)$$
$$y(t) = H(t)z(t) \tag{9}$$

where the dimension of $z(t)$ is $n_z < n$. Writing the weighting pattern in terms of both realizations gives

$$C(t)B(\sigma) = H(t)F(\sigma)$$

for all t and σ. This implies

$$C^T(t)C(t)B(\sigma)B^T(\sigma) = C^T(t)H(t)F(\sigma)B^T(\sigma)$$

for all t, σ. For any t_o and any $t_f > t_o$ we can integrate this expression with respect to t, and then with respect to σ, to obtain

$$M(t_o, t_f)W(t_o, t_f) = \int_{t_o}^{t_f} C^T(t)H(t) \, dt \int_{t_o}^{t_f} F(\sigma)B^T(\sigma) \, d\sigma \tag{10}$$

Since the right side is the product of an $n \times n_z$ matrix and an $n_z \times n$ matrix, it cannot be full rank, and thus (10) shows that $M(t_o, t_f)$ and $W(t_o, t_f)$ cannot both be invertible. Furthermore this argument holds regardless of t_o and $t_f > t_o$ so that the state equation (4), with $A(t)$ zero, cannot be both controllable and observable on any interval. Therefore sufficiency of the controllability/observability condition is established.

For the converse suppose (4) is a minimal realization of the weighting pattern $G(t, \sigma)$, again with $A(t) = 0$ for all t. To show that there is a t_o and $t_f > t_o$ such that

$$W(t_o, t_f) = \int_{t_o}^{t_f} B(t)B^T(t) \, dt$$

and

$$M(t_o, t_f) = \int_{t_o}^{t_f} C^T(t)C(t) \, dt$$

are invertible, we employ the following strategy. First we show that if either $W(t_o, t_f)$ or $M(t_o, t_f)$ is singular for all t_o and t_f with $t_f > t_o$, then minimality is contradicted. This gives existence of t_o^a, $t_f^a > t_o^a$, and t_o^b, $t_f^b > t_o^b$, such that $W(t_o^a, t_f^a)$ and $M(t_o^b, t_f^b)$ both are invertible. Then taking

$$t_o = \min[t_o^a, t_o^b], \quad t_f = \max[t_f^a, t_f^b]$$

the positive-definiteness properties of controllability and observability Gramians give that both $W(t_o, t_f)$ and $M(t_o, t_f)$ are invertible.

Embarking on this program, suppose that for all t_o and $t_f > t_o$ the matrix $W(t_o, t_f)$ is not invertible. Then given t_o and t_f there exists a nonzero $n \times 1$ vector x, depending on t_o and t_f, such that

$$0 = x^T W(t_o, t_f)x = \int_{t_o}^{t_f} x^T B(t)B^T(t)x \, dt$$

Of course this gives

$$x^T B(t) = 0, \quad t \in [t_o, t_f] \tag{11}$$

Now a contradiction argument will show that the vector x can be chosen to be independent of t_o and t_f. If this is not the case, then for some t_o^a, $t_f^a > t_o^a$ and some t_o^b, $t_f^b > t_o^b$, we must choose distinct nonzero vectors x_a and x_b to obtain the corresponding result (11) on the appropriate intervals. However taking $t_o = \min[t_o^a, t_o^b]$ and $t_f = \max[t_f^a, t_f^b]$, $W(t_o, t_f)$ is not invertible. Thus there exists a nonzero $n \times 1$ vector x such that (11) holds. But clearly this implies that the choice $x_a = x_b = x$ could have been made at the outset, contradicting the assumption above. This argument proves that there exists a nonzero vector x such that $x^T B(t) = 0$ for all t.

Now let P^{-1} be a constant, invertible, $n \times n$ matrix with bottom row x^T. Using P^{-1} as a change of state variables gives another minimal realization of the weighting pattern, with coefficient matrices

$$P^{-1}B(t) = \begin{bmatrix} \hat{B}_1(t) \\ 0_{1 \times m} \end{bmatrix}, \quad C(t)P = \begin{bmatrix} \hat{C}_1(t) & \hat{C}_2(t) \end{bmatrix}$$

where $\hat{B}_1(t)$ is $(n-1) \times m$, and $\hat{C}_1(t)$ is $p \times (n-1)$. Then an easy calculation gives

$$G(t, \sigma) = \hat{C}_1(t)\hat{B}_1(\sigma)$$

so that the linear state equation

$$\dot{z}(t) = \hat{B}_1(t)u(t)$$

$$y(t) = \hat{C}_1(t)z(t) \tag{12}$$

is a realization for $G(t, \sigma)$ of dimension $n-1$. This contradicts minimality of the original, dimension-n realization, so there must be at least one t_o^a and one $t_f^a > t_o^a$ such that $W(t_o^a, t_f^a)$ is invertible.

A similar argument shows that there exists at least one t_o^b and one $t_f^b > t_o^b$ such that $M(t_o^b, t_f^b)$ is invertible. Finally taking $t_o = min\,[t_o^a, t_o^b]$ and $t_f = max\,[t_f^a, t_f^b]$ shows that the minimal realization (4) is both controllable and observable on $[t_o, t_f]$.
□□□

Exercise 10.6 shows, in a somewhat indirect fashion, that all minimal realizations of a given weighting pattern are related by an invertible change of state variables. In the time-invariant setting, this result is proved in Theorem 10.10 by explicit construction of the state variable change. The important implication is that minimal realizations of a weighting pattern are unique in a meaningful sense. However it should be emphasized that, for time-varying realizations, properties of interest may not be shared by different minimal realizations. Example 10.5 provides a specific illustration.

Special Cases

An important issue in realization theory is characterization of realizability in terms of special classes of linear state equations. The cases of periodic and time-invariant linear state equations are addressed here. Of course by a T-periodic linear state equation we mean a state equation of the form (4) where $A(t)$, $B(t)$, and $C(t)$ all are periodic for the same T.

10.7 Theorem   A weighting pattern $G(t, \sigma)$ is realizable by a periodic linear state equation if and only if it is realizable, and there exists a finite positive constant T such that

$$G(t+T, \sigma+T) = G(t, \sigma) \tag{13}$$

for all t and σ. If these conditions hold, then there exists a *minimal* realization of $G(t, \sigma)$ that is periodic.

Proof   If $G(t, \sigma)$ has a periodic realization with period T, then obviously $G(t, \sigma)$ is realizable. Furthermore in terms of the realization we can write

$$G(t, \sigma) = C(t)\Phi_A(t, \sigma)B(\sigma)$$

and

$$G(t+T, \sigma+T) = C(t+T)\Phi_A(t+T, \sigma+T)B(\sigma+T)$$

In the proof of Property 5.11 it is shown that $\Phi_A(t+T, \sigma+T) = \Phi_A(t, \sigma)$ for T-periodic $A(t)$, so (13) follows easily.

Conversely suppose that $G(t, \sigma)$ is realizable and (13) holds. We assume that

$$\dot{x}(t) = B(t)u(t)$$

$$y(t) = C(t)x(t)$$

is a minimal realization of $G(t, \sigma)$ with dimension n. Then

$$G(t, \sigma) = C(t)B(\sigma) \tag{14}$$

and there exist finite times t_o and $t_f > t_o$ such that

$$W(t_o, t_f) = \int_{t_o}^{t_f} B(\sigma)B^T(\sigma)\, d\sigma$$

$$M(t_o, t_f) = \int_{t_o}^{t_f} C^T(t)C(t)\, dt$$

both are invertible. (Be careful in this proof not to confuse the transpose ^{T} and the constant T in (13).) Let

$$\hat{W}(t_o, t_f) = \int_{t_o}^{t_f} B(\sigma-T)B^T(\sigma)\, d\sigma$$

$$\hat{M}(t_o, t_f) = \int_{t_o}^{t_f} C^T(\sigma)C(\sigma+T)\, d\sigma$$

Then replacing σ by $\sigma-T$ in (13), and writing the result in terms of (14), leads to

$$C(t+T)B(\sigma) = C(t)B(\sigma-T) \tag{15}$$

for all t and σ. Postmultiplying this expression by $B^T(\sigma)$ and integrating with respect to σ from t_o to t_f gives

$$C(t+T) = C(t)\hat{W}(t_o, t_f)W^{-1}(t_o, t_f) \tag{16}$$

for all t. Similarly, premultiplying (15) by $C^T(t)$ and integrating with respect to t yields

$$B(\sigma-T) = M^{-1}(t_o, t_f)\hat{M}(t_o, t_f)B(\sigma) \tag{17}$$

for all σ. Substituting (16) and (17) back into (15), premultiplying and postmultiplying by $C^T(t)$ and $B^T(\sigma)$ respectively, and integrating with respect to both t and σ gives

$$M(t_o, t_f)\hat{W}(t_o, t_f)W^{-1}(t_o, t_f)W(t_o, t_f)$$

$$= M(t_o, t_f)M^{-1}(t_o, t_f)\hat{M}(t_o, t_f)W(t_o, t_f)$$

This implies

$$\hat{W}(t_o, t_f)W^{-1}(t_o, t_f) = M^{-1}(t_o, t_f)\hat{M}(t_o, t_f) \tag{18}$$

We denote by P the real $n \times n$ matrix in (18), and establish invertibility of P by a simple contradiction argument as follows. If P is not invertible, there exists a nonzero $n \times 1$ vector x such that $x^T P = 0$. Then (17) gives

$$x^T B(\sigma - T) = 0$$

for all σ. This implies

$$x^T \int_{t_o}^{t_f + T} B(\sigma - T)B^T(\sigma - T) \, d\sigma \, x = 0$$

and a change of integration variable yields $x^T W(t_o - T, t_f)x = 0$ But then $x^T W(t_o, t_f)x = 0$, which contradicts invertibility of $W(t_o, t_f)$.

Finally we use again the mathematical fact (see Exercise 5.17) that there exists a real $n \times n$ matrix A such that

$$P^2 = e^{A\,2T} \tag{19}$$

Letting

$$H(t) = C(t)e^{-At}$$

$$F(t) = e^{At}B(t)$$

it is easy to see from (14) that the state equation

$$\dot{z}(t) = Az(t) + F(t)u(t)$$

$$y(t) = H(t)z(t) \tag{20}$$

is a realization of $G(t, \sigma)$. Furthermore, using (16),

$$H(t+2T) = C(t+2T)e^{-A(t+2T)}$$

$$= C(t+T)Pe^{-A(t+2T)}$$

$$= C(t)P^2 e^{-A(t+2T)}$$

$$= H(t)$$

A similar demonstration for $F(t)$, using (17), shows that (20) is a $2T$-periodic realization for $G(t, \sigma)$. Also, since (20) has dimension n, it is a minimal realization.
□□□

Next we consider the characterization of weighting patterns that admit a time-invariant linear state equation

$$\dot{x}(t) = Ax(t) + Bu(t)$$

$$y(t) = Cx(t) \tag{21}$$

as a realization.

10.8 Theorem   A weighting pattern $G(t, \sigma)$ is realizable by a time-invariant linear state equation (21) if and only if $G(t, \sigma)$ is realizable, continuously differentiable with respect to both t and σ, and

$$G(t, \sigma) = G(t - \sigma, 0) \tag{22}$$

[margin note: should require $G(t,\sigma)$ to be analytic function of t & σ.]

for all t and σ. If these conditions hold, then there exists a *minimal* realization of $G(t, \sigma)$ that is time invariant.

Proof   If the weighting pattern has a time-invariant realization (21), then obviously it is realizable. Furthermore we can write

$$G(t, \sigma) = Ce^{A(t-\sigma)}B = Ce^{At}\, e^{-A\sigma}B$$

and continuous differentiability is clear, while verification of (22) is obvious.

For the converse, suppose the weighting pattern is realizable, continuously differentiable in both t and σ, and satisfies (22). Then $G(t, \sigma)$ has a minimal realization. Invoking a change of variables, assume that

$$\dot{x}(t) = B(t)u(t)$$

$$y(t) = C(t)x(t) \tag{23}$$

is an n-dimensional minimal realization, where both $C(t)$ and $B(t)$ are continuously differentiable. Also from Theorem 10.6 there exists a t_o and $t_f > t_o$ such that

$$W(t_o, t_f) = \int_{t_o}^{t_f} B(t)B^T(t)\, dt$$

$$M(t_o, t_f) = \int_{t_o}^{t_f} C^T(t)C(t)\, dt$$

both are invertible. These Gramians are deployed as follows to replace (23) by a time-invariant realization of the same dimension.

From (22), and the continuous-differentiability hypothesis,

$$\frac{\partial}{\partial t} G(t, \sigma) = -\frac{\partial}{\partial \sigma} G(t, \sigma)$$

for all t and σ. Writing this in terms of the minimal realization (23) and postmultiplying by $B^T(\sigma)$ yields

$$0 = \left[\frac{d}{dt} C(t) \right] B(\sigma)B^T(\sigma) + C(t) \left[\frac{d}{d\sigma} B(\sigma) \right] B^T(\sigma) \tag{24}$$

for all t, σ. Integrating both sides with respect to σ from t_o to t_f gives

$$0 = \left[\frac{d}{dt} C(t) \right] W(t_o, t_f) + C(t) \int_{t_o}^{t_f} \left[\frac{d}{d\sigma} B(\sigma) \right] B^T(\sigma) \, d\sigma \tag{25}$$

Now define a constant $n \times n$ matrix A by

$$A = -\int_{t_o}^{t_f} \left[\frac{d}{d\sigma} B(\sigma) \right] B^T(\sigma) \, d\sigma \, W^{-1}(t_o, t_f)$$

Then (25) can be rewritten as

$$\dot{C}(t) = C(t)A$$

This gives, as differentiation readily verifies,

$$C(t) = C(0)e^{At}$$

Therefore

$$G(t, \sigma) = C(t)B(\sigma) = C(t-\sigma)B(0)$$
$$= C(0)e^{A(t-\sigma)}B(0)$$

and the linear state equation

$$\dot{z}(t) = Az(t) + B(0)u(t)$$
$$y(t) = C(0)z(t) \tag{26}$$

is a time-invariant realization of $G(t, \sigma)$. Furthermore (26) has dimension n, and thus is a minimal realization.
□□□

  In the context of time-invariant linear state equations the weighting pattern (or impulse response) normally would be specified as a function of a single variable, say, $G(t)$. In this situation we can set $G_a(t, \sigma) = G(t-\sigma)$. Then (22) is satisfied automatically, and Theorem 10.4 can be applied to $G_a(t, \sigma)$.

10.9 Example   The weighting pattern

$$G(t, \sigma) = e^{t\sigma}$$

is realizable by Theorem 10.4, though the condition (22) for time-invariant realizability clearly fails. For the weighting pattern

$$G(t, \sigma) = e^{-t^2 + 2t\sigma - \sigma^2}$$

(22) is easy to verify:

$$G(t-\sigma, 0) = e^{-(t-\sigma)^2} = G(t, \sigma)$$

However it takes a bit of thought even in this simple case to see that by Theorem 10.4 the weighting pattern is not realizable.
□□□

Next we show that a minimal time-invariant realization of a weighting pattern is unique up to a change of state variables, and provide a formula for the variable change that relates any two minimal realizations. The finicky are asked to forgive a mild notational collision caused by yet another traditional use of the symbol G.

10.10 Theorem   Suppose the time-invariant, n-dimensional linear state equation (21) and

$$\dot{z}(t) = Fz(t) + Gu(t)$$

$$y(t) = Hz(t)$$

both are minimal realizations of a specified weighting pattern. Then there exists a unique, invertible $n \times n$ matrix P such that

$$F = P^{-1}AP , \quad G = P^{-1}B , \quad H = CP$$

Proof   To unclutter construction of the claimed P, let

$$C_a = [\, B \quad AB \quad \cdots \quad A^{n-1}B \,], \quad C_f = [\, G \quad FG \quad \cdots \quad F^{n-1}G \,]$$

$$O_a = \begin{bmatrix} C \\ CA \\ \vdots \\ CA^{n-1} \end{bmatrix}, \quad O_f = \begin{bmatrix} H \\ HF \\ \vdots \\ HF^{n-1} \end{bmatrix} \tag{27}$$

By hypothesis,

$$Ce^{At}B = He^{Ft}G \tag{28}$$

for all t. In particular, at $t = 0$,

$$CB = HG$$

Differentiating (28) with respect to t and evaluating at $t = 0$ gives

$$CAB = HFG$$

and repeating this process shows that

$$CA^kB = HF^kG , \quad k = 0, 1, \cdots \tag{29}$$

These equalities can be arranged in partitioned form to give

$$O_a C_a = O_f C_f \qquad (30)$$

Since a variable change P that relates the two linear state equations is such that

$$C_f = P^{-1} C_a, \quad O_f = O_a P$$

it is natural to construct the P of interest from these controllability and observability matrices. If $m = p = 1$, then C_f, C_a, O_f, and O_a all are invertible $n \times n$ matrices and definition of P is reasonably transparent. The general case is fussy.

By hypothesis the matrices in (27) all have (full) rank n, so a simple contradiction argument shows that the $n \times n$ matrices

$$C_a C_a^T, \quad C_f C_f^T, \quad O_a^T O_a, \quad O_f^T O_f$$

all are positive definite, hence invertible. Then the $n \times n$ matrices

$$P_c = C_a C_f^T (C_f C_f^T)^{-1}$$
$$P_o = (O_f^T O_f)^{-1} O_f^T O_a$$

are such that, applying (30),

$$P_o P_c = (O_f^T O_f)^{-1} O_f^T O_a \, C_a C_f^T (C_f C_f^T)^{-1}$$
$$= (O_f^T O_f)^{-1} O_f^T O_f \, C_f C_f^T (C_f C_f^T)^{-1}$$
$$= I$$

Therefore we can set $P = P_c$, and $P^{-1} = P_o$. Applying (30) again gives

$$P^{-1} C_a = (O_f^T O_f)^{-1} O_f^T O_a \, C_a = (O_f^T O_f)^{-1} O_f^T O_f C_f$$
$$= C_f$$
$$O_a P = O_a \, C_a C_f^T (C_f C_f^T)^{-1} = O_f C_f C_f^T (C_f C_f^T)^{-1}$$
$$= O_f \qquad (31)$$

Extracting the first m columns of the first relationship and the first p rows of the second gives

$$P^{-1} B = G, \quad CP = H$$

Finally another arrangement of (29) yields, in place of (30),

$$O_a A C_a = O_f F C_f$$

from which

$$P^{-1} A P = (O_f^T O_f)^{-1} O_f^T O_a \, A \, C_a C_f^T (C_f C_f^T)^{-1}$$
$$= (O_f^T O_f)^{-1} O_f^T O_f F C_f C_f^T (C_f C_f^T)^{-1}$$
$$= F \qquad (32)$$

Thus we have exhibited an invertible state variable change relating the two minimal realizations. Uniqueness of the variable change follows by noting that if $\hat{P}$ is another such variable change, then

$$HF^k = C\hat{P}(\hat{P}^{-1}A\hat{P})^k = CA^k\hat{P} , \quad k = 0, 1, \cdots$$

and thus

$$O_a\hat{P} = O_f$$

This gives, in conjunction with (31),

$$O_a(P - \hat{P}) = 0 \tag{33}$$

and since O_a has full rank n, $\hat{P} = P$.

Transfer Function Realizability

For the time-invariant case, realizability conditions on a weighting pattern $G(t)$ are addressed further as part of Chapter 11. Here we reconstitute the basic realizability criterion in Theorem 10.8 in terms of the Laplace transform of $G(t)$. That is, in place of the time-domain description of input-output behavior

$$y(t) = \int_0^t G(t-\tau)u(\tau)\, d\tau$$

the input-output relation is written as

$$Y(s) = G(s)U(s)$$

Of course

$$G(s) = \int_0^\infty G(t)e^{-st}\, dt$$

and, similarly, $Y(s)$ and $U(s)$ are the Laplace transforms of the output and input signals. Now the question of realizability is: Given a $p \times m$ transfer function $G(s)$, when does there exist a time-invariant linear state equation of the form (21) such that

$$C(sI - A)^{-1}B = G(s) \tag{34}$$

Recall from Chapter 5 that a rational function is *strictly proper* if the degree of the numerator polynomial is strictly less than the degree of the denominator polynomial.

10.11 Theorem   The transfer function $G(s)$ admits a time-invariant realization (21) if and only if each entry of $G(s)$ is a strictly-proper rational function of s.

Proof   If $G(s)$ has a time-invariant realization (21), then (34) holds. As argued in Chapter 5, each entry of $(sI - A)^{-1}$ is a strictly-proper rational function. Linear combinations of strictly-proper rational functions are strictly-proper rational functions, so $G(s)$ in (34) has entries that are strictly-proper rational functions.

Now suppose that each entry, $G_{ij}(s)$ is a strictly-proper rational function. We can assume that the denominator polynomial of each $G_{ij}(s)$ is *monic*, that is, the coefficient of the highest power of s is unity. Let

$$d(s) = s^r + d_{r-1}s^{r-1} + \cdots + d_0$$

be the (monic) least common multiple of these denominator polynomials. Then $d(s)G(s)$ can be written as a polynomial in s with coefficients that are $p \times m$ constant matrices:

$$d(s)G(s) = N_{r-1}s^{r-1} + \cdots + N_1 s + N_0 \tag{35}$$

From this data we will show that the mr-dimensional linear state equation specified by the partitioned coefficient matrices

$$A = \begin{bmatrix} 0_m & I_m & 0_m & \cdots & 0_m \\ 0_m & 0_m & I_m & \cdots & 0_m \\ \vdots & \vdots & \vdots & \vdots & \vdots \\ 0_m & 0_m & 0_m & \cdots & I_m \\ -d_0 I_m & -d_1 I_m & -d_2 I_m & \cdots & -d_{r-1}I_m \end{bmatrix}$$

$$B = \begin{bmatrix} 0_m \\ 0_m \\ \vdots \\ 0_m \\ I_m \end{bmatrix}, \quad C = \begin{bmatrix} N_0 & N_1 & \cdots & N_{r-1} \end{bmatrix}$$

is a realization of $G(s)$. Let

$$Z(s) = (sI - A)^{-1}B \tag{36}$$

and partition the $mr \times m$ matrix $Z(s)$ into r blocks $Z_1(s), \ldots, Z_r(s)$, each $m \times m$. Multiplying (36) by $(sI - A)$ and writing the result in terms of submatrices gives the set of relations

$$Z_{i+1}(s) = s Z_i(s), \quad i = 1, \ldots, r-1 \tag{37}$$

and

$$s \, Z_r(s) + d_0 Z_1(s) + d_1 Z_2(s) + \cdots + d_{r-1} Z_r(s) = I_m \tag{38}$$

Using (37) to rewrite (38) in terms of $Z_1(s)$ gives

$$Z_1(s) = \frac{1}{d(s)} I_m$$

Therefore, from (37) again,

$$Z(s) = \frac{1}{d(s)} \begin{bmatrix} I_m \\ s I_m \\ \vdots \\ s^{r-1} I_m \end{bmatrix}$$

Finally multiplying through by C yields

$$C(sI - A)^{-1} B = \frac{1}{d(s)} \left[N_0 + N_1 s + \cdots + N_{r-1} s^{r-1} \right]$$

$$= G(s)$$

□□□

The realization for $G(s)$ written down in this proof usually is far from minimal, though it is easy to show that it always is controllable. Construction of minimal realizations in both the time-varying and time-invariant cases is discussed in Chapter 11, along with further properties of minimal realizations.

10.12 Example For $m = p = 1$ the calculation in the proof of Theorem 10.11 simplifies to yield, in our customary notation, the result that the transfer function of the linear state equation

$$\dot{x}(t) = \begin{bmatrix} 0 & 1 & 0 & \cdots & 0 \\ 0 & 0 & 1 & \cdots & 0 \\ \vdots & \vdots & \vdots & \vdots & \vdots \\ 0 & 0 & 0 & \cdots & 1 \\ -a_0 & -a_1 & -a_2 & \cdots & -a_{n-1} \end{bmatrix} x(t) + \begin{bmatrix} 0 \\ 0 \\ \vdots \\ 0 \\ 1 \end{bmatrix} u(t)$$

$$y(t) = \begin{bmatrix} c_0 & c_1 & \cdots & c_{n-1} \end{bmatrix} x(t) \tag{39}$$

is given by

$$G(s) = \frac{c_{n-1}s^{n-1} + \cdots + c_1 s + c_0}{s^n + a_{n-1}s^{n-1} + \cdots + a_1 s + a_0} \tag{40}$$

Thus the controllable realization (39) can be written down by inspection of the numerator and denominator coefficients of the strictly-proper rational transfer function in (40). An easy drill in contradiction proofs shows that the linear state equation (39) is a minimal realization of the transfer function (40) (and thus also observable) if and only if the numerator and denominator polynomials in (40) have no roots in common. Arriving at this result in the multi-input, multi-output case takes additional work that is carried out in Chapters 16 and 17.

EXERCISES

Exercise 10.1   If F is $n \times n$ and $Ce^{At}B$ is $n \times n$, show that

$$G(t,\sigma) = e^{-Ft}Ce^{A(t-\sigma)}Be^{F\sigma}$$

has a time-invariant realization if and only if

$$FCA^jB = CA^jBF, \quad j = 0, 1, 2, \cdots$$

Exercise 10.2   Prove that the weighting pattern of the linear state equation

$$\dot{x}(t) = Ax(t) + e^{Ft}Bu(t)$$
$$y(t) = Ce^{-Ft}x(t)$$

admits a time-invariant realization if $AF = FA$. Under this condition, give one such realization.

Exercise 10.3   Show that if a scalar weighting pattern is a finite sum of terms of the form

$$\alpha_{ij}(t - \sigma)^i e^{\lambda_j(t-\sigma)}$$

where α_{ij} and λ_j are real constants, then it is realizable by a time-invariant linear state equation. What other types of terms can be added without destroying time-invariant realizability?

Exercise 10.4   Consider a scalar linear state equation with zero $A(t)$ and

$$B(t) = \begin{cases} \sin t, & t \in [0, 2\pi] \\ 0, & otherwise \end{cases}, \quad C(t) = \begin{cases} \sin t, & t \in [-2\pi, 0] \\ 0, & otherwise \end{cases}$$

Prove that this state equation is a minimal realization of its weighting pattern. What is the impulse response of the state equation (that is, $G(t, \sigma)$ for $t \geq \sigma$)? What is the dimension of a minimal realization of this impulse response?

Exercise 10.5   Given a weighting pattern $G(t,\sigma) = H(t)F(\sigma)$, where $H(t)$ is $p \times n$ and $F(\sigma)$ is $n \times m$, and a constant $n \times n$ matrix A, show how to find a realization of the form

$$\dot{x}(t) = Ax(t) + B(t)u(t)$$
$$y(t) = C(t)x(t)$$

Exercise 10.6   Suppose the linear state equations

$$\dot{x}(t) = B(t)u(t)$$
$$y(t) = C(t)x(t)$$

and

$$\dot{z}(t) = F(t)u(t)$$
$$y(t) = H(t)z(t)$$

both are minimal realizations of the weighting pattern $G(t, \sigma)$. Show that there exists a constant invertible matrix P such that $z(t) = Px(t)$. Conclude that any two minimal realizations of a given weighting pattern are related by a (time-varying) state variable change.

Exercise 10.7   Show that the weighting pattern $G(t, \sigma)$ admits a time-invariant realization if and only if $G(t, \sigma)$ is realizable, continuously differentiable with respect to both t and σ, and

$$G(t + \tau, \sigma + \tau) = G(t, \sigma)$$

for all t, σ, and τ.

Exercise 10.8   Suppose the weighting pattern $G(t, \sigma)$ admits a time-invariant realization. Show that the realization constructed in the proof of Theorem 10.7 turns out to be a time-invariant state equation.

Exercise 10.9   Using techniques from the proof of Theorem 10.8, prove that the only differentiable solutions of the $n \times n$ matrix functional equation

$$X(t + \sigma) = X(t)X(\sigma), \quad X(0) = I$$

are matrix exponentials.

Exercise 10.10   For

$$G(s) = \frac{1}{(s+1)^2}$$

provide time-invariant realizations that are, respectively, controllable and observable, controllable but not observable, observable but not controllable, and neither controllable nor observable.

Exercise 10.11   Suppose the $p \times m$ transfer function $G(s)$ has the partial fraction expansion

$$G(s) = \sum_{i=1}^{r} G_i \frac{1}{s+\lambda_i}$$

where $\lambda_1, \ldots, \lambda_r$ are distinct, and $G_1, \ldots, G_r$ are $p \times m$ matrices. Show that a minimal realization of $G(s)$ has dimension

$$n = \text{rank } G_1 + \cdots + \text{rank } G_r$$

(*Hint*: Write $G_i = C_i B_i$, and consider the corresponding diagonal-A realization of $G(s)$.)

NOTES

Note 10.1   In setting up the realizability question, we have circumvented important issues involving the generality of the input-output representation

$$y(t) = \int_{t_o}^{t} G(t, \sigma)u(\sigma) \, d\sigma$$

This can be defended on grounds that the integral representation suffices to describe the input-output behaviors that can be generated by a linear state equation, but leaves open the question of more general linear input-output behavior. Also the definition of concepts such as *causality* and *time invariance* for general linear input-output maps have been avoided. These issues call for a more sophisticated mathematical viewpoint, and are considered in

I.W. Sandberg, "Linear maps and impulse responses," *IEEE Transactions on Circuits and Systems,* Vol. 35, No. 2, pp. 201–206, 1988

I.W. Sandberg, "Integral representations for linear maps," *IEEE Transactions on Circuits and Systems,* Vol. 35, No. 5, pp. 536–544, 1988

Note 10.2   An important result we have omitted is the *Canonical Structure theorem.* Roughly this states that for a given linear state equation there exists a change of state variables that displays the new state equation in terms of four component state equations. These are, respectively, controllable and observable, controllable but not observable, observable but not controllable, and neither controllable nor observable. Furthermore the weighting pattern of the original state equation is identical to the weighting pattern of the controllable and observable part of the new state equation. Aside from structural insight, to compute a minimal realization we can start with any convenient realization, perform a state-variable change to display the controllable and observable part, and discard the other parts. This circle of ideas is discussed for the time-varying case in several papers, some dating from the heady period of setting foundations:

R.E. Kalman, "Mathematical description of linear dynamical systems," *SIAM Journal on Control and Optimization,* Vol. 1, No. 2, pp. 152–192, 1963

R.E. Kalman, "On the computation of the reachable/observable canonical form," *SIAM Journal on Control and Optimization,* Vol. 20, No. 2, pp. 258–260, 1982

D.C. Youla, "The synthesis of linear dynamical systems from prescribed weighting patterns," *SIAM Journal on Applied Mathematics,* Vol. 14, No. 3, pp. 527–549, 1966

L. Weiss, "On the structure theory of linear differential systems," *SIAM Journal on Control and Optimization,* Vol. 6, No. 4, pp. 659–680, 1968

P. D'Alessandro, A. Isidori, A. Ruberti, "A new approach to the theory of canonical decomposition of linear dynamical systems," *SIAM Journal on Control and Optimization,* Vol. 11, No. 1, pp. 148–158, 1973

For the time-invariant case there are many, many sources. Consult an original paper

E.G. Gilbert, "Controllability and observability in multivariable control systems," *SIAM Journal on Control and Optimization,* Vol. 1, No. 2, pp. 128–152, 1963

or the detailed textbook exposition, with variations, in Section 17 of

D.F. Delchamps, *State Space and Input-Output Linear Systems,* Springer-Verlag, New York, 1988

For a computational approach see

D. Boley, "Computing the Kalman decomposition: An optimal method," *IEEE Transactions on Automatic Control,* Vol. 29, No. 11, pp. 51–53, 1984 (Correction: Vol. 36, No. 11, p. 1341, 1991)

Finally some material in Chapter 13, including Exercise 13.14, is closely related to the canonical structure of time-invariant linear state equations.

Note 10.3   Subtleties regarding formulation of the realization question in terms of impulse responses versus formulation in terms of weighting patterns are discussed in Section 10.13 of

R.E. Kalman, P.L. Falb, M.A. Arbib, *Topics in Mathematical System Theory,* McGraw-Hill, New York, 1969

Note 10.4   An approach to the often-difficult problem of checking the realizability criterion in Theorem 10.4 is presented in

C. Bruni, A. Isidori, A. Ruberti, "A method of factorization of the impulse-response matrix," *IEEE Transactions on Automatic Control,* Vol. 13, No. 6, pp. 739–741, 1968

The hypotheses and constructions in this paper are related to those in Chapter 11.

Note 10.5   Further details and developments related to Exercise 10.9 can be found in

D. Kalman, A. Unger, "Combinatorial and functional identities in one-parameter matrices," *American Mathematical Monthly,* Vol. 94, No. 1, pp. 21–35, 1987

11

MINIMAL REALIZATION

We further consider the realization question introduced in Chapter 10, with two goals in mind. The first is to suitably strengthen the setting so that results can be obtained for realization of an *impulse response* rather than a weighting pattern. This is important because the impulse response in principle can be determined from input-output behavior of a physical system. The second goal is to obtain solutions of the minimal realization problem that are more constructive than those discussed in Chapter 10.

Assumptions

One adjustment we make to obtain a coherent minimal realization theory for impulse response representations is that the technical defaults are strengthened. It is assumed that a given $p \times m$ impulse response $G(t, \sigma)$, defined for all t, σ with $t \geq \sigma$, is such that any derivatives that appear in the development are continuous for all t, σ with $t \geq \sigma$. Similarly for the linear state equations considered in this chapter,

$$\dot{x}(t) = A(t)x(t) + B(t)u(t)$$

$$y(t) = C(t)x(t) \tag{1}$$

we assume $A(t)$, $B(t)$, and $C(t)$ are such that all derivatives that appear are continuous for all t. Imposing smoothness hypotheses in this way circumvents tedious counts and distracting lists of differentiability requirements.

Another adjustment is that strengthened forms of controllability and observability are used to characterize minimality of realizations. Recall from Definition 9.3 the $n \times m$ matrix functions

$$K_0(t) = B(t)$$

$$K_j(t) = -A(t)K_{j-1}(t) + \dot{K}_{j-1}(t), \quad j = 1, 2, \cdots \tag{2}$$

and for convenience let

$$W_k(t) = \left[K_0(t) \ K_1(t) \ \cdots \ K_{k-1}(t) \right], \quad k = 1, 2, \cdots \tag{3}$$

Similarly from Definition 9.9 recall the $p \times n$ matrices

$$L_0(t) = C(t)$$
$$L_j(t) = L_{j-1}(t)A(t) + \dot{L}_{j-1}(t), \quad j = 1, 2, \cdots \tag{4}$$

and let

$$M_k(t) = \begin{bmatrix} L_0(t) \\ L_1(t) \\ \vdots \\ L_{k-1}(t) \end{bmatrix}, \quad k = 1, 2, \cdots \tag{5}$$

We define new types of controllability and observability for (1) in terms of the matrices $W_n(t)$ and $M_n(t)$, where of course n is the dimension of the linear state equation (1). Unfortunately the terminology is not standard, though some justification for our selection can be found in Exercises 11.1 and 11.2.

11.1 Definition   The linear state equation (1) is called *instantaneously controllable* if *rank* $W_n(t) = n$ for every t, and *instantaneously observable* if *rank* $M_n(t) = n$ for every t.

If (1) is a realization of a given impulse response $G(t, \sigma)$, that is,

$$G(t, \sigma) = C(t)\Phi(t, \sigma)B(\sigma), \quad t \geq \sigma$$

then a straightforward calculation shows that

$$\frac{\partial^i}{\partial t^i} \frac{\partial^j}{\partial \sigma^j} G(t, \sigma) = L_i(t)\Phi(t, \sigma)K_j(\sigma); \quad i, j = 0, 1, \cdots \tag{6}$$

for all t, σ with $t \geq \sigma$. This motivates the appearance of the instantaneous controllability and instantaneous observability matrices in the realization problem, and leads directly to a sufficient condition for minimality of a realization.

11.2 Theorem   Suppose the linear state equation (1) is a realization of the impulse response $G(t, \sigma)$. Then (1) is a minimal realization of $G(t, \sigma)$ if it is instantaneously controllable and instantaneously observable.

Proof   Suppose $G(t, \sigma)$ has a dimension-n realization (1) that is instantaneously controllable and instantaneously observable, but is not minimal. Then we can assume that there is an $(n-1)$-dimensional realization

$$\dot{z}(t) = \widetilde{A}(t)z(t) + \widetilde{B}(t)u(t)$$
$$y(t) = \widetilde{C}(t)z(t) \tag{7}$$

and write

$$G(t, \sigma) = C(t)\Phi_A(t, \sigma)B(\sigma) = \widetilde{C}(t)\Phi_{\widetilde{A}}(t, \sigma)\widetilde{B}(\sigma)$$

for all t, σ with $t \geq \sigma$. Differentiating repeatedly with respect to both t and σ as in (6), evaluating at $\sigma = t$, and arranging the resulting identities in matrix form gives, using the obvious notation for instantaneous controllability and instantaneous observability matrices for (7),

$$M_n(t)W_n(t) = \widetilde{M}_n(t)\widetilde{W}_n(t)$$

Since $\widetilde{M}_n(t)$ has $n-1$ columns, and $\widetilde{W}_n(t)$ has $n-1$ rows, this equality shows that *rank* $[M_n(t)W_n(t)] \leq n-1$ for all t, which contradicts the hypotheses of instantaneous controllability and instantaneous observability of (1).
□□□

With slight modification the basic realizability criterion for weighting patterns, Theorem 10.4, applies to impulse responses. That is, an impulse response $G(t, \sigma)$ is realizable if and only if there exist continuous matrix functions $H(t)$ and $F(t)$ such that

$$G(t, \sigma) = H(t)F(\sigma)$$

for all t, σ with $t \geq \sigma$. However we will develop alternative realizability tests that lead to more effective methods for computing minimal realizations.

Time-Varying Realizations

The algebraic structure of the realization problem as well as connections to instantaneous controllability and instantaneous observability are captured in terms of properties of a set of matrices defined from the impulse response. For all t, σ such that $t \geq \sigma$, and positive integers i, j, define an $(ip) \times (jm)$ *behavior matrix* corresponding to $G(t, \sigma)$ as

$$\Gamma_{ij}(t, \sigma) = \begin{bmatrix} G(t, \sigma) & \dfrac{\partial}{\partial\sigma}G(t, \sigma) & \cdots & \dfrac{\partial^{j-1}}{\partial\sigma^{j-1}}G(t, \sigma) \\[2ex] \dfrac{\partial}{\partial t}G(t, \sigma) & \dfrac{\partial^2}{\partial t\partial\sigma}G(t, \sigma) & \cdots & \dfrac{\partial^j}{\partial t\partial\sigma^{j-1}}G(t, \sigma) \\[2ex] \vdots & \vdots & \vdots & \vdots \\[2ex] \dfrac{\partial^{i-1}}{\partial t^{i-1}}G(t, \sigma) & \dfrac{\partial^i}{\partial t^{i-1}\partial\sigma}G(t, \sigma) & \cdots & \dfrac{\partial^{i+j-2}}{\partial t^{i-1}\partial\sigma^{j-1}}G(t, \sigma) \end{bmatrix} \tag{8}$$

We use the set of behavior matrices for $G(t, \sigma)$ to develop a realizability test, and a construction for minimal realizations involving submatrices of $\Gamma_{ij}(t, \sigma)$.

A few observations might be helpful in digesting proofs involving behavior matrices. A *submatrix,* unlike a partition, need not be formed from adjacent rows and columns. For example one submatrix of a 3×3 matrix A is

$$\begin{bmatrix} a_{11} & a_{13} \\ a_{31} & a_{33} \end{bmatrix}$$

Matrix-algebra concepts associated with $\Gamma_{ij}(t, \sigma)$ in the sequel are applied pointwise in t and σ (with $t \geq \sigma$). For example linear independence of rows of $\Gamma_{ij}(t, \sigma)$ involves linear combinations of the rows using coefficients that are scalar functions of t and σ. To visualize the structure of behavior matrices it is useful to write (8) in more detail on a large sheet of paper, and use a sharp pencil to sketch various relationships developed in the proofs.

11.3 Theorem   Suppose for the impulse response $G(t, \sigma)$ there exist positive integers l, k, n such that $l, k \leq n$ and

$$\text{rank } \Gamma_{lk}(t, \sigma) = \text{rank } \Gamma_{l+1, k+1}(t, \sigma) = n \tag{9}$$

for all t, σ with $t \geq \sigma$. Also suppose there is a fixed $n \times n$ submatrix of $\Gamma_{lk}(t, \sigma)$ that is invertible for all t, σ with $t \geq \sigma$. Then $G(t, \sigma)$ is realizable and has a minimal realization of dimension n.

Proof   Assume (9) holds, and that $F(t, \sigma)$ is an $n \times n$ submatrix of $\Gamma_{lk}(t, \sigma)$ that is invertible for all t, σ with $t \geq \sigma$. Let $F_c(t, \sigma)$ be the $p \times n$ matrix comprising those columns of $\Gamma_{1k}(t, \sigma)$ that correspond to columns of $F(t, \sigma)$, and let

$$C_c(t, \sigma) = F_c(t, \sigma)F^{-1}(t, \sigma) \tag{10}$$

That is, the coefficients in the ith-row of $C_c(t, \sigma)$ specify the linear combination of rows of $F(t, \sigma)$ that gives the ith-row of $F_c(t, \sigma)$. Similarly let $F_r(t, \sigma)$ be the $n \times m$ matrix formed from those rows of $\Gamma_{l1}(t, \sigma)$ that correspond to rows of $F(t, \sigma)$, and let

$$B_r(t, \sigma) = F^{-1}(t, \sigma)F_r(t, \sigma) \tag{11}$$

The jth-column of $B_r(t, \sigma)$ specifies the linear combination of columns of $F(t, \sigma)$ that gives the jth-column of $F_r(t, \sigma)$. Then we claim that

$$G(t, \sigma) = C_c(t, \sigma)F(t, \sigma)B_r(t, \sigma) \tag{12}$$

for all t, σ with $t \geq \sigma$. This relationship holds because, by (9), any row (column) of $\Gamma_{lk}(t, \sigma)$ can be represented as a linear combination of those rows (columns) of $\Gamma_{lk}(t, \sigma)$ that correspond to rows (columns) of $F(t, \sigma)$. (Again, throughout this proof the linear combinations resulting from the rank property (9) have scalar coefficients that are functions of t and σ, defined for $t \geq \sigma$.)

In particular consider the single-input, single-output case. If $m = p = 1$, then $l = k = n$, $F(t, \sigma) = \Gamma_{nn}(t, \sigma)$, and $F_c(t, \sigma)$ is just the first row of $\Gamma_{nn}(t, \sigma)$. Therefore $C_c(t, \sigma) = e_1^T$, the first row of I_n. Similarly $B_r(t, \sigma) = e_1$, and (12) turns out to be the obvious

$$G(t, \sigma) = \Gamma_{11}(t, \sigma) = e_1^T \Gamma_{nn}(t, \sigma)e_1$$

(Throughout this proof consideration of the $m = p = 1$ case is a good way to gain understanding of the admittedly-complicated general situation.)

The next step is to show that $C_c(t, \sigma)$ is independent of σ. From (10), $F_c(t, \sigma) = C_c(t, \sigma)F(t, \sigma)$, and therefore

$$\frac{\partial}{\partial \sigma} F_c(t, \sigma) = \left[\frac{\partial}{\partial \sigma} C_c(t, \sigma) \right] F(t, \sigma) + C_c(t, \sigma)\frac{\partial}{\partial \sigma} F(t, \sigma) \qquad (13)$$

In $\Gamma_{l,k+1}(t, \sigma)$ each column of $(\partial F/\partial \sigma)(t, \sigma)$ occurs m columns to the right of the corresponding column of $F(t, \sigma)$, and the same holds for the relative locations of columns of $(\partial F_c/\partial \sigma)(t, \sigma)$ and $F_c(t, \sigma)$, By the rank property (9) the linear combination of the jth-column entries of $(\partial F/\partial \sigma)(t, \sigma)$ specified by the ith-row of $C_c(t, \sigma)$ gives precisely the entry that occurs m columns to the right of the i,j-entry of $F_c(t, \sigma)$. Of course this is the i,j-entry of $(\partial F_c/\partial \sigma)(t, \sigma)$. Therefore

$$\frac{\partial}{\partial \sigma} F_c(t, \sigma) = C_c(t, \sigma)\frac{\partial}{\partial \sigma} F(t, \sigma) \qquad (14)$$

Comparing (13) and (14), and using the invertibility of $F(t, \sigma)$, gives

$$\frac{\partial}{\partial \sigma} C_c(t, \sigma) = 0$$

for all t, σ with $t \geq \sigma$.

A similar argument can be used to show that $B_r(t, \sigma)$ in (11) is independent of t. Then with some abuse of notation we let

$$C_c(t) = F_c(t, t)F^{-1}(t, t)$$
$$B_r(\sigma) = F^{-1}(\sigma, \sigma)F_r(\sigma, \sigma)$$

and write (12) as

$$G(t, \sigma) = C_c(t)F(t, \sigma)B_r(\sigma) \qquad (15)$$

for all t, σ with $t \geq \sigma$.

The remainder of the proof involves reworking the factorization of the impulse response in (15) into a factorization of the type provided by a state equation realization. To this end the notation

$$F_s(t, \sigma) = \frac{\partial}{\partial t} F(t, \sigma)$$

is temporarily convenient. Clearly $F_s(t, \sigma)$ is an $n \times n$ submatrix of $\Gamma_{l+1,k+1}(t, \sigma)$ where each entry of $F_s(t, \sigma)$ occurs exactly p rows below the corresponding entry of

$F(t, \sigma)$. Therefore the rank condition (9) implies that each row of $F_s(t, \sigma)$ can be written as a linear combination of the rows of $F(t, \sigma)$. That is, collecting these linear combination coefficients into an $n \times n$ matrix $A(t, \sigma)$,

$$F_s(t, \sigma) = A(t, \sigma)F(t, \sigma) \tag{16}$$

Also each entry of $(\partial F/\partial \sigma)(t, \sigma)$ as a submatrix of $\Gamma_{l+1,k+1}(t, \sigma)$ occurs m columns to the right of the corresponding entry of $F(t, \sigma)$. But then the rank condition and the interchange of differentiation order permitted by the differentiability hypotheses give

$$\frac{\partial^2}{\partial t \partial \sigma} F(t, \sigma) = \frac{\partial}{\partial \sigma} F_s(t, \sigma) = A(t, \sigma) \frac{\partial}{\partial \sigma} F(t, \sigma) \tag{17}$$

This can be used as follows to show that $A(t, \sigma)$ is independent of σ. Differentiating (16) with respect to σ gives

$$\frac{\partial}{\partial \sigma} F_s(t, \sigma) = \left[\frac{\partial}{\partial \sigma} A(t, \sigma) \right] F(t, \sigma) + A(t, \sigma) \frac{\partial}{\partial \sigma} F(t, \sigma) \tag{18}$$

From (18) and (17), using the invertibility of $F(t, \sigma)$,

$$\frac{\partial}{\partial \sigma} A(t, \sigma) = 0$$

for all t, σ with $t \geq \sigma$. Thus $A(t, \sigma)$ depends only on t, and replacing the variable σ in (16) by a fixed τ we write

$$A(t) = F_s(t, \tau)F^{-1}(t, \tau)$$

Furthermore the transition matrix corresponding to $A(t)$ is given by

$$\Phi_A(t, \sigma) = F(t, \tau)F^{-1}(\sigma, \tau)$$

as is easily shown by verifying the relevant matrix differential equation, and the initial condition at $t = \sigma$. Again τ is a parameter that can be assigned any value.

To continue we similarly show that $F^{-1}(t, \tau)F(t, \sigma)$ is not a function of t since

$$\frac{\partial}{\partial t} \left[F^{-1}(t, \tau)F(t, \sigma) \right] = -F^{-1}(t, \tau) \left[\frac{\partial}{\partial t} F(t, \tau) \right] F^{-1}(t, \tau)F(t, \sigma)$$

$$+ F^{-1}(t, \tau) \frac{\partial}{\partial t} F(t, \sigma)$$

$$= -F^{-1}(t, \tau)A(t)F(t, \sigma) + F^{-1}(t, \tau)A(t)F(t, \sigma)$$

$$= 0$$

In particular this gives

$$F^{-1}(t, \tau)F(t, \sigma) = F^{-1}(\sigma, \tau)F(\sigma, \sigma)$$

that is,

$$F(t,\ \sigma) = F(t,\ \tau)F^{-1}(\sigma,\ \tau)F(\sigma,\ \sigma)$$

This means that the factorization (15) can be written as

$$G(t,\ \sigma) = C_c(t)F(t,\ \tau)F^{-1}(\sigma,\ \tau)F(\sigma,\ \sigma)B_r(\sigma) \tag{19}$$

$$= \left[F_c(t,\ t)F^{-1}(t,\ t) \right] \Phi_A(t,\ \sigma)\ F_r(\sigma,\ \sigma)$$

for all t, σ with $t \geq \sigma$. Now it is clear that an n-dimensional realization of $G(t,\ \sigma)$ is specified by

$$A(t) = F_s(t,\ t)F^{-1}(t,\ t)$$
$$B(t) = F_r(t,\ t)$$
$$C(t) = F_c(t,\ t)F^{-1}(t,\ t) \tag{20}$$

Finally since $l,\ k \leq n$, $\Gamma_{nn}(t,\ \sigma)$ has rank at least n for all t, σ such that $t \geq \sigma$. Therefore $\Gamma_{nn}(t,\ t)$ has rank at least n for all t. Evaluating (6) at $\sigma = t$ and forming $\Gamma_{nn}(t,\ t)$ gives $\Gamma_{nn}(t,\ t) = M_n(t)W_n(t)$, so that the realization we have constructed is instantaneously controllable and instantaneously observable, hence minimal. □□□

Another minimal realization of $G(t,\ \sigma)$ can be written from the factorization in (19), namely

$$\dot{z}(t) = F^{-1}(t,\ \tau)F_r(t,\ t)u(t)$$
$$y(t) = F_c(t,\ t)F^{-1}(t,\ t)F(t,\ \tau)z(t) \tag{21}$$

(with τ a fixed parameter). However it is easily shown that the realization specified by (20), unlike (21), has the desirable property that the coefficient matrices turn out to be constant if $G(t,\ \sigma)$ admits a time-invariant realization.

11.4 Example   Given the impulse response

$$G(t,\ \sigma) = e^{-t}\sin(t-\sigma)$$

the realization procedure in the proof of Theorem 11.3 begins with rank calculations. These show that, for all t, σ with $t \geq \sigma$,

$$\Gamma_{22}(t,\ \sigma) = \begin{bmatrix} e^{-t}\sin(t-\sigma) & -e^{-t}\cos(t-\sigma) \\ e^{-t}[\cos(t-\sigma)-\sin(t-\sigma)] & e^{-t}[\cos(t-\sigma)+\sin(t-\sigma)] \end{bmatrix}$$

has rank 2, while $det\ \Gamma_{33}(t,\ \sigma) = 0$. Thus the rank condition (9) is satisfied with $l = k = n = 2$, and we can take $F(t,\ \sigma) = \Gamma_{22}(t,\ \sigma)$. Then

$$F(t,\ t) = \begin{bmatrix} 0 & -e^{-t} \\ e^{-t} & e^{-t} \end{bmatrix}$$

Straightforward differentiation of $F(t, \sigma)$ with respect to t leads to

$$F_s(t, t) = \begin{bmatrix} e^{-t} & e^{-t} \\ -2e^{-t} & 0 \end{bmatrix} \tag{22}$$

Finally since $F_c(t, t)$ is the first row of $\Gamma_{22}(t, t)$, and $F_r(t, t)$ is the first column, the minimal realization specified by (20) is

$$\dot{x}(t) = \begin{bmatrix} 0 & 1 \\ -2 & -2 \end{bmatrix} x(t) + \begin{bmatrix} 0 \\ e^{-t} \end{bmatrix} u(t)$$

$$y(t) = \begin{bmatrix} 1 & 0 \end{bmatrix} x(t)$$

Time-Invariant Realizations

We now pursue the specialization and strengthening of Theorem 11.3 for the time-invariant case. A slight modification of Theorem 10.8 to fit the present setting gives that a realizable impulse response has a time-invariant realization if it can be written as $G(t - \sigma)$. For the remainder of this chapter we simply replace the difference $t - \sigma$ by t, and work with $G(t)$ for convenience. Of course $G(t)$ is defined for all $t \geq 0$, and there is no loss of generality in the time-invariant case in assuming that $G(t)$ is analytic. (Specifically a function of the form $Ce^{At}B$ is analytic, and thus a realizable impulse response must have this property.) Therefore we can differentiate $G(t)$ any number of times, and it is convenient to redefine the behavior matrices corresponding to $G(t)$ as

$$\Gamma_{ij}(t) = \begin{bmatrix} G(t) & \dfrac{d}{dt}G(t) & \cdots & \dfrac{d^{j-1}}{dt^{j-1}}G(t) \\[2ex] \dfrac{d}{dt}G(t) & \dfrac{d^2}{dt^2}G(t) & \cdots & \dfrac{d^j}{dt^j}G(t) \\[2ex] \vdots & \vdots & \vdots & \vdots \\[2ex] \dfrac{d^{i-1}}{dt^{i-1}}G(t) & \dfrac{d^i}{dt^i}G(t) & \cdots & \dfrac{d^{i+j-2}}{dt^{i+j-2}}G(t) \end{bmatrix} \tag{23}$$

where i, j are positive integers and $t \geq 0$. This differs from the definition of $\Gamma_{ij}(t, \sigma)$ in (8) in the sign of alternate block columns, though rank properties are unaffected. As a corresponding change, involving only signs of block columns in the instantaneous controllability matrix defined in (3), we will work with the customary controllability and observability matrices in the time-invariant case. Namely these matrices for the state equation

$$\dot{x}(t) = Ax(t) + Bu(t)$$

$$y(t) = Cx(t) \tag{24}$$

are given in the current notation by

$$W_n = \begin{bmatrix} B & AB & \cdots & A^{n-1}B \end{bmatrix}, \quad M_n = \begin{bmatrix} C \\ CA \\ \vdots \\ CA^{n-1} \end{bmatrix} \qquad (25)$$

Theorem 11.3, a sufficient condition for realizability, can be restated as a necessary and sufficient condition in the time-invariant case. The proof is strategically similar, employing linear-algebraic arguments applied pointwise in t.

11.5 Theorem   The analytic impulse response $G(t)$ admits a time-invariant realization (24) if and only if there exist positive integers l, k, n with l, $k \le n$ such that

$$\text{rank } \Gamma_{lk}(t) = \text{rank } \Gamma_{l+1,k+1}(t) = n, \quad t \ge 0 \qquad (26)$$

and there is a fixed $n \times n$ submatrix of $\Gamma_{lk}(t)$ that is invertible for all $t \ge 0$. If these conditions hold, then the dimension of a minimal realization of $G(t)$ is n.

Proof   Suppose (26) holds and $F(t)$ is an $n \times n$ submatrix of $\Gamma_{lk}(t)$ that is invertible for all $t \ge 0$. Let $F_c(t)$ be the $p \times n$ matrix comprising those columns of $\Gamma_{1k}(t)$ that correspond to columns of $F(t)$, and let $F_r(t)$ be the $n \times m$ matrix of rows of $\Gamma_{l1}(t)$ that correspond to rows of $F(t)$. Then

$$C_c(t) = F_c(t)F^{-1}(t)$$
$$B_r(t) = F^{-1}(t)F_r(t)$$

yields the preliminary factorization

$$G(t) = C_c(t)F(t)B_r(t), \quad t \ge 0 \qquad (27)$$

exactly as in the proof of Theorem 11.3.
        Next we show that $C_c(t)$ is a constant matrix by considering

$$\dot{C}_c(t) = \dot{F}_c(t)F^{-1}(t) - F_c(t)F^{-1}(t)\dot{F}(t)F^{-1}(t)$$
$$= \left[\dot{F}_c(t) - C_c(t)\dot{F}(t) \right]F^{-1}(t) \qquad (28)$$

In $\Gamma_{l,k+1}(t)$ each entry of $\dot{F}(t)$ occurs m columns to the right of the corresponding entry of $F(t)$. By the rank property (26) the linear combination of the jth-column entries of $\dot{F}(t)$ specified by the ith-row of $C_c(t)$ gives the entry that occurs m columns to the right of the i,j-entry of $F_c(t)$. This is precisely the i,j-entry of $\dot{F}_c(t)$, and so (28) shows that $\dot{C}_c(t) = 0$, $t \ge 0$. A similar argument shows that $\dot{B}_r(t) = 0$, $t \ge 0$. Therefore, with a familiar abuse of notation, we write these constant matrices as

$$C_c = F_c(0)F^{-1}(0)$$

$$B_r = F^{-1}(0)F_r(0) \tag{29}$$

Then (27) becomes

$$G(t) = C_c F(t)B_r , \quad t \geq 0 \tag{30}$$

The remainder of the proof involves further manipulations to obtain a factorization corresponding to a time-invariant realization of $G(t)$; that is, a three-part factorization with a matrix exponential in the middle. Preserving notation in the proof of Theorem 11.3, consider the submatrix $F_s(t) = \dot{F}(t)$ of $\Gamma_{l+1,k}(t)$. By (26) the rows of $F_s(t)$ must be expressible as a linear combination of the rows of $F(t)$ (with t-dependent scalar coefficients). That is, there is an $n \times n$ matrix $A(t)$ such that

$$F_s(t) = A(t)F(t) \tag{31}$$

However we can show that $A(t)$ is a constant matrix. From (31),

$$\dot{F}_s(t) = A(t)\dot{F}(t) + \dot{A}(t)F(t) \tag{32}$$

It is not difficult to check that $\dot{F}_s(t)$ is a submatrix of $\Gamma_{l+1,k+1}(t)$, and the rank condition gives

$$\dot{F}_s(t) = A(t)F_s(t) \tag{33}$$

Therefore from (32), (33), and the invertibility of $F(t)$ we conclude $\dot{A}(t) = 0$, $t \geq 0$. We simply write A for $A(t)$, and use, from (31),

$$A = F_s(0)F^{-1}(0)$$

Also from (31),

$$F(t) = e^{At}F(0) , \quad t \geq 0 \tag{34}$$

Putting together (29), (30), and (34) gives the factorization

$$G(t) = F_c(0)F^{-1}(0)e^{At}F_r(0)$$

from which we obtain an n-dimensional realization of the form (24) with coefficients

$$A = F_s(0)F^{-1}(0)$$

$$B = F_r(0)$$

$$C = F_c(0)F^{-1}(0) \tag{35}$$

Of course these coefficients are defined in terms of submatrices of $\Gamma_{l+1,k}(0)$, and bear a close resemblance to those specified by (20).

Extending the notation for controllability and observability matrices in (25), it is easy to verify that

$$\Gamma_{lk}(t) = M_l \, e^{At} \, W_k , \quad l, k = 1, 2, \cdots \tag{36}$$

and since

$$n \le \operatorname{rank} \Gamma_{lk}(0) \le \operatorname{rank} \Gamma_{nn}(0) \le \operatorname{rank} M_n W_n$$

the realization specified by (35) is controllable and observable. Therefore by Theorem 10.6 or by independent contradiction argument as in the proof of Theorem 11.2, we conclude that the realization specified by (35) is minimal.

For the converse argument suppose (24) is a minimal realization of $G(t)$. Then (36) and the Cayley-Hamilton theorem immediately imply that the rank condition (26) holds. Also there must exist invertible $n \times n$ submatrices F_o composed of linearly independent rows of M_n, and F_r composed of linearly independent columns of W_n. Consequently

$$F(t) = F_o e^{At} F_r$$

is an $n \times n$ submatrix of $\Gamma_{nn}(t)$ that has rank n for $t \ge 0$.

11.6 Example   Consider the impulse response

$$G(t) = \begin{bmatrix} 2e^{-t} & \alpha(e^t - e^{-t}) \\ e^{-t} & e^{-t} \end{bmatrix} \tag{37}$$

where α is a real parameter, inserted for illustration. Then $\Gamma_{11}(t) = G(t)$, and

$$\Gamma_{22}(t) = e^{-t} \begin{bmatrix} 2 & \alpha(e^{2t}-1) & -2 & \alpha(e^{2t}+1) \\ 1 & 1 & -1 & -1 \\ -2 & \alpha(e^{2t}+1) & 2 & \alpha(e^{2t}-1) \\ -1 & -1 & 1 & 1 \end{bmatrix}$$

For $\alpha = 0$,

$$\operatorname{rank} \Gamma_{11}(t) = \operatorname{rank} \Gamma_{22}(t) = 2 , \quad t \ge 0$$

so a minimal realization of $G(t)$ has dimension two. We can choose

$$F(t) = \Gamma_{11}(t) = e^{-t} \begin{bmatrix} 2 & 0 \\ 1 & 1 \end{bmatrix}$$

Then

$$F_s(t) = -F(t)$$
$$F_r(t) = F_c(t) = F(t)$$

and the prescription in (35) gives the minimal realization ($\alpha = 0$)

$$\dot{x}(t) = \begin{bmatrix} -1 & 0 \\ 0 & -1 \end{bmatrix} x(t) + \begin{bmatrix} 2 & 0 \\ 1 & 1 \end{bmatrix} u(t)$$

$$y(t) = \begin{bmatrix} 1 & 0 \\ 0 & 1 \end{bmatrix} x(t) \tag{38}$$

For the parameter value $\alpha = -2$, it is left as an exercise to show that minimal realizations again have dimension two. If $\alpha \neq 0, -2$, then matters are more interesting. Straightforward calculations verify

$$\text{rank } \Gamma_{22}(t) = \text{rank } \Gamma_{33}(t) = 3, \quad t \geq 0$$

The upper left 3×3 submatrix of $\Gamma_{22}(t)$ is not invertible, but selecting columns 1, 2, and 4 of the first three rows of $\Gamma_{22}(t)$ gives the invertible (for all $t \geq 0$) matrix

$$F(t) = e^{-t} \begin{bmatrix} 2 & \alpha(e^{2t}-1) & \alpha(e^{2t}+1) \\ 1 & 1 & -1 \\ -2 & \alpha(e^{2t}+1) & \alpha(e^{2t}-1) \end{bmatrix} \tag{39}$$

This specifies a minimal realization as follows. From $\dot{F}(t)$ we get

$$F_s(0) = \dot{F}(0) = \begin{bmatrix} -2 & 2\alpha & 0 \\ -1 & -1 & 1 \\ 2 & 0 & 2\alpha \end{bmatrix}$$

and, from $F(0)$,

$$F^{-1}(0) = \frac{1}{4\alpha(\alpha+2)} \begin{bmatrix} 2\alpha & 4\alpha^2 & -2\alpha \\ 2 & 4\alpha & 2\alpha+2 \\ 2\alpha+2 & -4\alpha & 2 \end{bmatrix}$$

Columns 1, 2 and 4 of $\Gamma_{12}(0)$ give

$$F_c(0) = \begin{bmatrix} 2 & 0 & 2\alpha \\ 1 & 1 & -1 \end{bmatrix}$$

and the first three rows of $\Gamma_{21}(0)$ provide

$$F_r(0) = \begin{bmatrix} 2 & 0 \\ 1 & 1 \\ -2 & 2\alpha \end{bmatrix}$$

Then a minimal realization is specified by ($\alpha \neq 0, -2$)

$$A = F_s(0)F^{-1}(0) = \begin{bmatrix} 0 & 0 & 1 \\ 0 & -1 & 0 \\ 1 & 0 & 0 \end{bmatrix}, \quad B = F_r(0) = \begin{bmatrix} 2 & 0 \\ 1 & 1 \\ -2 & 2\alpha \end{bmatrix}$$

$$C = F_c(0)F^{-1}(0) = \begin{bmatrix} 1 & 0 & 0 \\ 0 & 1 & 0 \end{bmatrix} \tag{40}$$

The skeptical observer might want to compute $Ce^{At}B$ to verify this realization, and check controllability and observability to confirm minimality.

Realization from Markov Parameters

There is an alternate formulation of the realization problem in the time-invariant case that often is used in place of Theorem 11.5. Again we restrict attention to impulse responses that are analytic for $t \geq 0$, since otherwise $G(t)$ is not realizable by a time-invariant linear state equation. Then the realization question can be cast in terms of coefficients in the power series expansion of $G(t)$ about $t = 0$. The sequence of $p \times m$ matrices

$$G_0, G_1, G_2, \cdots \qquad (41)$$

where

$$G_i = \frac{d^i}{dt^i} G(t) \bigg|_{t=0}, \quad i = 0, 1, \cdots$$

is called the *Markov parameter sequence* corresponding to the impulse response $G(t)$. Clearly if $G(t)$ has a realization (24), that is, $G(t) = Ce^{At}B$, then the Markov parameter sequence can be represented in the form

$$G_i = CA^iB, \quad i = 0, 1, \cdots \qquad (42)$$

This shows that the minimal realization problem in the time-invariant case can be viewed as the matrix-algebra problem of computing a minimal-dimension factorization of the form (42) for a given Markov parameter sequence.

The Markov parameter sequence also can be determined from a given transfer function representation $G(s)$. Since $G(s)$ is the Laplace transform of $G(t)$, the Initial Value theorem gives, assuming the indicated limits exist,

$$G_0 = \lim_{s \to \infty} sG(s)$$

$$G_1 = \lim_{s \to \infty} s[\, sG(s) - G_0\,]$$

$$G_2 = \lim_{s \to \infty} s[\, s^2G(s) - sG_0 - G_1\,]$$

and so on. Alternatively if $G(s)$ is a matrix of strictly-proper rational functions, as by Theorem 10.11 it must be if it is realizable, then this limit calculation can be implemented by polynomial division. For each entry of $G(s)$, dividing the denominator polynomial into the numerator polynomial produces a power series in s^{-1}. Arranging these power series in matrix form, the Markov parameter sequence appears as the sequence of matrix coefficients in the expression

$$G(s) = G_0 s^{-1} + G_1 s^{-2} + G_2 s^{-3} + \cdots$$

The time-invariant realization problem specified by a given Markov parameter sequence leads to consideration of the behavior matrix in (23) evaluated at $t = 0$. In this setup $\Gamma_{ij}(0)$ often is called a *block Hankel matrix* corresponding to $G(t)$, or $G(s)$, and is written as

$$\Gamma_{ij} = \begin{bmatrix} G_0 & G_1 & \cdots & G_{j-1} \\ G_1 & G_2 & \cdots & G_j \\ \vdots & \vdots & \vdots & \vdots \\ G_{i-1} & G_i & \cdots & G_{i+j-2} \end{bmatrix} \tag{43}$$

By repacking the data in (42) it is easy to verify that the controllability and observability matrices for a realization of a Markov parameter sequence are related to the block Hankel matrices by

$$\Gamma_{ij} = M_i W_j , \quad i, j = 1, 2, \cdots \tag{44}$$

In addition the pattern of entries in (43) captures essential algebraic features of the realization problem, and leads to a realization criterion and a method for computing minimal realizations.

11.7 Theorem   The analytic impulse response $G(t)$ admits a time-invariant realization (24) if and only if there exist positive integers l, k, n with l, $k \le n$ such that

$$\text{rank } \Gamma_{lk} = \text{rank } \Gamma_{l+1,k+j} = n , \quad j = 1, 2, \cdots \tag{45}$$

If this rank condition holds, then the dimension of a minimal realization of $G(t)$ is n.

Proof   Assuming l, k, and n are such that the rank condition (45) holds, we will construct a minimal realization for $G(t)$ of dimension n by a roughly similar construction as in preceding proofs. Again a large sketch of a block Hankel matrix is a useful scratch pad in deciphering the construction.

Let H_k denote the $n \times km$ submatrix formed from the first n linearly independent rows of Γ_{lk}, equivalently, the first n linearly independent rows of $\Gamma_{l+1,k}$. Also let H_k^s be another $n \times km$ submatrix defined as follows. The ith row of H_k^s is the row of $\Gamma_{l+1,k}$ that is p rows below the row of $\Gamma_{l+1,k}$ that is the ith-row of H_k. A realization of $G(t)$ can be constructed in terms of these submatrices. Let
(a) F be the invertible $n \times n$ matrix formed from the first n linearly independent columns of H_k,
(b) F_s be the $n \times n$ matrix occupying the same column positions in H_k^s as does F in H_k,
(c) F_c be the $p \times n$ matrix occupying the same column positions in Γ_{1k} as does F in H_k,
(d) F_r be the $n \times m$ matrix occupying the first m columns of H_k.

Specifically consider the coefficient matrices defined by

$$A = F_s F^{-1} , \quad B = F_r , \quad C = F_c F^{-1} \tag{46}$$

Since $F_s = AF$, entries in the ith-row of A give the linear combination of rows of F that results in the ith row of F_s. Therefore the ith-row of A also gives the linear combination of rows of H_k that yields the ith-row of H_k^s, that is, $H_k^s = AH_k$.

In fact a more general relationship holds. Let H_j be the extension or restriction of H_k in Γ_{1j}, $j = 1, 2, \cdots$, prescribed as follows. Each row of H_k, which is a row of Γ_{lk}, either is truncated (if $j < k$) or extended (if $j > k$) to match the corresponding row of Γ_{1j}. Similarly define H_j^s as the extension or restriction of H_k^s in $\Gamma_{l+1,j}$. Then (45) implies

$$H_j^s = AH_j , \quad j = 1, 2, \cdots \tag{47}$$

Also

$$H_j = \begin{bmatrix} F_r & H_{j-1}^s \end{bmatrix}, \quad j = 2, 3, \cdots \tag{48}$$

For example H_1 and H_2 are formed by the rows in

$$\begin{bmatrix} G_0 \\ G_1 \\ \vdots \\ G_{l-1} \end{bmatrix}, \quad \begin{bmatrix} G_0 & G_1 \\ G_1 & G_2 \\ \vdots & \vdots \\ G_{l-1} & G_l \end{bmatrix}$$

respectively, that correspond to the first n linearly independent rows in Γ_{lk}. But then H_1^s can be described as the rows of H_2 with the first m entries deleted, and from the definition of F_r it is immediate that $H_2 = [F_r \quad H_1^s]$.

Using (47) and (48) gives

$$H_j = \begin{bmatrix} F_r & AF_r & AH_{j-2}^s \end{bmatrix} \tag{49}$$

and, continuing,

$$H_j = \begin{bmatrix} F_r & AF_r & \cdots & A^{j-1}F_r \end{bmatrix}$$

$$= \begin{bmatrix} B & AB & \cdots & A^{j-1}B \end{bmatrix}, \quad j = 1, 2, \cdots$$

From (46) the ith-row of C specifies the linear combination of rows of F that gives the ith-row of F_c. But then the ith-row of C specifies the linear combination of rows of H_j that gives Γ_{1j}. Since every row of Γ_{1j} can be written as a linear combination of rows of H_j, it follows that

$$\Gamma_{1j} = CH_j = \begin{bmatrix} CB & CAB & \cdots & CA^{j-1}B \end{bmatrix}$$

$$= \begin{bmatrix} G_0 & G_1 & \cdots & G_{j-1} \end{bmatrix}, \quad j = 1, 2, \cdots$$

Therefore

$$G_j = CA^jB , \quad j = 0, 1, \cdots \tag{50}$$

and this shows that (46) specifies an n-dimensional realization for $G(t)$. Furthermore it is clear from a simple contradiction argument involving the rank condition (45), and (44), that this realization is minimal.

To prove the necessity portion of the theorem, suppose that $G(t)$ has a time-invariant realization. Then from (44) and the Cayley-Hamilton theorem there must exist integers l, k, n, with l, $k \le n$, such that the rank condition (45) holds.
□□□

It should be emphasized that the rank test (45) involves an infinite sequence of matrices, and this sequence cannot be truncated.

11.8 Example   The Markov parameter sequence for the impulse response

$$G(t) = e^t + \frac{t^{100}}{100!} \tag{51}$$

has 1's in the first 101 places. Yielding to temptation and truncating (45) at $l = k = n = 1$ would lead to a one-dimensional realization for $G(t)$ — a dramatically incorrect result. Since the transfer function corresponding to (51) is

$$G(s) = \frac{1}{s-1} + \frac{1}{s^{101}} = \frac{s^{101}+s-1}{s^{102}-s^{101}}$$

the observations in Example 10.12 lead to the conclusion that a minimal realization has dimension $n = 102$. As further illustration consider the Markov parameter sequence for $G(t) = exp(-t^2)$,

$$G_k = \begin{cases} \dfrac{(-1)^{k/2}k!}{k/2!}, & k \text{ even} \\ \\ 0, & k \text{ odd} \end{cases}$$

for $k = 0, 1, \cdots$. Pretending we don't know from Example 10.9 that $G(t)$ is not realizable, determination of realizability via rank properties of the corresponding Hankel matrix

$$\begin{bmatrix} 1 & 0 & -2 & 0 & \cdots \\ 0 & -2 & 0 & 12 & \cdots \\ -2 & 0 & 12 & 0 & \cdots \\ 0 & 12 & 0 & -120 & \cdots \\ 12 & 0 & -120 & 0 & \cdots \\ 0 & -120 & 0 & 1680 & \cdots \\ \vdots & \vdots & \vdots & \vdots & \vdots \end{bmatrix}$$

clearly is a precarious endeavor.
□□□

Suppose we know *a priori* that a given impulse response or transfer function has a realization of dimension no larger than some fixed number. Then the rank test (45) on an infinite number of block Hankel matrices can be truncated appropriately, and construction of a minimal realization can proceed. Specifically if there exists a realization of dimension n, then from (44), and the Cayley-Hamilton theorem applied to M_i and W_j,

$$\text{rank } \Gamma_{nn} = \text{rank } \Gamma_{n+i,n+j} \leq n , \quad i, j = 1, 2, \cdots \tag{52}$$

Therefore (45) need only be checked for $l,k < n$ and $k+j \leq n$. Further discussion of this issue is left to Note 11.2, except for an illustration.

11.9 Example   For the two-input, single-output transfer function

$$G(s) = \left[\frac{4s^2+7s+3}{s^3+4s^2+5s+2} \quad \frac{1}{s+1} \right] \tag{53}$$

a dimension-4 realization can be constructed by applying the procedure in Example 10.12 for each single-input, single-output component. This gives the realization

$$\dot{x}(t) = \begin{bmatrix} 0 & 1 & 0 & 0 \\ 0 & 0 & 1 & 0 \\ -2 & -5 & -4 & 0 \\ 0 & 0 & 0 & -1 \end{bmatrix} x(t) + \begin{bmatrix} 0 & 0 \\ 0 & 0 \\ 1 & 0 \\ 0 & 1 \end{bmatrix} u(t)$$

$$y(t) = \begin{bmatrix} 3 & 7 & 4 & 1 \end{bmatrix} x(t)$$

To check minimality and, if needed, construct a minimal realization, the first step is to divide each transfer function to obtain the corresponding Markov parameter sequence,

$$G_0 = \begin{bmatrix} 4 & 1 \end{bmatrix}, \quad G_1 = \begin{bmatrix} -9 & -1 \end{bmatrix}, \quad G_2 = \begin{bmatrix} 19 & 1 \end{bmatrix}, \quad G_3 = \begin{bmatrix} -39 & -1 \end{bmatrix},$$

$$G_4 = \begin{bmatrix} 79 & 1 \end{bmatrix}, \quad G_5 = \begin{bmatrix} -159 & -1 \end{bmatrix}, \quad G_6 = \begin{bmatrix} 319 & 1 \end{bmatrix}, \quad \cdots$$

Beginning application of the rank test,

$$\text{rank } \Gamma_{22} = \text{rank } \begin{bmatrix} 4 & 1 & -9 & -1 \\ -9 & -1 & 19 & 1 \end{bmatrix} = 2$$

$$\text{rank } \Gamma_{32} = \text{rank } \begin{bmatrix} 4 & 1 & -9 & -1 \\ -9 & -1 & 19 & 1 \\ 19 & 1 & -39 & -1 \end{bmatrix} = 2 \tag{54}$$

and continuing we find

$$\text{rank } \Gamma_{44} = 2$$

Thus by (52) the rank condition in (45) holds with $l = k = n = 2$, and the dimension of minimal realizations of $G(s)$ is two. Construction of a minimal realization can proceed on the basis of Γ_{22} and Γ_{32} in (54). The various submatrices

$$H_2 = \begin{bmatrix} 4 & 1 & -9 & -1 \\ -9 & -1 & 19 & 1 \end{bmatrix}, \quad H_2^s = \begin{bmatrix} -9 & -1 & 19 & 1 \\ 19 & 1 & -39 & -1 \end{bmatrix}$$

$$F = \begin{bmatrix} 4 & 1 \\ -9 & -1 \end{bmatrix}, \quad F_s = \begin{bmatrix} -9 & -1 \\ 19 & 1 \end{bmatrix}, \quad F_r = F, \quad F_c = \begin{bmatrix} 4 & 1 \end{bmatrix}$$

yield via (46) the minimal-realization coefficients

$$A = \begin{bmatrix} 0 & 1 \\ -2 & -3 \end{bmatrix}, \quad B = \begin{bmatrix} 4 & 1 \\ -9 & -1 \end{bmatrix}, \quad C = \begin{bmatrix} 1 & 0 \end{bmatrix}$$

The dimension reduction from 4 to 2 can be partly understood by writing the transfer function (53) in factored form as

$$G(s) = \begin{bmatrix} \dfrac{(4s+3)(s+1)}{(s+2)(s+1)^2} & \dfrac{1}{s+1} \end{bmatrix} \tag{55}$$

Canceling the common factor in the first entry and applying the approach from Example 10.12 yields a realization of dimension 3. The remaining dimension reduction to minimality is more subtle.

EXERCISES

Exercise 11.1   If the single-input linear state equation

$$\dot{x}(t) = A(t)x(t) + b(t)u(t)$$

is instantaneously controllable, show that at any time t_a an 'instantaneous' state transfer from $x(t_a) = 0$ to any desired state x_d can be made using an input of the form

$$u(t) = \sum_{k=0}^{n-1} \alpha_k \delta^{(k)}(t - t_a)$$

where $\delta^{(0)}(t)$ is the unit impulse, $\delta^{(1)}(t)$ is the unit doublet, and so on. *Hint*: Recall the sifting property

$$\int_{t_a^-}^{t_a^+} f(t)\delta^{(k)}(t-t_a)\, dt = (-1)^k \frac{d^k f}{dt^k}(t_a)$$

Exercise 11.2   If the linear state equation

$$\dot{x}(t) = A(t)x(t)$$
$$y(t) = C(t)x(t)$$

is instantaneously observable, show that at any time t_a the state $x(t_a)$ can be determined 'instantaneously' from a knowledge of the values of the output and its first $n-1$ derivatives at t_a.

Exercise 11.3   Show that instantaneous controllability and instantaneous observability are preserved under an invertible time-varying variable change (that has sufficiently many continuous derivatives).

Exercise 11.4   Is the linear state equation

$$\dot{x}(t) = \begin{bmatrix} -1 & 0 \\ 0 & -1 \end{bmatrix} x(t) + \begin{bmatrix} 1 \\ 1 \end{bmatrix} u(t)$$

$$y(t) = \begin{bmatrix} t^2 & 1 \end{bmatrix} x(t)$$

a minimal realization of its impulse response? If not, construct such a minimal realization.

Exercise 11.5   Show that

$$\dot{x}(t) = \begin{bmatrix} -1 & 0 \\ 0 & -2 \end{bmatrix} x(t) + \begin{bmatrix} 1 \\ 1 \end{bmatrix} u(t)$$

$$y(t) = \begin{bmatrix} t-3 & 1 \\ t & 5 \end{bmatrix} x(t)$$

is a minimal realization of its impulse response, yet the hypotheses of Theorem 11.3 are not satisfied.

Exercise 11.6   Construct a minimal realization for the impulse response

$$G(t, \sigma) = 1 + \tfrac{1}{2}e^{2t} + \tfrac{1}{2}e^{2\sigma} , \quad t \geq \sigma$$

Exercise 11.7   Show that the hypotheses of Theorem 11.3 imply that given any $i, j \geq 0$,

$$\text{rank } \Gamma_{l+i,k+j}(t, \sigma) = n$$

for all t, σ with $t \geq \sigma$.

Exercise 11.8   Prove that if an n-dimensional, time-varying linear state equation is instantaneously controllable and instantaneously observable, then

$$\text{rank } \Gamma_{nn}(t, \sigma) \leq \text{rank } \Gamma_{n+1,n+1}(t, \sigma) \leq n$$

for all t, σ such that $t \geq \sigma$.

Exercise 11.9   Show that the rank condition (45) implies

$$\text{rank } \Gamma_{l+i,k+j} = n ; \quad i, j = 1, 2, \cdots$$

Exercise 11.10   Compute a minimal realization corresponding to the Markov parameter sequence given by the *Fibonacci sequence*

$$0, 1, 1, 2, 3, 5, 8, 13, \cdots$$

(*Hint:* $f(k+2) = f(k+1) + f(k)$.)

Exercise 11.11   Compute a minimal realization corresponding to the Markov parameter sequence

$$1, 1, 1, 1, 1, 1, 1, 1, \cdots$$

Then compute a minimal realization corresponding to the 'truncated' sequence

$$1, 1, 1, 0, 0, 0, 0, \cdots$$

Exercise 11.12   For a scalar transfer function $G(s)$, suppose the infinite block Hankel matrix has rank n. Show that the first n columns are linearly independent, and that a minimal realization is given by

$$
A = \begin{bmatrix} G_1 & G_2 & \cdots & G_n \\ G_2 & G_3 & \cdots & G_{n+1} \\ \vdots & \vdots & \vdots & \vdots \\ G_n & G_{n+1} & \cdots & G_{2n-1} \end{bmatrix} \begin{bmatrix} G_0 & G_1 & \cdots & G_{n-1} \\ G_1 & G_2 & \cdots & G_n \\ \vdots & \vdots & \vdots & \vdots \\ G_{n-1} & G_n & \cdots & G_{2n-2} \end{bmatrix}^{-1} , \quad B = \begin{bmatrix} G_0 \\ G_1 \\ \vdots \\ G_{n-1} \end{bmatrix} , \quad C = \begin{bmatrix} 1 & 0 & \cdots & 0 \end{bmatrix}
$$

NOTES

Note 11.1   Our treatment of realization theory is based on

L.M. Silverman, "Representation and realization of time-variable linear systems," Technical Report No. 94, Department of Electrical Engineering, Columbia University, New York, 1966

L.M. Silverman, "Realization of linear dynamical systems," *IEEE Transactions on Automatic Control,* Vol. 16, No. 6, pp. 554–567, 1971

It can be shown that realization theory in the time-varying case can be founded on the single-variable matrix obtained by evaluating $\Gamma_{ij}(t, \sigma)$ at $\sigma = t$. Furthermore the assumption of a fixed invertible submatrix $F(t, \sigma)$ can be dropped. Using a more sophisticated algebraic framework, these extensions are discussed in

E.W. Kamen, "New results in realization theory for linear time-varying analytic systems," *IEEE Transactions on Automatic Control,* Vol. 24, No. 6, pp. 866–877, 1979

For the time-invariant case a different realization algorithm based on the block Hankel matrix is in

B.L. Ho, R.E. Kalman, "Effective construction of linear state variable models from input-output functions," *Regelungstechnik,* Vol. 14, pp. 545–548, 1966.

Note 11.2   A special type of realization where the controllability and observability Gramians are equal and diagonal, called a *balanced* realization, is studied for time-varying systems in

S. Shokoohi, L.M. Silverman, P.M. Van Dooren, "Linear time-variable systems: balancing and model reduction," *IEEE Transactions on Automatic Control,* Vol. 28, No. 8, pp. 810–822, 1983

E. Verriest, T. Kailath, "On generalized balanced realizations," *IEEE Transactions on Automatic Control,* Vol. 28, No. 8, pp. 833–844, 1983

The notion of balanced realization in the time-invariant case is introduced in

B.C. Moore, "Principal component analysis in linear systems: Controllability, observability, and model reduction," *IEEE Transactions on Automatic Control,* Vol. 26, No. 1, pp. 17–32, 1981

Note 11.3   In the time-invariant case the problem of realization from a finite number of Markov parameters is known as *partial realization*. Subtle issues arise in this problem, and these issues are studied in, for example,

R.E. Kalman, P.L. Falb, M.A. Arbib, *Topics in Mathematical System Theory,* Mc-Graw Hill, New York, 1969

R.E. Kalman, "On minimal partial realizations of a linear input/output map," in *Aspects of Network and System Theory,* R.E. Kalman and N. DeClaris, editors, Holt, Rinehart and Winston, New York, 1971

Note 11.4   The time-invariant realization problem can be based on information about the input-output behavior other than the Markov parameters. Realization based on the time-moments of the impulse response is discussed in

C. Bruni, A. Isidori, A. Ruberti, "A method of realization based on moments of the impulse-response matrix," *IEEE Transactions on Automatic Control,* Vol. 14, No. 2, pp. 203–204, 1969

The realization problem also can be formulated as an interpolation problem based on evaluations of the transfer function. Recent, in-depth studies can be found in the papers

A.C. Antoulas, B.D.O. Anderson, "On the scalar rational interpolation problem," *IMA Journal of Mathematical Control and Information,* Vol. 3, pp. 61–88, 1986

B.D.O. Anderson, A.C. Antoulas, "Rational interpolation and state-variable realizations," *Linear Algebra and its Applications,* Vol. 137/138, pp. 479–509, 1990

One motivation for the interpolation formulation is that certain types of transfer function evaluations in principle can be determined from input-output measurements on an unknown linear system. An elementary treatment is given in Chapter 6 of

W.J. Rugh, *Mathematical Description of Linear Systems,* Marcel-Dekker, New York, 1975

Also the realization problem can be based on arrangements of the Markov parameters other than the block Hankel matrix. See

A.A.H. Damen, P.M.J. Van den Hof, A.K. Hajdasinski, "Approximate realization based upon an alternative to the Hankel matrix: the Page matrix," *Systems & Control Letters,* Vol. 2, No. 4, pp. 202–208, 1982

12

INPUT-OUTPUT STABILITY

In this chapter we address stability properties appropriate to the input-output behavior (zero-state response) of the linear state equation

$$\dot{x}(t) = A(t)x(t) + B(t)u(t)$$

$$y(t) = C(t)x(t) \tag{1}$$

That is, the initial state is set to zero, and attention is focused on boundedness of the response to bounded inputs. There is no $D(t)u(t)$ term in (1) because a bounded $D(t)$ does not affect the treatment, while an unbounded $D(t)$ provides an unbounded response to an appropriate constant input. Of course the input-output behavior of (1) is specified by the impulse response

$$G(t, \sigma) = C(t)\Phi(t, \sigma)B(\sigma) , \quad t \ge \sigma \tag{2}$$

and stability results are characterized in terms of boundedness properties of $\|G(t, \sigma)\|$. (Notice in particular that the weighting pattern is not employed.) For the time-invariant case, input-output stability also is characterized in terms of the transfer function of the linear state equation.

Uniform Bounded-Input Bounded-Output Stability

Bounded-input, bounded-output stability is most simply discussed in terms of the largest value (over time) of the norm of the input signal, $\|u(t)\|$, in comparison to the largest value of the corresponding response norm $\|y(t)\|$. More precisely we use the standard notion of *supremum*. For example

$$\nu = \sup_{t \ge t_o} \|u(t)\|$$

is defined as the smallest constant such that $\|u(t)\| \le \nu$ for $t \ge t_o$. If no such bound exits, we write

$$\sup_{t \geq t_o} \| u(t) \| = \infty$$

The basic notion is that the zero-state response should exhibit finite 'gain' in terms of the input and output suprema.

12.1 Definition   The linear state equation (1) is called *uniformly bounded-input, bounded-output stable* if there exists a finite constant η such that for any t_o and any input signal $u(t)$ defined for $t \geq t_o$ the corresponding zero-state response satisfies

$$\sup_{t \geq t_o} \| y(t) \| \leq \eta \sup_{t \geq t_o} \| u(t) \| \qquad (3)$$

The adjective 'uniform' does double duty in this definition. It emphasizes the fact that the same η works for all values of t_o, and that the same η works for all input signals. An equivalent definition based on the pointwise norms of $u(t)$ and $y(t)$ is explored in Exercise 12.1. See Note 12.1 for discussion of related points, some quite subtle.

12.2 Theorem   The linear state equation (1) is uniformly bounded-input, bounded-output stable if and only if there exists a finite constant ρ such that for all t, τ with $t \geq \tau$,

$$\int_{\tau}^{t} \| G(t, \sigma) \| \, d\sigma \leq \rho \qquad (4)$$

Proof   Assume first that such a ρ exists. Then for any t_o and any input defined for $t \geq t_o$, the corresponding zero-state response of (1) satisfies

$$\| y(t) \| = \left\| \int_{t_o}^{t} C(t)\Phi(t, \sigma)B(\sigma)u(\sigma) \, d\sigma \right\|$$

$$\leq \int_{t_o}^{t} \| G(t, \sigma) \| \, \| u(\sigma) \| \, d\sigma , \quad t \geq t_o$$

Replacing $\| u(\sigma) \|$ by its supremum over $\sigma \geq t_o$, and using (4),

$$\| y(t) \| \leq \int_{t_o}^{t} \| G(t, \sigma) \| \, d\sigma \sup_{t \geq t_o} \| u(t) \|$$

$$\leq \rho \sup_{t \geq t_o} \| u(t) \| , \quad t \geq t_o$$

Therefore, taking the supremum of the left side over $t \geq t_o$, (3) holds with $\eta = \rho$, and the state equation is uniformly bounded-input, bounded-output stable.

Suppose now that (1) is uniformly bounded-input, bounded-output stable. Then there exists a constant η so that, in particular, the zero-state response for any t_o and any input signal such that

$$\|u(t)\| \le 1 , \quad t \ge t_o$$

satisfies

$$\|y(t)\| \le \eta , \quad t \ge t_o$$

To set up a contradiction argument, suppose no finite ρ exists that satisfies (4). In other words for any given constant ρ there exist τ_ρ and $t_\rho > \tau_\rho$ such that

$$\int_{\tau_\rho}^{t_\rho} \|G(t_\rho, \sigma)\| \, d\sigma > \rho$$

By Exercise 1.14 this implies, taking $\rho = \eta$, that there exist τ_η, $t_\eta > \tau_\eta$, and indices i, j such that the i,j-entry of the impulse response satisfies

$$\int_{\tau_\eta}^{t_\eta} |G_{ij}(t_\eta, \sigma)| \, d\sigma > \eta \tag{5}$$

With $t_o = \tau_\eta$ consider the $m \times 1$ input signal $u(t)$ defined for $t \ge t_o$ as follows. Set $u(t) = 0$ for $t > t_\eta$, and for $t \in [t_o, t_\eta]$ set every component of $u(t)$ to zero except for the jth component given by

$$u_j(t) = \begin{cases} 1 , & G_{ij}(t_\eta, t) > 0 \\ 0 , & G_{ij}(t_\eta, t) = 0 , \quad t \in [t_o, t_\eta] \\ -1 , & G_{ij}(t_\eta, t) < 0 \end{cases}$$

This input signal satisfies $\|u(t)\| \le 1$, for all $t \ge t_o$, but the ith component of the corresponding zero-state response satisfies, by (5),

$$y_i(t_\eta) = \int_{t_o}^{t_\eta} G_{ij}(t_\eta, \sigma) u_j(\sigma) \, d\sigma$$

$$= \int_{t_o}^{t_\eta} |G_{ij}(t_\eta, \sigma)| \, d\sigma$$

$$> \eta$$

Since $\|y(t_\eta)\| \ge |y_i(t_\eta)|$, a contradiction is obtained that completes the proof. □□□

For a time-invariant linear state equation, $G(t, \sigma) = G(t - \sigma)$, and thus the impulse response customarily is written as

$$G(t) = Ce^{At}B , \quad t \geq 0$$

Then a change of integration variable in (4) shows that a necessary and sufficient condition for uniform bounded-input, bounded-output stability is finiteness of the integral

$$\int_0^\infty \| G(t) \| \, dt \tag{6}$$

Relation to Uniform Exponential Stability

We now turn to establishing connections between uniform bounded-input, bounded-output stability and the property of uniform exponential stability of the zero-input response. This is not a trivial pursuit, as a simple example indicates.

12.3 Example   The time-invariant linear state equation

$$\dot{x}(t) = \begin{bmatrix} 0 & 1 \\ 1 & 0 \end{bmatrix} x(t) + \begin{bmatrix} 0 \\ 1 \end{bmatrix} u(t)$$

$$y(t) = [\, 1 \quad -1 \,] \, x(t) \tag{7}$$

is *not* uniformly exponentially stable, since the eigenvalues of A are $1, -1$. However the impulse response is given by $G(t) = -e^{-t}$, and therefore the state equation is uniformly bounded-input, bounded-output stable.
□□□

In the time-invariant setting of this example, a description of the key difficulty is that scalar exponentials appearing in e^{At} might be missing from $G(t)$. Again controllability and observability play key roles, since we are considering the relation between input-output (zero-state) and internal (zero-input) stability concepts.

In one direction the connection between input-output and internal stability is easy to establish, and a division of labor proves convenient.

12.4 Lemma   Suppose the linear state equation (1) is uniformly exponentially stable, and there exist finite constants β and μ such that for all t

$$\| B(t) \| \leq \beta , \quad \| C(t) \| \leq \mu \tag{8}$$

Then the state equation also is uniformly bounded-input, bounded-output stable.

Proof   Using the transition matrix bound implied by uniform exponential stability,

$$\int_\tau^t \| G(t, \sigma) \| \, d\sigma \leq \int_\tau^t \| C(t) \| \, \| \Phi(t, \sigma) \| \, \| B(\sigma) \| \, d\sigma$$

$$\leq \mu\beta \int_{\tau}^{t} \gamma e^{-\lambda(t-\sigma)} \, d\sigma$$

$$\leq \mu\beta\gamma/\lambda$$

for all τ, t with $t \geq \tau$. Therefore the state equation is uniformly bounded-input, bounded-output stable by Theorem 12.2.
□□□

That coefficient bounds as in (8) are needed to obtain the implication in Lemma 12.4 should be clear. However the simple proof might suggest that uniform exponential stability is a needlessly strong condition for uniform bounded-input, bounded-output stability. To dispel this notion we consider a variation of Example 6.11.

12.5 Example   The scalar linear state equation with bounded coefficients

$$\dot{x}(t) = \frac{-2t}{t^2 + 1} x(t) + u(t) , \quad x(t_o) = x_o$$

$$y(t) = x(t) \tag{9}$$

is not uniformly exponentially stable, as shown in Example 6.11. Since

$$\Phi(t, t_o) = \frac{t_o^2 + 1}{t^2 + 1}$$

it is easy to check that the state equation is uniformly stable, and that the zero-input response goes to zero for all initial states. However with $t_o = 0$ and the bounded input $u(t) = 1$ for $t \geq 0$, the zero-state response is unbounded:

$$y(t) = \int_{0}^{t} \frac{\sigma^2 + 1}{t^2 + 1} \, d\sigma = \frac{t^3/3 + t}{t^2 + 1}$$

□□□

In developing implications of uniform bounded-input, bounded-output stability for uniform exponential stability, we need to strengthen the usual controllability and observability properties. Specifically we will assume these properties are uniform in time in a special way. For simplicity, admittedly a commodity in short supply for the next few pages, the development is subdivided into two parts. First we deal with linear state equations where the output is precisely the state vector ($C(t)$ is the $n \times n$ identity). In this instance the natural terminology is *uniform bounded-input, bounded-state stability*.

Recall from Chapter 9 the controllability Gramian

$$W(t_o, t_f) = \int_{t_o}^{t_f} \Phi(t_o, t)B(t)B^T(t)\Phi^T(t_o, t) \, dt$$

12.6 Theorem   Suppose for the linear state equation

$$\dot{x}(t) = A(t)x(t) + B(t)u(t)$$

$$y(t) = x(t)$$

there exist finite positive constants α, β, ε, and δ, such that for all t

$$\|A(t)\| \le \alpha, \quad \|B(t)\| \le \beta, \quad \varepsilon I \le W(t-\delta, t) \tag{10}$$

Then the state equation is uniformly bounded-input, bounded-state stable if and only if it is uniformly exponentially stable.

Proof   One direction of proof is supplied by Lemma 12.4, so assume the linear state equation (1) is uniformly bounded-input, bounded-state stable. Applying Theorem 12.2 with $C(t) = I$, there exists a finite constant ρ such that

$$\int_{\tau}^{t} \|\Phi(t, \sigma)B(\sigma)\| \, d\sigma \le \rho \tag{11}$$

for all t, τ such that $t \ge \tau$. We next show that this implies existence of a finite constant ψ such that

$$\int_{\tau}^{t} \|\Phi(t, \sigma)\| \, d\sigma \le \psi$$

for all t, τ such that $t \ge \tau$, and thus conclude uniform exponential stability by Theorem 6.8.

We need to use some elementary facts from earlier exercises. First, since $A(t)$ is bounded, corresponding to the constant δ in (10) there exists a finite constant κ such that

$$\|\Phi(t, \sigma)\| \le \kappa, \quad |t - \sigma| \le \delta \tag{12}$$

(See Exercise 6.2.) Second, the lower bound on the controllability Gramian in (10) together with Exercise 1.10 gives

$$W^{-1}(t-\delta, t) \le \frac{1}{\varepsilon} I$$

for all t, and therefore

$$\|W^{-1}(t-\delta, t)\| \le \frac{1}{\varepsilon}$$

for all t. In particular these bounds show that

$$\|B^T(\gamma)\Phi^T(\sigma-\delta, \gamma)W^{-1}(\sigma-\delta, \sigma)\| \le \|B^T(\gamma)\| \, \|\Phi^T(\sigma-\delta, \gamma)\| \, \|W^{-1}(\sigma-\delta, \sigma)\|$$

$$\le \frac{\beta\kappa}{\varepsilon} \tag{13}$$

for all σ, γ satisfying $|\sigma-\delta-\gamma| \le \delta$. Therefore writing

$$\Phi(t,\ \sigma-\delta) = \Phi(t,\ \sigma-\delta)W(\sigma-\delta,\ \sigma)W^{-1}(\sigma-\delta,\ \sigma)$$

$$= \int_{\sigma-\delta}^{\sigma} \Phi(t,\ \gamma)B(\gamma)B^{T}(\gamma)\Phi^{T}(\sigma-\delta,\ \gamma)\ W^{-1}(\sigma-\delta,\ \sigma)\ d\gamma$$

we obtain, since $\sigma-\delta \le \gamma \le \sigma$ implies $|\sigma-\delta-\gamma| \le \delta$,

$$\|\Phi(t,\ \sigma-\delta)\| \le \frac{\beta\kappa}{\varepsilon} \int_{\sigma-\delta}^{\sigma} \|\Phi(t,\ \gamma)B(\gamma)\|\ d\gamma$$

Then

$$\int_{\tau}^{t} \|\Phi(t,\ \sigma-\delta)\|\ d(\sigma-\delta) \le \frac{\beta\kappa}{\varepsilon} \int_{\tau}^{t} \left[\int_{\sigma-\delta}^{\sigma} \|\Phi(t,\ \gamma)B(\gamma)\|\ d\gamma \right] d(\sigma-\delta) \qquad (14)$$

The proof can be completed by showing that the right side of (14) is bounded for all t, τ such that $t \ge \tau$.

In the inside integral on the right side of (14), change the integration variable from γ to $\xi = \gamma-\sigma+\delta$, and then interchange the order of integration to write the right side of (14) as

$$\frac{\beta\kappa}{\varepsilon} \int_{0}^{\delta} \left[\int_{\tau}^{t} \| \Phi(t,\ \xi+\sigma-\delta)B(\xi+\sigma-\delta)\|\ d(\sigma-\delta) \right] d\xi$$

In the inside integral in this expression, change the integration variable from $\sigma-\delta$ to $\zeta = \xi+\sigma-\delta$ to obtain

$$\frac{\beta\kappa}{\varepsilon} \int_{0}^{\delta} \left[\int_{\tau+\xi}^{t+\xi} \|\Phi(t,\ \zeta)B(\zeta)\|\ d\zeta \right] d\xi \qquad (15)$$

Since $0 \le \xi \le \delta$ we can use (11) and (12) with the composition property to bound the inside integral in (15) as

$$\int_{\tau+\xi}^{t+\xi} \|\Phi(t,\ \zeta)B(\zeta)\|\ d\zeta \le \|\Phi(t,\ t+\xi)\| \int_{\tau+\xi}^{t+\xi} \|\Phi(t+\xi,\ \zeta)B(\zeta)\|\ d\zeta$$

$$\le \kappa\rho$$

Therefore (14) becomes

$$\int_{\tau}^{t} \|\Phi(t,\ \sigma-\delta)\|\ d(\sigma-\delta) \le \frac{\beta\kappa}{\varepsilon} \int_{0}^{\delta} \kappa\rho\ d\xi$$

$$\le \frac{\beta\kappa^{2}\rho\delta}{\varepsilon}$$

This holds for all t, τ such that $t \ge \tau$, so uniform exponential stability of the linear state equation with $C(t) = I$ follows from Theorem 6.8.
□□□

To address the general case, where $C(t)$ is not an identity matrix, recall that the observability Gramian for the state equation (1) is defined by

$$M(t_o, t_f) = \int_{t_o}^{t_f} \Phi^T(t, t_o) C^T(t) C(t) \Phi(t, t_o) \, dt \tag{16}$$

12.7 Theorem   Suppose that for the linear state equation (1) there exist finite positive constants α, β, μ, ε_1, δ_1, ε_2, and δ_2, such that

$$\|A(t)\| \le \alpha, \quad \|B(t)\| \le \beta, \quad \|C(t)\| \le \mu,$$

$$\varepsilon_1 I \le W(t - \delta_1, t), \quad \varepsilon_2 I \le M(t, t + \delta_2) \tag{17}$$

for all t. Then the state equation is uniformly bounded-input, bounded-output stable if and only if it is uniformly exponentially stable.

Proof   Again uniform exponential stability implies uniform bounded-input, bounded-output stability by Lemma 12.4. So suppose that (1) is uniformly bounded-input, bounded-output stable, and η is such that the zero-state response satisfies

$$\sup_{t \ge t_o} \|y(t)\| \le \eta \sup_{t \ge t_o} \|u(t)\| \tag{18}$$

for all inputs $u(t)$. We will show that the associated state equation with $C(t) = I$, namely,

$$\dot{x}(t) = A(t)x(t) + B(t)u(t)$$

$$y_a(t) = x(t) \tag{19}$$

also is uniformly bounded-input, bounded-state stable. To set up a contradiction argument, assume the negation. Then for the positive constant $\sqrt{\eta^2 \delta_2 / \varepsilon_2}$ there exists a t_o, $t_a > t_o$, and bounded input signal $u(t)$ such that

$$\|y(t_a)\| = \|x(t_a)\| > \sqrt{\eta^2 \delta_2 / \varepsilon_2} \sup_{t \ge t_o} \|u(t)\| \tag{20}$$

Applying $u(t)$ to (1), keeping the same initial time t_o, the zero-state response satisfies

$$\delta_2 \sup_{t_a \le t \le t_a + \delta_2} \|y(t)\|^2 \ge \int_{t_a}^{t_a + \delta_2} \|y(t)\|^2 \, dt$$

$$= \int_{t_a}^{t_a + \delta_2} x^T(t_a) \Phi^T(t, t_a) C^T(t) C(t) \Phi(t, t_a) x(t_a) \, dt$$

$$= x^T(t_a) M(t_a, t_a + \delta_2) x(t_a)$$

Invoking the hypothesis on the observability Gramian, and then (20),

$$\delta_2 \sup_{t_a \le t \le t_a + \delta_2} \|y(t)\|^2 \ge \varepsilon_2 \|x(t_a)\|^2$$

$$> \eta^2 \delta_2 \left[\sup_{t \ge t_o} \|u(t)\| \right]^2$$

Using elementary properties of the supremum, including

$$\left[\sup_{t_a \le t \le t_a + \delta_2} \|y(t)\| \right]^2 = \sup_{t_a \le t \le t_a + \delta_2} \|y(t)\|^2$$

yields

$$\sup_{t \ge t_o} \|y(t)\| > \eta \sup_{t \ge t_o} \|u(t)\| \tag{21}$$

Thus we have shown that the bounded input $u(t)$ is such that the bound (18) for uniform bounded-input, bounded-output stability of (1) is violated. This contradiction implies (19) is uniformly bounded-input, bounded-state stable. Then by Theorem 12.7 the state equation (19) is uniformly exponentially stable, and hence (1) also is uniformly exponentially stable.

Time-Invariant Case

Seemingly contrived manipulations in the proofs of Theorem 12.6 and Theorem 12.7 motivate separate consideration of the time-invariant case. In this setting the simpler characterizations of stability properties, and of controllability and observability, yield more straightforward proofs. For the linear state equation

$$\dot{x}(t) = Ax(t) + Bu(t)$$

$$y(t) = Cx(t) \tag{22}$$

the main task in proving an analog of Theorem 12.7 is to show that controllability, observability, and finiteness of

$$\int_0^\infty \|Ce^{At}B\| \, dt \tag{23}$$

imply finiteness of

$$\int_0^\infty \|e^{At}\| \, dt$$

12.8 Theorem   Suppose the time-invariant linear state equation (22) is controllable and observable. Then the state equation is uniformly bounded-input, bounded-output stable if and only if it is exponentially stable.

Proof   Clearly exponential stability implies uniform bounded-input, bounded-output stability since

$$\int_0^\infty \| Ce^{At}B \| \, dt \le \| C \| \, \| B \| \int_0^\infty \| e^{At} \| \, dt$$

Conversely suppose (2) is uniformly bounded-input, bounded-output stable. Then (23) is finite, and this implies

$$\lim_{t \to \infty} Ce^{At}B = 0 \tag{24}$$

Using Property 5.8 we can write the impulse response in the form

$$Ce^{At}B = \sum_{k=1}^{m} \sum_{j=1}^{\sigma_k} G_{kj} \frac{t^{j-1}}{(j-1)!} \, e^{\lambda_j t}$$

where $\lambda_1, \ldots, \lambda_m$ are the distinct eigenvalues of A, and the G_{kj} are $p \times m$ constant matrices. Then

$$\frac{d}{dt} Ce^{At}B = \sum_{k=1}^{m} \sum_{j=1}^{\sigma_k} G_{kj} \left[\frac{\lambda_j t^{j-1}}{(j-1)!} + \frac{t^{j-2}}{(j-2)!} \right] e^{\lambda_j t}$$

where we abuse notation by ignoring the negative-power-of-t terms occurring for $j = 1$. A contradiction argument based on these representations proves that (24) implies

$$\lim_{t \to \infty} \left[\frac{d}{dt} Ce^{At}B \right] = 0$$

That is,

$$\lim_{t \to \infty} CAe^{At}B = \lim_{t \to \infty} Ce^{At}AB = 0$$

This reasoning can be repeated to show that any time derivative of the impulse response goes to zero as $t \to \infty$. Explicitly,

$$\lim_{t \to \infty} CA^i e^{At} A^j B = 0 \, ; \quad i, j = 0, 1, \, \cdots$$

Arranging this data in matrix form,

$$\lim_{t \to \infty} \begin{bmatrix} C \\ CA \\ \vdots \\ CA^{n-1} \end{bmatrix} e^{At} \begin{bmatrix} B & AB & \cdots & A^{n-1}B \end{bmatrix} = 0 \tag{25}$$

Using the controllability and observability hypotheses, select n linearly independent columns of the controllability matrix to form an invertible matrix W_a, and n linearly independent rows of the observability matrix to form an invertible M_a. Then, from (25),

$$\lim_{t \to \infty} M_a e^{At} W_a = 0$$

Therefore

$$\lim_{t \to \infty} e^{At} = 0$$

and exponential stability follows as in the proof of Theorem 6.10.
□□□

For some purposes it is useful to express the condition for uniform bounded-input, bounded-output stability of (22) in terms of the transfer function $G(s) = C(sI - A)^{-1}B$. We use the familiar terminology that a *pole* of $G(s)$ is a (complex, in general) value of s, say s_o, such that for some i, j, $|G_{ij}(s_o)| = \infty$.

If each entry of $G(s)$ has negative-real-part poles, then the partial-fraction-expansion representation developed for the matrix exponential in Chapter 5 shows that each entry of $G(t)$ has a 'sum of exponentials' form, with negative-real-part exponents. Therefore

$$\int_0^\infty \| G(t) \| \, dt \qquad (26)$$

is finite, and any realization of $G(s)$ is uniformly bounded-input, bounded-output stable. On the other hand if (26) is finite, then the exponential terms in any entry of $G(t)$ must have negative real parts. (Write a general entry in terms of distinct exponentials, and use a contradiction argument.) But then every entry of $G(s)$ has negative-real-part poles.

Supplying this reasoning with a little more specificity proves a standard result.

12.9 Theorem   The time-invariant linear state equation (22) is uniformly bounded-input, bounded-output stable if and only if all poles of the transfer function $G(s) = C(sI - A)^{-1}B$ have negative real parts.

For the time-invariant linear state equation (22), the relation between input-output stability and internal stability depends on whether all distinct eigenvalues of A appear as poles of $G(s) = C(sI - A)^{-1}B$. (Review Example 12.3 from a transfer-function perspective.) Controllability and observability guarantee that this is the case. (Unfortunately, eigenvalues of A sometimes are called 'poles of A,' a loose terminology that at best obscures delicate distinctions.)

EXERCISES

Exercise 12.1   Show that the linear state equation

$$\dot{x}(t) = A(t)x(t) + B(t)u(t)$$

$$y(t) = C(t)x(t)$$

is uniformly bounded-input, bounded output stable if and only if given any finite constant δ there exists a finite constant ε such that the following property holds for any t_o. If the the input signal satisfies

$$\|u(t)\| \le \delta, \quad t \ge t_o$$

then the corresponding zero-state response satisfies

$$\|y(t)\| \le \varepsilon, \quad t \ge t_o$$

(Note that ε depends only on δ, not on the particular input signal, nor on t_o.)

Exercise 12.2   Is the state equation below uniformly bounded-input, bounded-output stable? Is it uniformly exponentially stable?

$$\dot{x}(t) = \begin{bmatrix} 1/2 & 1 & 0 \\ 0 & -1 & 0 \\ 0 & 0 & -1 \end{bmatrix} x(t) + \begin{bmatrix} 0 \\ 1 \\ 0 \end{bmatrix} u(t)$$

$$y(t) = \begin{bmatrix} 1 & 0 & 0 \end{bmatrix} x(t)$$

Exercise 12.3   Determine whether the state equation below is uniformly bounded-input, bounded-output stable, and whether it is uniformly exponentially stable.

$$\dot{x}(t) = \begin{bmatrix} 0 & 1 \\ 2 & -1 \end{bmatrix} x(t) + \begin{bmatrix} 0 \\ 1 \end{bmatrix} u(t)$$

$$y(t) = \begin{bmatrix} 0 & 1 \end{bmatrix} x(t)$$

Exercise 12.4   Determine whether the state equation given below is uniformly exponentially stable, and whether it is uniformly bounded-input, bounded-output stable.

$$\dot{x}(t) = \begin{bmatrix} -1 & 0 \\ 0 & e^t \end{bmatrix} x(t) + \begin{bmatrix} 1 \\ 0 \end{bmatrix} u(t)$$

$$y(t) = [\, 1 \quad 0 \,] x(t)$$

Exercise 12.5   Find a linear state equation that satisfies all the hypotheses of Theorem 12.8 except for existence of ε_2 and δ_2, and is uniformly exponentially stable but not uniformly bounded-input, bounded-output stable.

Exercise 12.6   Devise a simple example of a linear state equation that is uniformly stable but not uniformly bounded-input, bounded-output stable. Can you give simple conditions on $B(t)$ and $C(t)$ under which the positive implication holds?

Exercise 12.7   Show that a time-invariant linear state equation is controllable if and only if there exist positive constants δ and ε such that for all t

$$\varepsilon I \le W(t-\delta, t)$$

Find a time-varying linear state equation that does not satisfy this condition but is controllable on $[t-\delta, t]$ for all t and some positive constant δ.

Exercise 12.8 Show that if the input signal to a uniformly bounded-input, bounded-output linear state equation goes to zero as $t \to \infty$, then the corresponding zero-state response also goes to zero as $t \to \infty$.

Exercise 12.9 With the obvious definition of *uniform bounded-input, bounded-state stable,* give proofs or counterexamples to the following claims.
(a) A linear state equation that is uniformly bounded-input, bounded-state stable also is uniformly bounded-input, bounded-output stable.
(b) A linear state equation that is uniformly bounded-input, bounded-output stable also is uniformly bounded-input, bounded-state stable.

Exercise 12.10 Suppose the linear state equation

$$\dot{x}(t) = A(t)x(t)$$

with $A(t)$ bounded, satisfies the following *total stability* property. Given $\varepsilon > 0$ there exist $\delta_1(\varepsilon), \delta_2(\varepsilon) > 0$ such that if $\|z_o\| < \delta_1$, and the continuous function $g(z, t)$ satisfies $\|g(z, t)\| < \delta_2$ for all z and t, then the solution of

$$\dot{z}(t) = A(t)z(t) + g(z(t), t), \quad z(t_o) = z_o$$

satisfies

$$\|z(t)\| < \varepsilon, \quad t \geq t_o$$

for any t_o. Show that the state equation $\dot{x}(t) = A(t)x(t)$ is uniformly exponentially stable.

Exercise 12.11 Consider a uniformly bounded-input, bounded-output stable, single-input, time-invariant linear state equation with transfer function $G(s)$. If λ and η are positive constants, show that the zero-state response $y(t)$ to

$$u(t) = e^{-\lambda t}, \quad t \geq 0$$

satisfies

$$\int_0^\infty y(t)e^{-\eta t}\, dt = \frac{1}{\lambda + \eta}\, G(\eta)$$

Under what conditions can such a relationship hold if the state equation is not uniformly bounded-input, bounded-output stable?

Exercise 12.12 Show that the single-input, single-output, linear state equations

$$\dot{x}(t) = Ax(t) + bu(t)$$
$$y(t) = cx(t) + u(t)$$

and

$$\dot{x}(t) = [A - bc]x(t) + bu(t)$$
$$y(t) = -cx(t) + u(t)$$

are inverses for each other in the sense that the product of their transfer functions is unity. If the first state equation is uniformly bounded-input, bounded-output stable, what is implied about input-output stability of the second?

Exercise 12.13   For the linear state equation

$$\dot{x}(t) = Ax(t) + Bu(t), \quad x(0) = x_o$$

$$y(t) = Cx(t)$$

suppose $m = p$, and CB is invertible. Let $P = I - B(CB)^{-1}C$ and consider the state equation

$$\dot{z}(t) = APz(t) + AB(CB)^{-1}v(t), \quad z(0) = x_o$$

$$w(t) = -(CB)^{-1}CAPz(t) - (CB)^{-1}CAB(CB)^{-1}v(t) + (CB)^{-1}\dot{v}(t)$$

Show that if $v(t) = y(t)$ for $t \geq 0$, then $w(t) = u(t)$ for $t \geq 0$. That is, show that the second state equation is an inverse for the first. If the first state equation is uniformly bounded-input, bounded-output stable, what is implied about input-output stability of the second? If the first is exponentially stable, what is implied about internal stability of the second?

NOTES

Note 12.1   By introduction of suprema in Definition 12.1, we surreptitiously employ a function-space norm, rather than the customary pointwise-in-time norm. See Exercise 12.1 for an equivalent definition in terms of pointwise norms. A more economical definition is that a linear state equation is *bounded-input, bounded-output stable* if a bounded input yields a bounded zero-state response. More precisely, given a t_o and $u(t)$ satisfying $\|u(t)\| \leq \delta$ for $t \geq t_o$, where δ is a finite positive constant, there is a finite positive constant ε such that the corresponding zero-state response satisfies $\|y(t)\| \leq \varepsilon$ for $t \geq t_o$. Obviously the requisite ε depends on δ, but also ε can depend on t_o, or on the particular input signal $u(t)$. Compare this to Exercise 12.1, where ε depends only on δ. Perhaps surprisingly, bounded-input, bounded-output stability is *equivalent* to Definition 12.1, and to the definition in Exercise 12.1, though the proof is difficult. See the papers:

C.A. Desoer, A.J. Thomasian, "A note on zero-state stability of linear systems," *Proceedings of the First Allerton Conference on Circuit and System Theory,* University of Illinois, Urbana, Illinois, 1963

D.C. Youla, "On the stability of linear systems," *IEEE Transactions on Circuits and Systems,* Vol. 10, No. 2, pp. 276–279, 1963

By this equivalence Theorem 12.2 is valid for the superficially weaker property of bounded-input, bounded-output stability, though the proof is less simple.

Note 12.2   The proof of Theorem 12.7 is based on

L.M. Silverman, B.D.O. Anderson, "Controllability, observability, and stability of linear systems," *SIAM Journal on Control and Optimization,* Vol. 6, No. 1, pp. 121–130, 1968

This paper contains a number of related results, and citations to earlier literature. See also

B.D.O. Anderson, J.B. Moore, "New results in linear system stability," *SIAM Journal on Control and Optimization,* Vol. 7, No. 3, pp. 398–414, 1969

Note 12.3   Exercises 12.12 and 12.13 are examples of *inverse system* calculations, a notion that is connected to several aspects of linear system theory. A general treatment for time-varying linear state equations is in

L.M. Silverman, "Inversion of multivariable linear systems," *IEEE Transactions on Automatic Control,* Vol. 14, No. 3, pp. 270–276, 1969

Further developments and a more general formulation for the time-invariant case can be found in

L.M. Silverman, H.J. Payne, "Input-output structure of linear systems with application to the decoupling problem," *SIAM Journal on Control and Optimization,* Vol. 9, No. 2, pp. 199–233, 1971

P.J. Moylan, "Stable inversion of linear systems," *IEEE Transactions on Automatic Control,* Vol. 22, No. 1, pp. 74–78, 1977

E. Soroka, U. Shaked, "On the geometry of the inverse system," *IEEE Transactions on Automatic Control,* Vol. 31, No. 8, pp. 751–754, 1986

These papers presume a linear state equation with fixed initial state. A somewhat different formulation is discussed in

H.L. Weinert, "On the inversion of linear systems," *IEEE Transactions on Automatic Control,* Vol. 29, No. 10, pp. 956–958, 1984

13

CONTROLLER AND OBSERVER FORMS

Though much of the material in Chapters 14 and 15 pertains to time-varying linear systems, our attention in the sequel increasingly turns to properties of the time-invariant case. Some of these properties rest on special techniques for time-invariant linear systems, for example the Laplace transform representation. Other properties simply have not been developed for time-varying systems, or the development is so tentative, or complicated, or requires such restrictive hypotheses that potential utility is unclear. Thus we begin to explore in more detail time-invariant linear state equations.

Even in the time-invariant case multi-input, multi-output linear state equations have a remarkably complicated algebraic structure. One approach to coping with this complexity is to apply a state variable change yielding a special form for the state equation that displays the structure. We adopt this approach and consider variable changes related to the controllability and observability structure of time-invariant linear state equations. Additional criteria for controllability and observability are obtained in the course of this development. A second approach, adopting an abstract geometric viewpoint that subordinates algebraic detail to a larger view, is explored in Chapter 18.

The standard notation

$$\dot{x}(t) = Ax(t) + Bu(t)$$
$$y(t) = Cx(t) \tag{1}$$

is continued for an n-dimensional, time-invariant, linear state equation with m inputs and p outputs. Recall that if two such state equations are related by a (constant) state variable change, then the $n \times nm$ controllability matrices for the two state equations have the same rank. Also the two $np \times n$ observability matrices have the same rank.

194

Controllability

We begin by showing that there is a state variable change for (1) that displays the 'controllable part' of the state equation. This result is of interest in itself, and it is used to develop new criteria for controllability.

13.1 Theorem   Suppose the controllability matrix for the linear state equation (1) satisfies

$$\text{rank} \begin{bmatrix} B & AB & \cdots & A^{n-1}B \end{bmatrix} = q \tag{2}$$

where $0 < q < n$. Then there exists an invertible $n \times n$ matrix P such that

$$P^{-1}AP = \begin{bmatrix} \hat{A}_{11} & \hat{A}_{12} \\ 0_{(n-q) \times q} & \hat{A}_{22} \end{bmatrix}, \quad P^{-1}B = \begin{bmatrix} \hat{B}_{11} \\ 0_{(n-q) \times m} \end{bmatrix} \tag{3}$$

where $\hat{A}_{11}$ is $q \times q$, $\hat{B}_{11}$ is $q \times m$, and

$$\text{rank} \begin{bmatrix} \hat{B}_{11} & \hat{A}_{11}\hat{B}_{11} & \cdots & \hat{A}_{11}^{q-1}\hat{B}_{11} \end{bmatrix} = q$$

Proof   The state variable change matrix P is constructed as follows. Pick q linearly independent columns, $p_1, \ldots, p_q$ from the controllability matrix for (1), that is, pick a basis for the range space of the controllability matrix. Then let $p_{q+1}, \ldots, p_n$ be additional $n \times 1$ vectors such that

$$P = \begin{bmatrix} p_1 & \cdots & p_q & p_{q+1} & \cdots & p_n \end{bmatrix}$$

is invertible. Define $G = P^{-1}B$, equivalently, $PG = B$. The jth column of B is given by postmultiplication of P by the jth column of G, in other words, by a linear combination of columns of P with coefficients given by the jth column of G. Since the jth column of B can be written as a linear combination of $p_1, \ldots, p_q$, and the columns of P are linearly independent, the last $n-q$ entries of the jth column of G must be zero. This argument applies for $j = 1, \ldots, m$, and therefore $G = P^{-1}B$ has the claimed form.

Now let $F = P^{-1}AP$ so that

$$PF = \begin{bmatrix} Ap_1 & Ap_2 & \cdots & Ap_n \end{bmatrix} \tag{4}$$

Since each column of A^kB, $k \geq 0$, can be written as a linear combination of $p_1, \ldots, p_q$, the column vectors $Ap_1, \ldots, Ap_q$ can be written as linear combinations of $p_1, \ldots, p_q$. Thus an argument similar to the argument for G gives that the first q columns of F must have zeros as the last $n-q$ entries. Therefore F has the claimed form. To complete the proof multiply the rank-q controllability matrix by the invertible matrix P^{-1} to obtain

$$P^{-1}\begin{bmatrix} B & AB & \cdots & A^{n-1}B \end{bmatrix} = \begin{bmatrix} P^{-1}B & P^{-1}AB & \cdots & P^{-1}A^{n-1}B \end{bmatrix}$$

$$= \begin{bmatrix} G & FG & \cdots & F^{n-1}G \end{bmatrix}$$

$$= \begin{bmatrix} \hat{B}_{11} & \hat{A}_{11}\hat{B}_{11} & \cdots & \hat{A}_{11}^{n-1}\hat{B}_{11} \\ 0 & 0 & \cdots & 0 \end{bmatrix} \tag{5}$$

The rank is preserved at each step in (5), and applying again the Cayley Hamilton theorem shows that

$$\text{rank} \begin{bmatrix} \hat{B}_{11} & \hat{A}_{11}\hat{B}_{11} & \cdots & \hat{A}_{11}^{q-1}\hat{B}_{11} \end{bmatrix} = q \tag{6}$$

□□□

An interpretation of this result is shown in Figure 13.2. Writing the variable change as

$$\begin{bmatrix} z_c(t) \\ z_{nc}(t) \end{bmatrix} = P^{-1}x(t)$$

where the partition $z_c(t)$ is $q \times 1$, yields a linear state equation that can be written in the decomposed form

$$\dot{z}_c(t) = \hat{A}_{11}z_c(t) + \hat{A}_{12}z_{nc}(t) + \hat{B}_{11}u(t)$$

$$\dot{z}_{nc}(t) = \hat{A}_{22}z_{nc}(t)$$

Clearly $z_{nc}(t)$ is not influenced by the input signal. Thus the second component state equation is not controllable, while by (6) the first component is controllable.

13.2 Figure   A state equation decomposition related to controllability.

The character of the decomposition aside, Theorem 13.1 is an important technical device in the proof of a new characterization of controllability.

13.3 Theorem   The linear state equation (1) is controllable if and only if for every complex scalar λ the only complex $n \times 1$ vector p that satisfies

$$p^T A = \lambda p^T , \quad p^T B = 0 \tag{7}$$

is $p = 0$.

Proof   The strategy is to show that (7) can be satisfied for some λ and some $p \neq 0$ if and only if the state equation is not controllable. If there exists a nonzero, complex, $n \times 1$ vector p and a complex scalar λ such that (7) is satisfied, then

$$p^T \begin{bmatrix} B & AB & \cdots & A^{n-1}B \end{bmatrix} = \begin{bmatrix} p^T B & p^T AB & \cdots & p^T A^{n-1}B \end{bmatrix}$$
$$= \begin{bmatrix} p^T B & p^T \lambda B & \cdots & p^T \lambda^{n-1} B \end{bmatrix}$$
$$= 0$$

Therefore the n rows of the controllability matrix are linearly dependent, and thus the state equation is not controllable.

On the other hand suppose the linear state equation (1) is not controllable. Then by Theorem 13.1 there exists an invertible P such that (3) holds. Let $p^T = [\, 0 \quad p_q^T \,]P^{-1}$, where the zero is $1 \times q$, and p_q is a left eigenvector for $\hat{A}_{22}$. That is, for some complex scalar λ,

$$p_q^T \hat{A}_{22} = \lambda p_q^T, \quad p_q \neq 0$$

Then $p \neq 0$, and

$$p^T B = \begin{bmatrix} 0 & p_q^T \end{bmatrix} \begin{bmatrix} \hat{B}_{11} \\ 0 \end{bmatrix} = 0$$

$$p^T A = \begin{bmatrix} 0 & p_q^T \end{bmatrix} \begin{bmatrix} \hat{A}_{11} & \hat{A}_{12} \\ 0 & \hat{A}_{22} \end{bmatrix} P^{-1} = \begin{bmatrix} 0 & \lambda p_q^T \end{bmatrix} P^{-1} = \lambda p^T$$

This completes the proof.
□□□

A solution λ, p of (7) with $p \neq 0$ must be an eigenvalue and left eigenvector for A. Thus a quick paraphrase of the condition in Theorem 13.3 is: "there is no left eigenvector of A that is orthogonal to the columns of B." Phrasing aside, the result can be used to obtain another controllability criterion that appears as a rank condition.

13.4 Theorem   The linear state equation (1) is controllable if and only if

$$\text{rank} \begin{bmatrix} sI - A & B \end{bmatrix} = n \tag{8}$$

for every complex scalar s.

Proof   Again we show equivalence of the negation of the claim and the negation of the condition. By Theorem 13.3 the state equation is not controllable if and only if there is a nonzero, complex, $n \times 1$ vector p and complex scalar λ such that (7) holds. That is, if and only if

$$p^T \begin{bmatrix} \lambda I - A & B \end{bmatrix} = 0, \quad p \neq 0$$

But this condition is equivalent to

$$\text{rank} \begin{bmatrix} \lambda I - A & B \end{bmatrix} < n$$

that is, equivalent to the negation of the condition in (8).
□□□

Observe from the proof that the rank test in (8) need only be applied for those values of s that are eigenvalues of A. However in many instances it is just as easy to argue that the rank condition holds for all complex scalars, thereby avoiding the chore of computing eigenvalues.

Controller Form

A special form for a controllable linear state equation (1) that can be obtained by a change of state variables is discussed next. The derivation of this form is intricate, but the result is important in revealing the structure of multi-input, multi-output, linear state equations. The special form is used in Chapter 14 to treat eigenvalue placement by linear state feedback, and in Chapter 17 where the minimal realization problem is revisited for time-invariant systems.

To avoid fussy and uninteresting complications, we assume that

$$\text{rank } B = m \qquad (9)$$

in addition to controllability. Of course if $rank\ B < m$, then the input components do not independently affect the state vector, and the state equation can be recast with a lower-dimensional input. For notational convenience the kth column of B is written as B_k. Then the controllability matrix for the state equation (1) can be displayed in column-partitioned form as

$$\begin{bmatrix} B_1 & \cdots & B_m & AB_1 & \cdots & AB_m & \cdots & A^{n-1}B_1 & \cdots & A^{n-1}B_m \end{bmatrix} \qquad (10)$$

To begin construction of the desired variable change, we search the columns of (10) from left to right to select a set of n linearly independent columns. This search is made easier by the following fact. If $A^q B_r$ is linearly dependent on columns to its left in (10), namely, the columns in

$$B, AB, \ldots, A^{q-1}B \; ; \; A^q B_1, A^q B_2, \ldots, A^q B_{r-1}$$

then $A^{q+1}B_r$ is linearly dependent on the columns in

$$AB, A^2 B, \ldots, A^q B \; ; \; A^{q+1}B_1, A^{q+1}B_2, \ldots, A^{q+1}B_{r-1}$$

That is, $A^{q+1}B_r$ is linearly dependent on columns to *its* left in (10). This means that, in the left-to-right search of (10), once a dependent column involving a product of a power of A and the column B_r is found, all columns that are products of higher powers of A and B_r can be ignored.

13.5 Definition   For $j = 1, \ldots, m$, the jth *controllability index* ρ_j for the controllable linear state equation (1) is the least integer such that column vector $A^{\rho_j}B_j$ is linearly dependent on column vectors occurring to the left of it in the controllability matrix (10).

The columns to the left of $A^{\rho_j}B_j$ in (10) can be listed as

$$B_1, \ldots, A^{\rho_j-1}B_1, \ldots, B_m, \ldots, A^{\rho_j-1}B_m ; A^{\rho_j}B_1, \ldots, A^{\rho_j}B_{j-1} \qquad (11)$$

where, compared to (10), a different arrangement of columns has been adopted to display the columns defining the controllability index ρ_j. For use in the sequel it is convenient to express $A^{\rho_j}B_j$ as a linear combination of only the linearly independent columns in (11). From the discussion above,

$$B_1, AB_1, \ldots, A^{\rho_1-1}B_1, \ldots, B_m, AB_m, \ldots, A^{\rho_m-1}B_m \qquad (12)$$

form a linearly independent set of columns in (10). In particular this is the linearly independent set obtained from a complete left-to-right search. Therefore any column to the left of the semicolon in (11) not included in (12) is linearly dependent. Thus $A^{\rho_j}B_j$ can be written as a linear combination of linearly independent columns to its left in (10):

$$A^{\rho_j}B_j = \sum_{r=1}^{m} \sum_{q=1}^{\min[\rho_j, \rho_r]} \alpha_{jrq}A^{q-1}B_r + \sum_{\substack{r=1 \\ \rho_j < \rho_r}}^{j-1} \beta_{jr}A^{\rho_j}B_r \qquad (13)$$

Additional facts to remember about this setup are that $\rho_1, \ldots, \rho_m \geq 1$ by (9), and $\rho_1 + \cdots + \rho_m = n$ by the assumption that (1) is controllable. Also it is easy to show that the controllability indices for (1) remain the same under a change of state variables (Exercise 13.10).

Now consider the invertible $n \times n$ matrix defined column-wise by

$$M^{-1} = \begin{bmatrix} B_1 & AB_1 & \cdots & A^{\rho_1-1}B_1 & \cdots & B_m & AB_m & \cdots & A^{\rho_m-1}B_m \end{bmatrix}$$

and partition the inverse matrix by rows as

$$M = \begin{bmatrix} M_1 \\ M_2 \\ \vdots \\ M_n \end{bmatrix}$$

The change of state variables we use is constructed from rows $\rho_1, \rho_1+\rho_2, \ldots,$ $\rho_1 + \cdots + \rho_m = n$ of M by setting

$$P = \begin{bmatrix} P_1 \\ P_2 \\ \vdots \\ P_m \end{bmatrix}, \quad P_i = \begin{bmatrix} M_{\rho_1+\cdots+\rho_i} \\ M_{\rho_1+\cdots+\rho_i}A \\ \vdots \\ M_{\rho_1+\cdots+\rho_i}A^{\rho_i-1} \end{bmatrix} \qquad (14)$$

13.6 Lemma   The $n \times n$ matrix P in (14) is invertible.

Proof   Suppose there is a linear combination of the rows of P that yields zero,

$$\sum_{i=1}^{m} \sum_{q=1}^{\rho_i} \gamma_{i,q} M_{\rho_1 + \cdots + \rho_i} A^{q-1} = 0 \tag{15}$$

Then the scalar coefficients in this linear combination can be shown to be zero as follows. From $MM^{-1} = I$, in particular rows $\rho_1, \rho_1 + \rho_2, \ldots, \rho_1 + \cdots + \rho_m = n$ of this identity, we have, for $i = 1, \ldots, m$,

$$M_{\rho_1 + \cdots + \rho_i} \begin{bmatrix} B_1 \; AB_1 \; \cdots \; A^{\rho_1 - 1} B_1 \; \cdots \; B_m \; AB_m \; \cdots \; A^{\rho_m - 1} B_m \end{bmatrix}$$

$$= \begin{bmatrix} 0 \; \cdots \; 0 \; \underset{\rho_1 + \cdots + \rho_i}{1} \; 0 \; \cdots \; 0 \end{bmatrix}$$

This can be rewritten as the set of identities

$$M_{\rho_1 + \cdots + \rho_i} A^{q-1} B_j = \begin{cases} 0, & q = 1, \ldots, \rho_j - 1 \\ 0, & j \neq i, \quad q = \rho_j \\ 1, & j = i, \quad q = \rho_i \end{cases} \tag{16}$$

Now suppose the columns $B_{j_1}, \ldots, B_{j_s}$ of B correspond to the largest controllability-index value $\rho_{j_1} = \cdots = \rho_{j_s}$. Multiplying the linear combination in (15) on the right by any one of these columns, say B_{j_r}, gives

$$\sum_{i=1}^{m} \sum_{q=1}^{\rho_i} \gamma_{i,q} M_{\rho_1 + \cdots + \rho_i} A^{q-1} B_{j_r} = 0 \tag{17}$$

The highest power of A in this expression is $\rho_i - 1 \leq \rho_{j_r} - 1$. Therefore, using (16), the only nonzero coefficient of a γ on the left side of (17) corresponds to indices $i = j_r, q = \rho_{j_r}$, and this gives

$$0 = \gamma_{j_r, \rho_{j_r}} M_{\rho_1 + \cdots + \rho_{j_r}} A^{\rho_{j_r} - 1} B_{j_r} = \gamma_{j_r, \rho_{j_r}} \tag{18}$$

Of course this argument shows that (18) holds for $r = 1, \ldots, s$. Now repeat the calculation with the columns of B corresponding to the next-largest controllability index, and so on. At the end of this process it will have been shown that

$$\gamma_{i, \rho_i} = 0, \quad i = 1, \ldots, m$$

Therefore the linear combination in (15) can be written as

$$\sum_{i=1}^{m} \sum_{q=1}^{\rho_i - 1} \gamma_{i,q} M_{\rho_1 + \cdots + \rho_i} A^{q-1} = 0 \tag{19}$$

where it is understood that values of i for which $\rho_i = 1$ are absent.

Again working with B_{j_r}, a column of B corresponding to the largest controllability-index value, multiply (19) on the right by AB_{j_r} to obtain

$$\sum_{i=1}^{m} \sum_{q=1}^{\rho_i-1} \gamma_{i,q} M_{\rho_1+\cdots+\rho_i} A^q B_{j_r} = 0 \tag{20}$$

From (16) the only nonzero γ-coefficient on the left side of (20) is the one with indices $i = j_r$, $q = \rho_{j_r}-1$, and therefore

$$\gamma_{j_r,\rho_{j_r}-1} = 0 \tag{21}$$

Again (21) holds for $r = 1, \ldots, s$. Proceeding with the columns of B corresponding to the next largest controllability index, and so on, gives

$$\gamma_{i,\rho_i-1} = 0, \quad i = 1, \ldots, m$$

That is, the $q = \rho_i-1$ term in the linear combination (20) can be removed, and we proceed by multiplying by $A^2 B_{j_r}$, and repeating the argument. Clearly this leads to the conclusion that all the γ-scalars in the linear combination in (15) are zero. Thus the n rows of P are linearly independent, and P is invertible. (To appreciate the importance of proceeding in decreasing order of controllability-index values, consider Exercise 13.6.)
□ □ □

To ease description of the special form obtained by changing state variables via P, we introduce a special notation.

13.7 Definition   Given a set of k positive integers $\alpha_1, \ldots, \alpha_k$, with $\alpha_1 + \cdots + \alpha_k - n$, the corresponding *integrator coefficient matrices* are defined by

$$A_o = \text{block diagonal} \left\{ \begin{bmatrix} 0 & 1 & 0 & \cdots & 0 \\ 0 & 0 & 1 & \cdots & 0 \\ \vdots & \vdots & \vdots & & \vdots \\ 0 & 0 & 0 & \cdots & 1 \\ 0 & 0 & 0 & \cdots & 0 \end{bmatrix}_{(\alpha_i \times \alpha_i)} , \quad i = 1, \ldots, k \right\}$$

$$B_o = \text{block diagonal} \left\{ \begin{bmatrix} 0 \\ \vdots \\ 0 \\ 1 \end{bmatrix}_{(\alpha_i \times 1)} , \quad i = 1, \ldots, k \right\} \tag{22}$$

The dimensional subscripts in (22) emphasize the diagonal-block sizes, while overall A_o is $n \times n$, and B_o is $n \times k$. The terminology in this definition is descriptive in that the n-dimensional state equation specified by (22) represents k parallel chains of integrators, with α_i integrators in the ith chain, as shown in Figure 13.8. Moreover (22) provides a useful notation for our special form for controllable state equations. Namely the core of the special form is the set of integrator chains specified by the controllability indices.

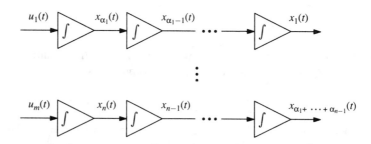

13.8 Figure   State variable diagram for the integrator-coefficient state equation.

For convenience of definition we have inverted our customary state variable change notation. That is, $z(t) = Px(t)$, and the resulting coefficient matrices are PAP^{-1}, PB, and CP^{-1}.

13.9 Theorem   Suppose the time-invariant linear state equation (1) satisfies *rank B = m*, and is controllable with controllability indices $\rho_1, \ldots, \rho_m$. Then the change of state variables $z(t) = Px(t)$, with P as in (14), yields the *controller form* state equation

$$\dot{z}(t) = \left[A_o + B_o UP^{-1} \right] z(t) + B_o R \ u(t)$$

$$y(t) = CP^{-1}z(t) \tag{23}$$

where A_o and B_o are the integrator coefficient matrices corresponding to $\rho_1, \ldots, \rho_m$, and where the $m \times n$ coefficient matrix U and the $m \times m$ invertible coefficient matrix R are given by

$$U = \begin{bmatrix} M_{\rho_1} A^{\rho_1} \\ M_{\rho_1 + \rho_2} A^{\rho_2} \\ \vdots \\ M_n A^{\rho_m} \end{bmatrix}, \quad R = \begin{bmatrix} M_{\rho_1} A^{\rho_1 - 1} B \\ M_{\rho_1 + \rho_2} A^{\rho_2 - 1} B \\ \vdots \\ M_n A^{\rho_m - 1} B \end{bmatrix} \tag{24}$$

Proof   The relation

$$PAP^{-1} = A_o + B_o \begin{bmatrix} M_{\rho_1}A^{\rho_1} \\ M_{\rho_1+\rho_2}A^{\rho_2} \\ \vdots \\ M_nA^{\rho_m} \end{bmatrix} P^{-1}$$

can be verified by easy inspection after multiplying on the right by P and writing out terms using the special forms of P, A_o and B_o. For example the ith-block of ρ_i rows in the resulting expression is

$$\begin{bmatrix} M_{\rho_1+\cdots+\rho_i}A \\ \vdots \\ M_{\rho_1+\cdots+\rho_i}A^{\rho_i-1} \\ M_{\rho_1+\cdots+\rho_i}A^{\rho_i} \end{bmatrix} = \begin{bmatrix} M_{\rho_1+\cdots+\rho_i}A \\ \vdots \\ M_{\rho_1+\cdots+\rho_i}A^{\rho_i-1} \\ 0 \end{bmatrix} + \begin{bmatrix} 0 \\ \vdots \\ 0 \\ M_{\rho_1+\cdots+\rho_i}A^{\rho_i} \end{bmatrix}$$

Unfortunately it takes more work to verify

$$PB = B_oR \tag{25}$$

However invertibility of R will be clear once this is established, since P is invertible and *rank* B_o = *rank* $B = m$. Writing (25) in terms of the special forms of P, B_o and R gives, for the ith-block of ρ_i rows,

$$\begin{bmatrix} M_{\rho_1+\cdots+\rho_i}B \\ \vdots \\ M_{\rho_1+\cdots+\rho_i}A^{\rho_i-2}B \\ M_{\rho_1+\cdots+\rho_i}A^{\rho_i-1}B \end{bmatrix} = \begin{bmatrix} 0 \\ \vdots \\ 0 \\ M_{\rho_1+\cdots+\rho_i}A^{\rho_i-1}B \end{bmatrix}$$

Therefore we must show that

$$M_{\rho_1+\cdots+\rho_i}A^{q-1}B_j = 0 , \quad q = 1, \ldots, \rho_i-1 \tag{26}$$

for $i, j = 1, \ldots, m$. First note that if $i = j$, or if $i \neq j$ and $\rho_i \leq \rho_j+1$, then (26) follows directly from (16). So suppose $i \neq j$, and $\rho_i = \rho_j + \kappa$, where $\kappa \geq 2$. Then we need to prove that

$$M_{\rho_1+\cdots+\rho_i}A^{q-1}B_j = 0, \quad i \ne j, \quad q = 1, \ldots, \rho_i - 1 = \rho_j + \kappa - 1$$

Again using (16), it remains only to show that

$$M_{\rho_1+\cdots+\rho_i}A^{q-1}B_j = 0, \quad i \ne j, \quad q = \rho_j + 1, \ldots, \rho_j + \kappa - 1 \qquad (27)$$

To set up an induction proof it is convenient to write (27) as

$$M_{\rho_1+\cdots+\rho_i}A^{\rho_j+k}B_j = 0, \quad i \ne j, \quad k = 0, \ldots, \kappa - 2 \qquad (28)$$

where, again, $\kappa \ge 2$. To establish (28) for $k = 0$, we use (13), which is repeated here for convenience:

$$A^{\rho_j}B_j = \sum_{r=1}^{m} \sum_{q=1}^{\min[\rho_j, \rho_r]} \alpha_{jrq}A^{q-1}B_r + \sum_{\substack{r=1 \\ \rho_j < \rho_r}}^{j-1} \beta_{jr}A^{\rho_j}B_r \qquad (13)$$

Replacing ρ_j by $\rho_i - \kappa$ on the right side, and multiplying through by $M_{\rho_1+\cdots+\rho_i}$ gives

$$M_{\rho_1+\cdots+\rho_i}A^{\rho_j}B_j = \sum_{r=1}^{m} \sum_{q=1}^{\min[\rho_i-\kappa, \rho_r]} \alpha_{jrq}M_{\rho_1+\cdots+\rho_i}A^{q-1}B_r$$

$$+ \sum_{\substack{r=1 \\ \rho_i-\kappa < \rho_r}}^{j-1} \beta_{jr}M_{\rho_1+\cdots+\rho_i}A^{\rho_i-\kappa}B_r \qquad (29)$$

In the first expression on the right side, all summands can be shown to be zero (ignoring the scalar coefficients). For $r = i$ the summands are those corresponding to

$$M_{\rho_1+\cdots+\rho_i}B_i, \ldots, M_{\rho_1+\cdots+\rho_i}A^{\rho_i-\kappa-1}B_i$$

and these terms are zero by (16) and the fact that $\kappa \ge 2$. For $r \ne i$ the summands are those corresponding to

$$M_{\rho_1+\cdots+\rho_i}B_r, \ldots, M_{\rho_1+\cdots+\rho_i}A^{\min[\rho_i-\kappa, \rho_r]-1}B_r, \quad r \ne i$$

and again these are zero by (16). For the second expression on the right side of (29), the $r = i$ term, if present (that is, if $i < j$), corresponds to

$$M_{\rho_1+\cdots+\rho_i}A^{\rho_i-\kappa}B_i$$

Again this is zero by (16) and $\kappa \ge 2$. Any term with $r \ne i$ that is present has the form

$$M_{\rho_1+\cdots+\rho_i}A^{\rho_i-\kappa}B_r, \quad r \ne i$$

and since $\rho_i - \kappa \le \rho_r - 1$, this term is zero by (16). Thus (28) has been established for $k = 0$.

Now assume that (28) holds for $k = 0, \ldots, K$, where $K < \kappa - 2$. Then for $k = K+1$, we multiply (13) by $M_{\rho_1+\cdots+\rho_i}A^{K+1}$, and replace ρ_j by $\rho_i - \kappa$ on the right side, to obtain

$$M_{\rho_1+\cdots+\rho_i}A^{\rho_j+K+1}B_j = \sum_{r=1}^{m}\sum_{q=1}^{\min[\rho_i-\kappa,\,\rho_r]}\alpha_{jrq}M_{\rho_1+\cdots+\rho_i}A^{K+q}B_r$$

$$+\sum_{\substack{r=1\\ \rho_i-\kappa<\rho_r}}^{j-1}\beta_{jr}M_{\rho_1+\cdots+\rho_i}A^{K+1+\rho_i-\kappa}B_r \qquad (30)$$

In the first expression on the right side of (30), the summands for $r = i$ correspond to

$$M_{\rho_1+\cdots+\rho_i}A^{K+1}B_i,\ldots, M_{\rho_1+\cdots+\rho_i}A^{K+\rho_i-\kappa}B_i$$

Since $K+\rho_i-\kappa < \kappa-2+\rho_i-\kappa = \rho_i-2$, these terms are zero by (16). The summands for $r \neq i$ involve

$$M_{\rho_1+\cdots+\rho_i}A^{K+1}B_r,\ldots, M_{\rho_1+\cdots+\rho_i}A^{K+\min[\rho_i-\kappa,\,\rho_r]}B_r, \quad r \neq i \qquad (31)$$

But no power of A in (31) is greater than ρ_r+K, so by the inductive hypothesis all terms in (31) are zero.

Finally, for the second expression on the right side of (30), the $r = i$ term, if present, is

$$M_{\rho_1+\cdots+\rho_i}A^{K+1+\rho_i-\kappa}B_i$$

Since $\kappa > K+2$, this term is zero by (16). For $r \neq i$ the power of A present in the summand is $K+1+\rho_i-\kappa < K+1+\rho_r$, that is, $K+1+\rho_i-\kappa \le K+\rho_r$. Therefore the inductive hypothesis gives that such a term is zero since $r \neq i$. In summary this induction establishes (27), and thus completes the proof.
□□□

Additional investigation of the matrix R in (23) yields a further simplification of the controller form.

13.10 Proposition Under the hypotheses of Theorem 13.9, the invertible $m \times m$ matrix R defined in (24) is an upper-triangular matrix with unity diagonal entries.

Proof The (i, j)-entry of R is $M_{\rho_1+\cdots+\rho_i}A^{\rho_i-1}B_j$, and for $i = j$ this is unity by the identities in (16). For entries below the diagonal it must be shown that

$$M_{\rho_1+\cdots+\rho_i}A^{\rho_i-1}B_j = 0, \quad i > j \qquad (32)$$

To do this the identities in (26), established in the proof of Theorem 13.7, are used. Specifically (26) can be written as

$$M_{\rho_1+\cdots+\rho_i}B_j = \cdots = M_{\rho_1+\cdots+\rho_i}A^{\rho_i-2}B_j = 0; \quad i, j = 1,\ldots, m \qquad (33)$$

To begin an induction proof fix $j = 1$, and suppose $i > 1$. If $\rho_i \le \rho_1$, then (32) follows from (16). So suppose $\rho_i = \rho_1 + \kappa$, where $\kappa \ge 1$. Then (13) gives, after multiplying through by $M_{\rho_1+\cdots+\rho_i}A^{\kappa-1}$,

$$M_{\rho_1 + \cdots + \rho_i} A^{\rho_i - 1} B_1 = M_{\rho_1 + \cdots + \rho_i} A^{\kappa - 1} A^{\rho_i} B_1$$

$$= \sum_{r=1}^{m} \sum_{q=1}^{\min[\rho_1, \rho_r]} \alpha_{1rq} M_{\rho_1 + \cdots + \rho_i} A^{q+\kappa-2} B_r$$

Since the highest power of A among the summands is no greater than $\rho_1 + \kappa - 2 = \rho_i - 2$, all the summands are zero by (33).

Now suppose (32) has been established for $j = 1, \ldots, J$. To show the case $j = J+1$, first note that if $i \geq J+2$ and $\rho_i \leq \rho_{J+1}$, then (32) is zero by (16). So suppose $i \geq J+2$ and $\rho_i = \rho_{J+1} + \kappa$, where $\kappa \geq 1$. Using (13) again gives

$$M_{\rho_1 + \cdots + \rho_i} A^{\rho_i - 1} B_{J+1} = M_{\rho_1 + \cdots + \rho_i} A^{\kappa - 1} A^{\rho_{J+1}} B_{J+1}$$

$$= \sum_{r=1}^{m} \sum_{q=1}^{\min[\rho_{J+1}, \rho_r]} \alpha_{J+1,rq} M_{\rho_1 + \cdots + \rho_i} A^{q+\kappa-2} B_r$$

$$+ \sum_{\substack{r=1 \\ \rho_{J+1} < \rho_r}}^{J} \beta_{J+1,r} M_{\rho_1 + \cdots + \rho_i} A^{\rho_{J+1} + \kappa - 1} B_r$$

In the first expression on the right side, the highest power of A is no greater than $\rho_{J+1} + \kappa - 2 = \rho_i - 2$. Therefore (33) can be used to show that the first expression is zero. For the second expression on the right side, any term that appears has the form (ignoring the scalar coefficient)

$$M_{\rho_1 + \cdots + \rho_i} A^{\rho_{J+1} + \kappa - 1} B_r = M_{\rho_1 + \cdots + \rho_i} A^{\rho_i - 1} B_r , \quad r \leq J$$

and these terms are zero by the inductive hypothesis. Therefore the proof is complete.
□□□

While the special structure of the controller form state equation in (23) is not immediately transparent, it emerges on contemplating a few specific cases. It also becomes obvious that the special form of R revealed in Proposition 13.10 plays an important role in the structure of $B_o R$.

13.11 Example   For the case $n = 6$, $m = 2$, $\rho_1 = 4$, and $\rho_2 = 2$, (23) takes the form

$$\dot{z}(t) = \begin{bmatrix} 0 & 1 & 0 & 0 & 0 & 0 \\ 0 & 0 & 1 & 0 & 0 & 0 \\ 0 & 0 & 0 & 1 & 0 & 0 \\ \times & \times & \times & \times & \times & \times \\ 0 & 0 & 0 & 0 & 0 & 1 \\ \times & \times & \times & \times & \times & \times \end{bmatrix} z(t) + \begin{bmatrix} 0 & 0 \\ 0 & 0 \\ 0 & 0 \\ 1 & \times \\ 0 & 0 \\ 0 & 1 \end{bmatrix} u(t)$$

$$y(t) = CP^{-1} z(t) \tag{34}$$

where "×" denotes entries that are not necessarily either zero or one. (The output equation has no special structure, and simply is repeated from (23).)
□□□

The controller form for a linear state equation is useful in the sequel for addressing the multi-input, multi-output minimal realization problem, and the capabilities of linear state feedback. Of course controller form when $m = 1$, $\rho_1 = n$ is familiar from Example 2.5, and Example 10.12.

Observability

Next we address concepts related to observability, though proofs can be left as errant exercises since they are so similar to corresponding proofs in the controllability case.

13.12 Theorem   Suppose the observability matrix for the linear state equation (1) satisfies

$$
\text{rank} \begin{bmatrix} C \\ CA \\ \vdots \\ CA^{n-1} \end{bmatrix} = l
$$

where $0 < l < n$. Then there exists an invertible $n \times n$ matrix Q such that

$$
Q^{-1}AQ = \begin{bmatrix} \hat{A}_{11} & 0 \\ \hat{A}_{21} & \hat{A}_{22} \end{bmatrix}, \quad CQ = [\, \hat{C}_{11} \quad 0 \,] \tag{35}
$$

where $\hat{A}_{11}$ is $l \times l$, $\hat{C}_{11}$ is $p \times l$, and

$$
\text{rank} \begin{bmatrix} \hat{C}_{11} \\ \hat{C}_{11}\hat{A}_{11} \\ \vdots \\ \hat{C}_{11}\hat{A}_{11}^{l-1} \end{bmatrix} = l
$$

The state variable change in Theorem 13.12 is constructed by choosing $n - l$ vectors in the nullspace of the observability matrix, and preceding them by l vectors that yield a set of n linearly independent vectors. The linear state equation resulting from $z(t) = Q^{-1}x(t)$ can be written as

$$\dot{z}_o(t) = \hat{A}_{11} z_o(t)$$

$$\dot{z}_{no}(t) = \hat{A}_{21} z_o(t) + \hat{A}_{22} z_{no}(t)$$

$$y(t) = \hat{C}_{11} z_o(t)$$

and is shown in Figure 13.13.

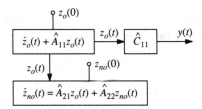

13.13 Figure   Observable and unobservable subsystems displayed by (35).

13.14 Theorem   The linear state equation (1) is observable if and only if for every complex scalar λ the only complex $n \times 1$ vector p that satisfies

$$Ap = \lambda p , \quad Cp = 0$$

is $p = 0$.

A compact locution for Theorem 13.14 is that observability is equivalent to nonexistence of a right eigenvector of A that is orthogonal to the rows of C.

13.15 Theorem   The linear state equation (1) is observable if and only if

$$\text{rank} \begin{bmatrix} C \\ sI - A \end{bmatrix} = n \tag{36}$$

for every complex scalar s.

Exactly as in the corresponding controllability test, the rank condition in (36) need only be applied for those values of s that are eigenvalues of A.

Observer Form

To develop a special form for linear state equations that is related to the concept of observability, we assume that (1) is observable, and that *rank* $C = p$. Then the observability matrix for (1) can be written in row-partitioned form, where the ith-block of p rows is

$$\begin{bmatrix} C_1 A^{i-1} \\ C_2 A^{i-1} \\ \vdots \\ C_p A^{i-1} \end{bmatrix}, \quad i = 1, \ldots, n$$

and C_j denotes the jth-row of C.

13.16 Definition   For $j = 1, \ldots, p$, the jth *observability index* η_j for the observable linear state equation (1) is the least integer such that row vector $C_j A^{\eta_j}$ is linearly dependent on vectors occurring above it in the observability matrix.

Specifically for each j, η_j is the least integer for which there exist scalars α_{jrq} and β_{jr} such that

$$C_j A^{\eta_j} = \sum_{r=1}^{p} \sum_{q=1}^{\min[\eta_j, \, \eta_r]} \alpha_{jrq} C_r A^{q-1} + \sum_{\substack{r=1 \\ \eta_j < \eta_r}}^{j-1} \beta_{jr} C_r A^{\eta_j} \tag{37}$$

As in the controllability case, our formulation is such that $\eta_1, \ldots, \eta_p \geq 1$, and $\eta_1 + \cdots + \eta_p = n$. Also it can be shown that the observability indices are unaffected by a change of state variables.

Consider the invertible $n \times n$ matrix N^{-1} defined in row-partitioned form with the ith-block containing the η_i rows

$$\begin{bmatrix} C_i \\ C_i A \\ \vdots \\ C_i A^{\eta_i - 1} \end{bmatrix}, \quad i = 1, \ldots, p$$

Partition the inverse of N^{-1} by columns as

$$N = \begin{bmatrix} N_1 & N_2 & \cdots & N_n \end{bmatrix}$$

Then the change of state variables of interest is specified by

$$Q = \begin{bmatrix} N_{\eta_1} & \cdots & A^{\eta_1 - 1} N_{\eta_1} & N_{\eta_1 + \eta_2} & \cdots & A^{\eta_2 - 1} N_{\eta_1 + \eta_2} \\ & & & \cdots & N_n & \cdots & A^{\eta_p - 1} N_n \end{bmatrix} \tag{38}$$

On verification that Q is invertible, a computation much in the style of the proof of Lemma 13.6, the main result can be stated as follows.

13.17 Theorem   Suppose the time-invariant linear state equation (1) satisfies *rank* $C = p$, and is observable with observability indices $\eta_1, \ldots, \eta_p$. Then the change of state variables $z(t) = Q^{-1}x(t)$, with Q as in (38), yields the *observer form* state equation

$$\dot{z}(t) = \left[A_o^T + Q^{-1}VB_o^T \right] z(t) + Q^{-1}Bu(t)$$

$$y(t) = SB_o^T z(t) \tag{39}$$

where A_o and B_o are the integrator coefficient matrices corresponding to $\eta_1, \ldots, \eta_p$, and where the $n \times p$ coefficient matrix V and the $p \times p$ invertible coefficient matrix S are given by

$$V = \left[A^{\eta_1} N_{\eta_1} \quad A^{\eta_2} N_{\eta_1+\eta_2} \quad \cdots \quad A^{\eta_p} N_n \right]$$

$$S = \left[CA^{\eta_1-1} N_{\eta_1} \quad CA^{\eta_2-1} N_{\eta_1+\eta_2} \quad \cdots \quad CA^{\eta_p-1} N_n \right] \tag{40}$$

13.18 Proposition   Under the hypotheses of Theorem 13.17, the invertible $p \times p$ matrix S defined in (40) is lower triangular with unity diagonal entries.

13.19 Example   The special structure of an observer form state equation becomes apparent in specific cases. With $n = 7$, $p = 3$, $\eta_1 = \eta_2 = 3$, and $\eta_3 = 1$, (39) takes the form

$$\dot{z}(t) = \begin{bmatrix} 0 & 0 & \times & 0 & 0 & \times & \times \\ 1 & 0 & \times & 0 & 0 & \times & \times \\ 0 & 1 & \times & 0 & 0 & \times & \times \\ 0 & 0 & \times & 0 & 0 & \times & \times \\ 0 & 0 & \times & 1 & 0 & \times & \times \\ 0 & 0 & \times & 0 & 1 & \times & \times \\ 0 & 0 & \times & 0 & 0 & \times & \times \end{bmatrix} z(t) + Q^{-1}Bu(t)$$

$$y(t) = \begin{bmatrix} 0 & 0 & 1 & 0 & 0 & 0 & 0 \\ 0 & 0 & \times & 0 & 0 & 1 & 0 \\ 0 & 0 & \times & 0 & 0 & \times & 1 \end{bmatrix} z(t)$$

where $\times$ denotes entries that are not necessarily zero or one. Note that a unity observability index renders nonspecial a corresponding portion of the structure.

EXERCISES

Exercise 13.1   Show that a single-input linear state equation of dimension $n = 2$,

$$\dot{x}(t) = Ax(t) + bu(t)$$

is controllable for every nonzero vector b if and only if the eigenvalues of A are complex. (For the hearty a more strenuous exercise is to show that a single-input linear state equation of dimension $n > 1$ is controllable for every nonzero b if and only if $n = 2$ and the eigenvalues of A are complex.)

Exercise 13.2   Prove that the linear state equation

$$\dot{x}(t) = Ax(t) + Bu(t)$$

is controllable if and only if the only $n \times n$ matrix X that satisfies

$$XA = AX , \quad XB = 0$$

is $X = 0$. (*Hint*: Employ right and left eigenvectors of A.)

Exercise 13.3   Suppose the linear state equations

$$\dot{x}_a(t) = A_a x_a(t) + B_a u(t)$$
$$y(t) = C_a x_a(t)$$

and

$$\dot{x}_b(t) = A_b x_b(t) + B_b u(t)$$

are controllable, with $p_a = m_b$. Show that if

$$\text{rank} \begin{bmatrix} sI - A_a & B_a \\ C_a & 0 \end{bmatrix} = n_a + p_a$$

for each s that is an eigenvalue of A_b, then

$$\dot{x}(t) = \begin{bmatrix} A_a & 0 \\ B_b C_a & A_b \end{bmatrix} x(t) + \begin{bmatrix} B_a \\ 0 \end{bmatrix} u(t)$$

is controllable. What does the last state equation represent?

Exercise 13.4   Show that if the time-invariant linear state equation

$$\dot{x}(t) = Ax(t) + Bu(t)$$
$$y(t) = Cx(t) + Du(t)$$

with $m \geq p$ is controllable, and

$$\text{rank} \begin{bmatrix} A & B \\ C & D \end{bmatrix} = n + p$$

then the state equation

$$\dot{z}(t) = \begin{bmatrix} A & 0 \\ C & 0 \end{bmatrix} z(t) + \begin{bmatrix} B \\ D \end{bmatrix} u(t)$$

is controllable. Also prove the converse.

Exercise 13.5   Consider a Jordan form state equation

$$\dot{x}(t) = Jx(t) + Bu(t)$$

in the case where J has a single eigenvalue of multiplicity n. That is, J is block diagonal, and each block has the form

$$\begin{bmatrix} \lambda & 1 & \cdots & 0 \\ 0 & \lambda & \cdots & 0 \\ \vdots & \vdots & \vdots & \vdots \\ 0 & 0 & \cdots & 1 \\ 0 & 0 & \cdots & \lambda \end{bmatrix}$$

with the same λ. Determine conditions on B that are necessary and sufficient for controllability. Does your answer lead to a controllability criterion for general Jordan form state equations?

Exercise 13.6   In the proof of Lemma 13.6, show why it is important to proceed in order of decreasing controllability indices by considering the case $n = 3$, $m = 2$, $\rho_1 = 2$ and $\rho_2 = 1$. Write out the proof twice: first beginning with B_1, and then beginning with B_2.

Exercise 13.7   Determine the form of the matrix R in Theorem 13.10 for the case $\rho_1 = 1$, $\rho_2 = 3$, $\rho_3 = 2$. In particular, which entries above the diagonal are nonzero?

Exercise 13.8   Prove that if the controllability indices for a linear state equation satisfy $1 \leq \rho_1 \leq \rho_2 \leq \cdots \leq \rho_m$, then the matrix R in Theorem 13.10 is the identity matrix.

Exercise 13.9   By considering the example

$$A = \begin{bmatrix} 0 & 0 & 0 & 0 \\ 0 & 0 & 0 & 0 \\ 1 & 0 & -1 & 0 \\ 2 & -2 & 0 & 0 \end{bmatrix}, \quad B = \begin{bmatrix} 1 & 0 & 0 \\ 0 & 1 & 0 \\ 0 & 0 & 1 \\ 1/2 & 0 & 0 \end{bmatrix}$$

show that in general the controllability indices cannot be placed in nondecreasing order by relabeling input components.

Exercise 13.10   If P is an invertible $n \times n$ matrix, and G is an invertible $m \times m$ matrix, show that the controllability indices for

$$\dot{x}(t) = Ax(t) + Bu(t)$$

(with *rank* $B = m$) are identical to the controllability indices for

$$\dot{z}(t) = P^{-1}APx(t) + P^{-1}Bu(t)$$

and are the same, up to reordering, as the controllability indices for

$$\dot{x}(t) = Ax(t) + BGu(t)$$

(*Hint*: Write, for example

$$\begin{bmatrix} BG & ABG \end{bmatrix} = \begin{bmatrix} B & AB \end{bmatrix} \begin{bmatrix} G & 0 \\ 0 & G \end{bmatrix}$$

and show that the number of linearly dependent columns in $A^k B$ that arise in the left-to-right search of $[\, B \quad AB \quad \cdots \quad A^{n-1}B\,]$ is the same as the number of linearly dependent columns in $A^k BG$ that arise in the left-to-right search of $[\, BG \quad ABG \quad \cdots \quad A^{n-1}BG\,]$.)

Exercise 13.11   Suppose the linear state equation

$$\dot{x}(t) = Ax(t) + Bu(t)$$

is controllable. If K is $m \times n$, prove that

$$\dot{z}(t) = [A + BK]z(t) + Bv(t)$$

is controllable. Repeat the problem for the time-varying case, where the original state equation is assumed to be controllable on $[t_o, \, t_f]$. (*Hint*: While an explicit argument can be used in the time-invariant case, apparently a clever, indirect argument is required in the time-varying case.)

Exercise 13.12   Use controller form to show the following. If the m-input linear state equation

$$\dot{x}(t) = Ax(t) + Bu(t)$$

is controllable (and *rank B* $= m$), then there exists an $m \times n$ matrix K and an $m \times 1$ vector b such that the single-input linear state equation

$$\dot{x}(t) = [\, A + BK\,]x(t) + Bbu(t)$$

is controllable. (*Hint*: Review Example 10.12.) Give an example to show that this cannot be accomplished in general with the choice $K = 0$.

Exercise 13.13   For a linear state equation

$$\dot{x}(t) = Ax(t) + Bu(t)$$

$$y(t) = Cx(t)$$

define the *controllability index* ρ as the least nonnegative integer such that

$$\text{rank} \left[B \quad AB \quad \cdots \quad A^{\rho-1}B \right] = \text{rank} \left[B \quad AB \quad \cdots \quad A^{\rho}B \right]$$

Prove that
(a) for any $k \geq \rho$,

$$\text{rank} \left[B \quad AB \quad \cdots \quad A^{\rho-1}B \right] = \text{rank} \left[B \quad AB \quad \cdots \quad A^{k}B \right]$$

(b) if *rank B* $= r > 0$, then $1 \leq \rho \leq n - r + 1$,
(c) the controllability index is invariant under invertible state variable changes.
State the corresponding results for the apparent notion of an *observability index* η for the state equation.

Exercise 13.14   Continuing Exercise 13.13, show that if

$$\text{rank} \left\{ \begin{bmatrix} C \\ CA \\ \vdots \\ CA^{\eta-1} \end{bmatrix} \left[B \quad AB \quad \cdots \quad A^{\rho-1}B \right] \right\} = s$$

then there is an invertible $n \times n$ matrix P such that

$$P^{-1}AP = \begin{bmatrix} \hat{A}_{11} & 0 & \hat{A}_{13} \\ \hat{A}_{21} & \hat{A}_{22} & \hat{A}_{23} \\ 0 & 0 & \hat{A}_{33} \end{bmatrix} , \quad P^{-1}B = \begin{bmatrix} \hat{B}_{11} \\ \hat{B}_{21} \\ 0 \end{bmatrix}$$

$$CP = \begin{bmatrix} \hat{C}_{11} & 0 & \hat{C}_{13} \end{bmatrix}$$

where the s-dimensional state equation

$$\dot{z}(t) = \hat{A}_{11}z(t) + \hat{B}_{11}u(t)$$

$$y(t) = \hat{C}_{11}z(t)$$

is controllable, observable, and has the same input-output behavior as the original n-dimensional linear state equation.

Exercise 13.15   Consider the n-dimensional linear state equation

$$\dot{x}(t) = \begin{bmatrix} A_{11} & A_{12} \\ A_{21} & A_{22} \end{bmatrix} x(t) + \begin{bmatrix} B_{11} \\ 0 \end{bmatrix} u(t)$$

where A_{11} is $q \times q$ and B_{11} is $q \times m$ with rank q. Prove that this state equation is controllable if and only if the $(n-q)$-dimensional linear state equation

$$\dot{z}(t) = A_{22}z(t) + A_{21}v(t)$$

is controllable.

NOTES

Note 13.1   The state-variable changes yielding the block triangular forms in Theorem 13.1 and Theorem 13.12 can be combined (in a nonobvious way) into a variable change that displays a linear state equation in terms of 4 component state equations that are, respectively, controllable and observable, controllable but not observable, observable but not controllable, and neither controllable nor observable. References for this *Canonical Structure theorem* are cited in Note 10.2. (Note that Exercise 13.14 involves a rather obvious combination of the variable changes.)

Note 13.2   The eigenvector test for controllability in Theorem 13.3 is attributed to W. Hahn in

R.E. Kalman, "Lectures on controllability and observability," Centro Internazionale Matematico Estivo Seminar Notes, Bologna, Italy, 1968

The rank and eigenvector tests for controllability and observability are sometimes called "PBH tests" because original sources include

V.M. Popov, *Hyperstability of Control Systems*, Springer-Verlag, Berlin, 1973 (translation of a 1966 version in Rumanian)

V. Belevitch, *Classical Network Theory*, Holden-Day, San Francisco, 1968

M.L.J. Hautus, "Controllability and observability conditions for linear autonomous systems," *Proceedings of the Koninklijke Akademie van Wetenschappen, Serie A*, Vol. 72, pp. 443–448, 1969

Note 13.3   Controller form is based on

D.G. Luenberger, "Canonical forms for linear multivariable systems," *IEEE Transactions on Automatic Control,* Vol. 12, pp. 290–293, 1967

Our different notation is intended to facilitate explicit, detailed derivation. (In most sources on the subject, phrases such as 'tedious but straightforward calculations show' appear, perhaps for humanitarian reasons.) When $m = 1$ the transformation to controller form is unique, but in general it is not. Also various $\times$'s in a particular case, say Example 13.11, are guaranteed to be zero, depending on inequalities among the controllability indices, and the specific vectors that appear in the linear-dependence relation (13). Thus, in technical terms, controller form is not a *canonical form* for controllable linear state equations (unless $m = p = 1$). Extensive discussion of these issues, including the precise mathematical meaning of canonical form, can be found in Chapter 6 of

T. Kailath, *Linear Systems,* Prentice Hall, New York, 1980

See also

V.M. Popov, "Invariant description of linear, time-invariant controllable systems," *SIAM Journal on Control and Optimization,* Vol. 10, No. 2, pp. 252–264, 1972

Of course similar remarks apply to observer form.

Note 13.4   Controller and observer forms are convenient, elementary theoretical tools for exploring the algebraic structure of linear state equations and linear feedback problems, and we apply them several times in the sequel. However, dispensing with any technical gloss, the numerical properties of such forms can be miserable. Even in single-input or single-output cases. Consult

C. Kenney, A.J. Laub, "Controllability and stability radii for companion form systems," *Mathematics of Control, Signals, and Systems,* Vol. 1, No. 3, pp. 239–256, 1988

Note 13.5   Standard forms analogous to controller and observer forms are available for time-varying linear state equations. The basic assumptions involve strong types of controllability and observability, much like the instantaneous controllability and instantaneous observability of Chapter 11. Consult, for a start,

L.M. Silverman, "Transformation of time-variable systems to canonical (phase-variable) form," *IEEE Transactions on Automatic Control,* Vol. 11, pp. 300–303, 1966

R.S. Bucy, "Canonical forms for multivariable systems," *IEEE Transactions on Automatic Control,* Vol. 13, No. 5, pp. 567–569, 1968

K. Ramar, B. Ramaswami, "Transformation of time-variable multi-input systems to a canonical form," *IEEE Transactions on Automatic Control,* Vol. 16, No. 4, pp. 371–374, 1971

A. Ilchmann, "Time-varying linear systems and invariants of system equivalence," *International Journal of Control,* Vol. 42, No. 4, pp. 759–790, 1985

14

LINEAR FEEDBACK

The theory of linear systems provides the basis for *linear control theory*. In this chapter we introduce concepts and results of linear control theory for time-varying linear state equations. In addition the controller form in Chapter 13 is applied to prove the celebrated eigenvalue assignment capability of linear feedback in the time-invariant case.

Linear control theory involves modification of the behavior of a given m-input, p-output, n-dimensional linear state equation

$$\dot{x}(t) = A(t)x(t) + B(t)u(t)$$

$$y(t) = C(t)x(t) \tag{1}$$

in this context often called the *plant* or *open-loop state equation,* by applying linear feedback. As shown in Figure 14.1, linear *state feedback* replaces the plant input $u(t)$ by an expression of the form

$$u(t) = K(t)x(t) + N(t)r(t) \tag{2}$$

where $r(t)$ is the new name for the $m \times 1$ input signal. Convenient default assumptions are that the $m \times n$ matrix $K(t)$ and the $m \times m$ matrix $N(t)$ are defined and continuous for all t. Substituting (2) into (1) gives a new linear state equation, called the *closed-loop state equation,* described by

$$\dot{x}(t) = [A(t) + B(t)K(t)]x(t) + B(t)N(t)r(t)$$

$$y(t) = C(t)x(t) \tag{3}$$

Similarly linear *output feedback* takes the form

$$u(t) = L(t)y(t) + N(t)r(t) \tag{4}$$

216

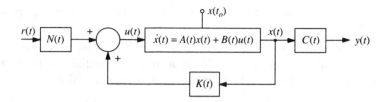

14.1 Figure   Structure of linear state feedback.

where again coefficients are assumed to be defined and continuous for all t. Output feedback, clearly a special case of state feedback, is diagramed in Figure 14.2. The resulting closed-loop state equation is described by

$$\dot{x}(t) = [A(t) + B(t)L(t)C(t)]x(t) + B(t)N(t)r(t)$$

$$y(t) = C(t)x(t) \tag{5}$$

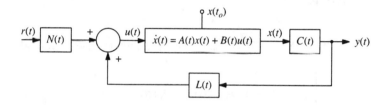

14.2 Figure   Structure of linear output feedback.

One important (if obvious) feature of either type of linear feedback is that the closed-loop state equation remains a linear state equation. If the coefficient matrices in (2) or (4) are constant, then the feedback is called *time invariant*. In any case the feedback is called *static* because at any t the value of $u(t)$ depends only on the values of $r(t)$ and $x(t)$ or $y(t)$ at that same time. Dynamic feedback where $u(t)$ is the output of a linear state equation with inputs $r(t)$ and $x(t)$ or $y(t)$ is considered in Chapter 15.

Effects of Feedback

We begin the discussion by considering the relationship between the closed-loop state equation and the plant. This is the initial step in describing what can be achieved by feedback. The available answers turn out to be disappointingly complicated for the general case in that a convenient, explicit relationship is not obtained. However matters are more encouraging in the time-invariant case.

First the effect of state feedback on the transition matrix is considered.

14.3 Theorem   If $\Phi_A(t, \tau)$ is the transition matrix for the open-loop state equation (1) and $\Phi_{A+BK}(t, \tau)$ is the transition matrix for the closed-loop state equation (3) resulting from state feedback (2), then

$$\Phi_{A+BK}(t,\ \tau) = \Phi_A(t,\ \tau) + \int_\tau^t \Phi_A(t,\ \sigma)B(\sigma)K(\sigma)\Phi_{A+BK}(\sigma,\ \tau)\,d\sigma \qquad (6)$$

If the open-loop state equation and state feedback both are time-invariant, then the Laplace transform of the closed-loop matrix exponential can be expressed in terms of the Laplace transform of the open-loop matrix exponential as

$$(sI - A - BK)^{-1} = [I - (sI - A)^{-1}BK]^{-1}(sI - A)^{-1} \qquad (7)$$

Proof   To verify (6), suppose τ is arbitrary but fixed. Then evaluation of the right side of (6) at $t = \tau$ yields the identity matrix. Furthermore differentiation of the right side of (6) with respect to t yields

$$\frac{d}{dt}\left[\Phi_A(t,\ \tau) + \int_\tau^t \Phi_A(t,\ \sigma)B(\sigma)K(\sigma)\Phi_{A+BK}(\sigma,\ \tau)\,d\sigma\right]$$

$$= A(t)\Phi_A(t,\ \tau)$$

$$+ \Phi_A(t,\ t)B(t)K(t)\Phi_{A+BK}(t,\ \tau) + \int_\tau^t A(t)\Phi_A(t,\ \sigma)B(\sigma)K(\sigma)\Phi_{A+BK}(\sigma,\ \tau)\,d\sigma$$

$$= A(t)\left[\Phi_A(t,\ \tau) + \int_\tau^t \Phi_A(t,\ \sigma)B(\sigma)K(\sigma)\Phi_{A+BK}(\sigma,\ \tau)\,d\sigma\right] + B(t)K(t)\Phi_{A+BK}(t,\ \tau)$$

Therefore the right side of (6) satisfies the matrix differential equation that uniquely characterizes $\Phi_{A+BK}(t,\ \tau)$, and this argument applies for any value of τ.

For a time-invariant linear state equation, rewriting (6) in terms of matrix exponentials, with $\tau = 0$, gives

$$e^{(A+BK)t} = e^{At} + \int_0^t e^{A(t-\sigma)}BKe^{(A+BK)\sigma}\,d\sigma$$

Taking Laplace transforms, using in particular the convolution property, yields

$$(sI - A - BK)^{-1} = (sI - A)^{-1} + (sI - A)^{-1}BK(sI - A - BK)^{-1} \qquad (8)$$

an expression that easily rearranges to (7).
□□□

A result similar to Theorem 14.3 holds for static, linear output feedback upon replacing $K(t)$ by $L(t)C(t)$. For output feedback, a relation between the input-output representations for the plant and closed-loop state equation also can be obtained. Again the relation is implicit, in general, though convenient formulas can be derived in the time-invariant case. (It is left as an exercise to show for state feedback that (6) and (7) yield only cumbersome expressions involving the open-loop and closed-loop weighting patterns or transfer functions.)

14.4 Theorem   If $G(t, \tau)$ is the weighting pattern of the open-loop state equation (1) and $\hat{G}(t, \tau)$ is the weighting pattern of the closed-loop state equation (5) resulting from static output feedback (4), then

$$\hat{G}(t, \tau) = G(t, \tau)N(\tau) + \int_{\tau}^{t} G(t, \sigma)L(\sigma)\hat{G}(\sigma, \tau) \, d\sigma \tag{9}$$

If the open-loop state equation and output feedback are time invariant, then the transfer function of the closed-loop state equation can be expressed in terms of the transfer function of the open-loop state equation by

$$\hat{G}(s) = [\, I - G(s)L \,]^{-1}G(s)N \tag{10}$$

Proof   In (6), we can replace $K(\sigma)$ by $L(\sigma)C(\sigma)$ to reflect output feedback. Then premultiplying by $C(t)$ and postmultiplying by $B(\tau)N(\tau)$ gives (9). Specializing (9) to the time-invariant case, with $\tau = 0$, Laplace transformation of the resulting impulse-response relation gives

$$\hat{G}(s) = G(s)N + G(s)L\hat{G}(s)$$

from which (10) follows easily.
□□□
An alternate expression for $\hat{G}(s)$ in (10) can be derived from the time-invariant version of the diagram in Figure 14.2. Using Laplace transforms we can write

$$[\, I - LG(s) \,]U(s) = NR(s)$$

$$Y(s) = G(s)U(s)$$

This gives

$$\hat{G}(s) = G(s)[\, I - LG(s) \,]^{-1}N \tag{11}$$

Of course in the single-input, single-output case, both (10) and (11) collapse to

$$\hat{G}(s) = \frac{G(s)}{1 - G(s)L} \, N$$

In a different notation, with different sign conventions for feedback, this is a familiar formula in elementary control systems.

State Feedback Stabilization

One of the first specific objectives that arises in considering the capabilities of feedback involves stabilization of a given plant. The basic problem is that of choosing a state feedback gain $K(t)$ such that the resulting closed-loop state equation is uniformly exponentially stable. (In addressing uniform exponential stability, the input gain $N(t)$ plays no role and is ignored.) Then appropriate boundedness assumptions on the coefficient matrices yield uniform bounded-input, bounded-output stability, as discussed in Chapter 12. Despite the complicated, implicit relation between the open- and closed-loop transition matrices, it turns out that exhibiting a control law to accomplish stabilization under reasonable hypotheses is simple in theory, though admittedly complicated in calculation.

Actually somewhat more than uniform exponential stability can be achieved, and for this purpose we slightly refine Definition 6.5 on uniform exponential stability by attaching a lower bound on the decay rate.

14.5 Definition   The linear state equation (1) is called *uniformly exponentially stable with rate* λ, where λ is a positive constant, if there exists a constant γ such that for any t_o and x_o the corresponding solution of (1) satisfies

$$\|x(t)\| \le \gamma e^{-\lambda(t-t_o)} \|x_o\| , \quad t \ge t_o$$

14.6 Lemma   The linear state equation (1) is uniformly exponentially stable with rate $\lambda + \alpha$, where λ and α are positive constants, if the linear state equation

$$\dot{z}(t) = [A(t) + \alpha I]z(t)$$

is uniformly exponentially stable with rate λ.

Proof   It is easy to show by differentiation that $x(t)$ satisfies

$$\dot{x}(t) = A(t)x(t) , \quad x(t_o) = x_o$$

if and only if $z(t) = e^{\alpha(t-t_o)}x(t)$ satisfies

$$\dot{z}(t) = [A(t) + \alpha I]z(t) , \quad z(t_o) = x_o \tag{12}$$

Now assume there is a γ such that for any x_o and t_o the resulting solution of (12) satisfies

$$\|z(t)\| \le \gamma e^{-\lambda(t-t_o)} \|x_o\| , \quad t \ge t_o$$

Then, substituting for $z(t)$,

$$\|e^{\alpha(t-t_o)}x(t)\| = e^{\alpha(t-t_o)} \|x(t)\|$$
$$\le \gamma e^{-\lambda(t-t_o)} \|x_o\|$$

and this immediately implies that (1) is uniformly exponentially stable with rate $\lambda + \alpha$.
□□□

The following stabilization result relies on a strengthened form of controllability assumption for the state equation (1). Recalling from Chapter 9 the controllability Gramian

$$W(t_o, t_f) = \int_{t_o}^{t_f} \Phi(t_o, \sigma)B(\sigma)B^T(\sigma)\Phi^T(t_o, \sigma) \, d\sigma \tag{13}$$

we use also the related notation

$$W_\alpha(t_o, t_f) = \int_{t_o}^{t_f} 2e^{4\alpha(t_o-\sigma)}\Phi(t_o, \sigma)B(\sigma)B^T(\sigma)\Phi^T(t_o, \sigma)\, d\sigma \tag{14}$$

for $\alpha > 0$.

14.7 Theorem For the linear state equation (1) suppose there exist positive constants δ, ε_1, and ε_2 such that

$$\varepsilon_1 I \le W(t, t+\delta) \le \varepsilon_2 I \tag{15}$$

for all t. Then given a positive constant α the state feedback gain

$$K(t) = -B^T(t)W_\alpha^{-1}(t, t+\delta) \tag{16}$$

is such that the resulting closed-loop state equation is uniformly exponentially stable with rate α.

Proof Comparing the quadratic forms $x^T W_\alpha(t, t+\delta)x$ and $x^T W(t, t+\delta)x$, using the definitions (13) and (14), yields

$$2e^{-4\alpha\delta}W(t, t+\delta) \le W_\alpha(t, t+\delta) \le 2W(t, t+\delta)$$

for all t. Therefore (15) implies

$$2\varepsilon_1 e^{-4\alpha\delta}I \le W_\alpha(t, t+\delta) \le 2\varepsilon_2 I \tag{17}$$

for all t, and in particular existence of the inverse in (16) is obvious. Next we show that the linear state equation

$$\dot{z}(t) = \left[A(t) - B(t)B^T(t)W_\alpha^{-1}(t, t+\delta) + \alpha I \right] z(t) \tag{18}$$

is uniformly exponentially stable by applying Theorem 7.4 with the choice

$$Q(t) = W_\alpha^{-1}(t, t+\delta) \tag{19}$$

Obviously $Q(t)$ is symmetric and continuously differentiable. From (17),

$$\frac{1}{2\varepsilon_2}I \le Q(t) \le \frac{e^{4\alpha\delta}}{2\varepsilon_1}I \tag{20}$$

for all t. Therefore it remains only to show that there is a positive constant ν such that

$$[A(t) - B(t)B^T(t)Q(t) + \alpha I]^T Q(t)$$
$$+ Q(t)[A(t) - B(t)B^T(t)Q(t) + \alpha I] + \dot{Q}(t) \le -\nu I \tag{21}$$

for all t. Using the formula for derivative of an inverse,

$$\dot{Q}(t) = -Q(t)\left[\frac{d}{dt}W_\alpha(t,\ t+\delta)\right]Q(t)$$

$$= -Q(t)\left[2e^{-4\alpha\delta}\Phi(t,\ t+\delta)B(t+\delta)B^T(t+\delta)\Phi^T(t,\ t+\delta) - 2B(t)B^T(t)\right.$$

$$\left. + 4\alpha Q^{-1}(t) + A(t)Q^{-1}(t) + Q^{-1}(t)A^T(t)\right]Q(t)$$

Substituting this expression into (21) shows that the left side of (21) is bounded above (in the matrix sign-definite sense) by $-2\alpha Q(t)$. Using (20) then gives that an appropriate choice for ν is α/ε_2. Thus uniform exponential stability of (18) (at some positive rate) is established. Invoking Lemma 14.6 completes the proof
□□□

For a time-invariant linear state equation,

$$\dot{x}(t) = Ax(t) + Bu(t)$$

$$y(t) = Cx(t) \tag{22}$$

it is not difficult to specialize Theorem 14.7 to obtain a time-varying linear state feedback gain that stabilizes. However a profitable alternative is available by applying algebraic results related to constant-Q Lyapunov functions that are the bases for some exercises in earlier chapters. Furthermore this alternative directly yields a constant state-feedback gain. For blithe spirits who have not worked exercises cited in the proof, another argument is outlined in Exercise 14.5.

14.8 Theorem   Suppose the time-invariant linear state equation (22) is controllable, and let

$$\alpha_m = \|A\|$$

Then for any $\alpha > \alpha_m$ the constant state feedback gain

$$K = -B^T Q^{-1} \tag{23}$$

where Q is the positive definite solution of

$$(A + \alpha I)Q + Q(A + \alpha I)^T = BB^T \tag{24}$$

is such that the resulting closed-loop state equation is exponentially stable with rate α.

Proof   Suppose $\alpha > \alpha_m$ is fixed. We first show that the state equation

$$\dot{z}(t) = -(A + \alpha I)z(t) + Bv(t) \tag{25}$$

is exponentially stable. But this follows easily from Theorem 7.4 with the choice $Q(t) = I$. Indeed the easy calculation

$$-(A + \alpha I)^T Q - Q(A + \alpha I) = -2\alpha I - A - A^T$$

$$\leq -2\alpha I + 2\alpha_m I$$

shows that an appropriate choice for v is $2(\alpha - \alpha_m)$.

Therefore, using Exercise 9.7 to conclude that (25) also is controllable, Exercise 9.8 gives that there exists a symmetric, positive-definite Q such that (24) is satisfied. Then $(A + \alpha I - BB^T Q^{-1})$ satisfies

$$(A + \alpha I - BB^T Q^{-1})Q + Q(A + \alpha I - BB^T Q^{-1})^T$$

$$= (A + \alpha I)Q + Q(A + \alpha I)^T - 2BB^T$$

$$= -BB^T$$

By Exercise 13.11 the linear state equation

$$\dot{z}(t) = (A + \alpha I - BB^T Q^{-1})z(t) + Bv(t) \tag{26}$$

is controllable also, and thus by Exercise 9.9 we have that (26) is exponentially stable. Finally Lemma 14.6 gives that the state equation

$$\dot{x}(t) = (A - BB^T Q^{-1})x(t)$$

is exponentially stable with rate α, and of course this is the closed-loop state equation resulting from the gain (23) and $N(t) = 0$.

Eigenvalue Assignment

Stabilization in the time-invariant case can be developed in several directions to further show what can be accomplished by state feedback. Summoning controller form from Chapter 13, we quickly provide one celebrated result as an illustration. Given a set of desired eigenvalues, the objective is to compute a constant state feedback gain K such that the closed-loop state equation

$$\dot{x}(t) = (A + BK)x(t) \tag{27}$$

has precisely these eigenvalues. Of course in almost all situations eigenvalues are specified to have negative real parts for exponential stability. The capability of assigning specific values for the real parts directly influences the rate of decay of the zero-input response component, and assigning imaginary parts influences the frequencies of oscillation that occur.

Because of the minor, fussy issue that eigenvalues of a real-coefficient state equation must occur in complex-conjugate pairs, it is convenient to specify, instead of eigenvalues, a real-coefficient, degree-n characteristic polynomial for (27).

14.9 Theorem   Suppose the the time-invariant linear state equation (22) is controllable and *rank* $B = m$. Given any monic degree-n polynomial $p(\lambda)$ there is a constant feedback gain K such that *det* $(\lambda I - A - BK) = p(\lambda)$.

Proof   First suppose that the controllability indices of (22) are $\rho_1, \ldots, \rho_m$, and the state variable change to controller form described in Theorem 13.9 has been applied. Then the controller-form coefficient matrices are

$$PAP^{-1} = A_o + B_o UP^{-1} , \quad PB = B_o R$$

and given $p(\lambda) = \lambda^n + p_{n-1}\lambda^{n-1} + \cdots + p_0$ a feedback gain K_{CF} for the new state equation can be computed as follows. Clearly

$$PAP^{-1} + PBK_{CF} = A_o + B_o UP^{-1} + B_o RK_{CF}$$

$$= A_o + B_o [UP^{-1} + RK_{CF}] \tag{28}$$

Reviewing the form of the integrator coefficient matrices A_o and B_o, the ith-row of $UP^{-1} + RK_{CF}$ becomes row $\rho_1 + \cdots + \rho_i$ of $PAP^{-1} + PBK_{CF}$. With this observation there are several ways to proceed. One is to set

$$K_{CF} = -R^{-1}UP^{-1} + R^{-1}
\begin{bmatrix}
e_{\rho_1+1} \\
e_{\rho_1+\rho_2+1} \\
\vdots \\
e_{\rho_1+\cdots+\rho_{m-1}+1} \\
-p_0 \ -p_1 \ \cdots \ -p_{n-1}
\end{bmatrix}$$

where e_j denotes the jth-row of the $n \times n$ identity matrix. Then from (28),

$$PAP^{-1} + PBK_{CF} = A_o + B_o
\begin{bmatrix}
e_{\rho_1+1} \\
e_{\rho_1+\rho_2+1} \\
\vdots \\
e_{\rho_1+\cdots+\rho_{m-1}+1} \\
-p_0 \ -p_1 \ \cdots \ -p_{n-1}
\end{bmatrix}$$

$$=
\begin{bmatrix}
0 & 1 & 0 & \cdots & 0 \\
0 & 0 & 1 & \cdots & 0 \\
\vdots & \vdots & \vdots & \vdots & \vdots \\
0 & 0 & 0 & \cdots & 1 \\
-p_0 & -p_1 & -p_2 & \cdots & -p_{n-1}
\end{bmatrix}$$

Either by straightforward calculation or review of Example 10.12 it can be shown that $PAP^{-1} + PBK_{CF}$ has the desired characteristic polynomial. Finally note that the characteristic polynomial of $A + BK_{CF}P$ is the same as the characteristic polynomial of

$$P[A + BK_{CF}P]P^{-1} = PAP^{-1} + PBK_{CF} \tag{29}$$

Therefore the choice $K = K_{CF}P$ is such that the characteristic polynomial of $A + BK$ is $p(\lambda)$.
□ □ □

The input gain $N(t)$ has not participated in stabilization or eigenvalue placement, obviously because these objectives pertain to the zero-input response of the closed-loop state equation. The gain $N(t)$ becomes important when zero-state response behavior is an issue. One illustration is provided by Exercise 2.8, and another occurs in the next section.

Noninteracting Control

The stabilization and eigenvalue placement problems employ linear state feedback to change the dynamical behavior of a given plant — asymptotic character of the zero-input response, overall speed of response, and so on. Another capability of feedback is that structural features of the zero-state response of the closed-loop state equation can be changed. As an illustration we consider a plant of the form (1) with the additional assumption that $p = m$, and discuss the problem of *noninteracting control*. This problem involves using linear state feedback to achieve two input-output objectives on a specified time interval $[t_o, t_f]$. First the closed-loop state equation (3) should be such that for $i \neq j$ the jth-input component $r_j(t)$ has no effect on the ith-output component $y_i(t)$ for all $t \in [t_o, t_f]$. The second objective, imposed in part to avoid a trivial solution where all output components are uninfluenced by any input component, is that the closed-loop state equation should be output controllable in the sense of Exercise 9.10.

It is clear from the problem statement that the zero-input response plays no role in noninteracting control, so we assume for simplicity that $x(t_o) = 0$. Then the first objective is equivalent to the requirement that the closed-loop impulse response

$$\hat{G}(t, \sigma) = C(t)\Phi_{A+BK}(t, \sigma)B(\sigma)N(\sigma)$$

be a diagonal matrix for all t and σ such that $t_f \geq t \geq \sigma \geq t_o$. A closed-loop state equation with this property can be viewed from an input-output perspective as a collection of m independent, single-input, single-output linear systems. This simplifies the output controllability objective: from Exercise 9.10 output controllability is achieved if each diagonal entry of $\hat{G}(t, \sigma)$ is not identically zero for $t_f \geq t \geq \sigma \geq t_o$. (This condition also is necessary for output controllability if *rank* $C(t_f) = m$.)

To further simplify analysis the input-output representation can be deconstructed to exhibit each output component. Let $C_1(t), \ldots, C_m(t)$ denote the rows of the $m \times n$ matrix $C(t)$. Then the ith-row of $\hat{G}(t, \sigma)$ can be written as

$$\hat{G}_i(t, \sigma) = C_i(t)\Phi_{A+BK}(t, \sigma)B(\sigma)N(\sigma) \tag{30}$$

and the ith-output component is described by

$$y_i(t) = \int_{t_o}^{t} \hat{G}_i(t,\ \sigma)u(\sigma)\ d\sigma$$

In this format the objective of noninteracting control is that the rows of $\hat{G}(t,\ \sigma)$ have the form

$$\hat{G}_i(t,\ \sigma) = g_i(t,\ \sigma)e_i\ ,\quad i = 1,\ \dots,\ m \tag{31}$$

for $t_f \geq t \geq \sigma \geq t_o$, where each scalar function $g_i(t,\ \sigma)$ is not identically zero, and e_i denotes the ith-row of I_m.

Solvability of the noninteracting control problem involves smoothness assumptions stronger than our default continuity. To unclutter the development we proceed as in Chapters 9 and 11, and simply assume every derivative that appears is endowed with existence and continuity. After digesting the proofs, the fastidious will find it satisfyingly easy to sum the continuous-differentiability requirements.

An existence condition for solution of the noninteracting control problem can be phrased in terms of the matrices $L_0(t)$, $L_1(t)$, $\cdots$ introduced in the context of observability in Definition 9.9. However a somewhat different notation is both convenient and traditional. Define a linear operator that maps $1 \times n$ time functions, for example $C_i(t)$, into $1 \times n$ time functions according to

$$L_A[C_i](t) = C_i(t)A(t) + \dot{C}_i(t) \tag{32}$$

In this notation a superscript denotes composition of linear operators,

$$L_A^{j+1}[C_i](t) = L_A\left[L_A^j[C_i](t) \right](t)$$

$$= L_A^j[C_i](t)A(t) + \frac{d}{dt}\ L_A^j[C_i](t)\ ,\quad j = 1,\ 2,\ \cdots$$

and, by definition,

$$L_A^0[C_i](t) = C_i(t)$$

An analogous notation is used in relation to the closed-loop linear state equation:

$$L_{A+BK}[C_i](t) = C_i(t)[\ A(t) + B(t)K(t)\ ] + \dot{C}_i(t)$$

It is easy to prove by induction that

$$L_{A+BK}^j[C_i](t)\Phi_{A+BK}(t,\ \sigma) = \frac{\partial^j}{\partial t^j}\left[C_i(t)\Phi_{A+BK}(t,\ \sigma) \right],\quad j = 0,\ 1,\ \cdots \tag{33}$$

an expression that on evaluation at $\sigma = t$ and translation of notation recalls equation (20) of Chapter 9. Going further, (30) and (33) give

$$\frac{\partial^j}{\partial t^j}\ \hat{G}_i(t,\ \sigma) = L_{A+BK}^j[C_i](t)\Phi_{A+BK}(t,\ \sigma)B(\sigma)N(\sigma)\ ,\quad j = 0,\ 1,\ \cdots \tag{34}$$

A basic structural concept for the linear state equation (1) can be introduced in terms of this notation. The underlying calculation is repeated differentiation of the ith-component of the zero-state response of (1) until the input $u(t)$ appears with a coefficient that is not identically zero. For example

$$\dot{y}_i(t) = \dot{C}_i(t)x(t) + C_i(t)\dot{x}(t)$$

$$= [\ \dot{C}_i(t) + C_i(t)A(t)\ ]x(t) + C_i(t)B(t)u(t)$$

In continuing this calculation the coefficient of $u(t)$ in the jth-derivative is

$$L_A^{j-1}[C_i](t)B(t)$$

at least up to and including the derivative where the coefficient of the input is nonzero. The number of output derivatives until the input appears with nonzero coefficient is of main interest, and a key assumption is that this number not change with time.

14.10 Definition   The linear state equation (1) is said to have *constant relative degree* $\kappa_1, \ldots, \kappa_m$ on $[t_o, t_f]$ if $\kappa_1, \ldots, \kappa_m$ are finite positive integers such that

$$L_A^j[C_i](t)B(t) = 0\ ,\quad t \in [t_o, t_f]\ ,\ \ j = 0, \ldots, \kappa_i-2$$

$$L_A^{\kappa_i-1}[C_i](t)B(t) \neq 0\ ,\quad t \in [t_o, t_f] \tag{35}$$

for $i = 1, \ldots, m$.

We emphasize that the same *constant* κ_i must be such that the relations (35) hold at *every* t in the interval. Straightforward application of the definition, left as a small exercise, provides a useful identity relating open-loop and closed-loop operators.

14.11 Lemma   Suppose the linear state equation (1) has constant relative degree $\kappa_1, \ldots, \kappa_m$ on $[t_o, t_f]$. Then for any state feedback gain $K(t)$, and $i = 1, \ldots, m$,

$$L_{A+BK}^j[C_i](t) = L_A^j[C_i](t)\ ,\quad j = 0, \ldots, \kappa_i-1\ ,\ \ t \in [t_o, t_f] \tag{36}$$

Existence conditions for solution of the noninteracting control problem on a specified time interval $[t_o, t_f]$ rely on intricate but elementary calculations. A slight complication is that $N(t)$ could be singular (even zero) on subintervals of $[t_o, t_f]$, so that the closed-loop state equation ignores portions of the reference input, yet is output controllable on $[t_o, t_f]$. We circumvent this impracticality by considering only the case where $N(t)$ is invertible at each $t \in [t_o, t_f]$. In a similar vein note that the following existence condition cannot be satisfied unless

$$\text{rank } B(t) = m\ ,\quad t \in [t_o, t_f]$$

14.12 Theorem   Suppose the linear state equation (1) with $p = m$ has constant relative degree $\kappa_1, \ldots, \kappa_m$ on $[t_o, t_f]$. Then there exist feedback gains $K(t)$ and $N(t)$ that achieve noninteracting control on $[t_o, t_f]$, with $N(t)$ invertible at each $t \in [t_o, t_f]$, if and only if the $m \times m$ matrix

$$\Delta(t) = \begin{bmatrix} L_A^{\kappa_1-1}[C_1](t)B(t) \\ \vdots \\ L_A^{\kappa_m-1}[C_m](t)B(t) \end{bmatrix} \tag{37}$$

is invertible at each $t \in [t_o, t_f]$.

Proof   To streamline the presentation we compute for a general value of index i, $i = 1, \ldots, m$, and neglect repetitive display of the argument range $t_f \geq t \geq \sigma \geq t_o$. The first step is to develop via basic calculus a representation for $\hat{G}_i(t, \sigma)$ in terms of its own derivatives. This permits characterizing the objective of noninteracting control in terms of $L_A[C_i](t)$ by (34).

For any σ the $1 \times m$ matrix function $\hat{G}_i(t, \sigma)$ can be written as

$$\hat{G}_i(t, \sigma) = \hat{G}_i(t, \sigma)\bigg|_{t=\sigma} + \int_\sigma^t \frac{\partial}{\partial\sigma_1}\hat{G}_i(\sigma_1, \sigma)\, d\sigma_1 \tag{38}$$

Similarly we can write

$$\frac{\partial}{\partial\sigma_1}\hat{G}_i(\sigma_1, \sigma) = \frac{\partial}{\partial\sigma_1}\hat{G}_i(\sigma_1, \sigma)\bigg|_{\sigma_1=\sigma} + \int_\sigma^{\sigma_1}\frac{\partial^2}{\partial\sigma_2^2}\hat{G}_i(\sigma_2, \sigma)\, d\sigma_2$$

and substitute into (38) to obtain

$$\hat{G}_i(t, \sigma) = \hat{G}_i(t, \sigma)\bigg|_{t=\sigma} + \int_\sigma^t\left[\frac{\partial}{\partial\sigma_1}\hat{G}_i(\sigma_1, \sigma)\bigg|_{\sigma_1=\sigma}\right]d\sigma_1 + \int_\sigma^t\int_\sigma^{\sigma_1}\frac{\partial^2}{\partial\sigma_2^2}\hat{G}_i(\sigma_2, \sigma)\, d\sigma_2 d\sigma$$

$$= \hat{G}_i(\sigma, \sigma) + \left[\frac{\partial}{\partial\sigma_1}\hat{G}_i(\sigma_1, \sigma)\bigg|_{\sigma_1=\sigma}\right](t-\sigma) + \int_\sigma^t\int_\sigma^{\sigma_1}\frac{\partial^2}{\partial\sigma_2^2}\hat{G}_i(\sigma_2, \sigma)\, d\sigma_2 d\sigma_1 \tag{39}$$

Next write

$$\frac{\partial^2}{\partial\sigma_2}\hat{G}_i(\sigma_2, \sigma) = \frac{\partial^2}{\partial\sigma_2}\hat{G}_i(\sigma_2, \sigma)\bigg|_{\sigma_2=\sigma} + \int_\sigma^{\sigma_2}\frac{\partial^3}{\partial\sigma_3^3}\hat{G}_i(\sigma_3, \sigma)\, d\sigma_3$$

and substitute into (39). Repeating this process κ_i-1 times yields the representation

$$\hat{G}_i(t, \sigma) = \hat{G}_i(\sigma, \sigma) + \left[\frac{\partial}{\partial\sigma_1}\hat{G}_i(\sigma_1, \sigma)\bigg|_{\sigma_1=\sigma}\right](t-\sigma)$$

$$+ \cdots + \left[\frac{\partial^{\kappa_i-1}}{\partial\sigma_{\kappa_i-1}^{\kappa_i-1}}\hat{G}_i(\sigma_{\kappa_i-1}, \sigma)\bigg|_{\sigma_{\kappa_i-1}=\sigma}\right]\frac{(t-\sigma)^{\kappa_i-1}}{(\kappa_i-1)!}$$

$$+ \int_\sigma^t\int_\sigma^{\sigma_1}\cdots\int_\sigma^{\sigma_{\kappa_i-1}}\frac{\partial^{\kappa_i}}{\partial\sigma_{\kappa_i}^{\kappa_i}}\hat{G}_i(\sigma_{\kappa_i}, \sigma)\, d\sigma_{\kappa_i}\cdots d\sigma_1$$

Using (34) gives

$$\hat{G}_i(t, \sigma) = L^0_{A+BK}[C_i](\sigma)B(\sigma)N(\sigma) + L_{A+BK}[C_i](\sigma)B(\sigma)N(\sigma)(t-\sigma)$$

$$+ \cdots + L^{\kappa_i-1}_{A+BK}[C_i](\sigma)B(\sigma)N(\sigma)\frac{(t-\sigma)^{\kappa_i-1}}{(\kappa_i-1)!}$$

$$+ \int_\sigma^t \int_\sigma^{\sigma_1} \cdots \int_\sigma^{\sigma_{\kappa_i-1}} L^{\kappa_i}_{A+BK}[C_i](\sigma_{\kappa_i})\Phi_{A+BK}(\sigma_{\kappa_i}, \sigma)B(\sigma)N(\sigma)\, d\sigma_{\kappa_i} \cdots d\sigma_1$$

Then from (35) and (36) we obtain

$$\hat{G}_i(t, \sigma) = L^{\kappa_i-1}_A[C_i](\sigma)B(\sigma)N(\sigma)\frac{(t-\sigma)^{\kappa_i-1}}{(\kappa_i-1)!}$$

$$+ \int_\sigma^t \int_\sigma^{\sigma_1} \cdots \int_\sigma^{\sigma_{\kappa_i-1}} L^{\kappa_i}_{A+BK}[C_i](\sigma_{\kappa_i})\Phi_{A+BK}(\sigma_{\kappa_i}, \sigma)B(\sigma)N(\sigma)\, d\sigma_{\kappa_i} \cdots d\sigma_1 \quad (40)$$

In terms of this representation for the rows of the impulse response, noninteracting control is achieved if and only if for each i there exist a pair of scalar functions $g_i(\sigma)$ and $f_i(\sigma_{\kappa_i}, \sigma)$, not both identically zero, such that

$$L^{\kappa_i-1}_A[C_i](\sigma)B(\sigma)N(\sigma) = g_i(\sigma)e_i \quad (41)$$

and

$$L^{\kappa_i}_{A+BK}[C_i](\sigma_{\kappa_i})\Phi_{A+BK}(\sigma_{\kappa_i}, \sigma)B(\sigma)N(\sigma) = f_i(\sigma_{\kappa_i}, \sigma)e_i \quad (42)$$

For the sufficiency portion of the proof we need to choose gains $K(t)$ and $N(t)$ to satisfy (41) and (42) for $i = 1, \ldots, m$. Surprisingly clever choices can be made. The assumed invertibility of $\Delta(t)$ at each t permits the gain selection

$$N(t) = \Delta^{-1}(t) \quad (43)$$

Then

$$L^{\kappa_i-1}_A[C_i](\sigma)B(\sigma)N(\sigma) = L^{\kappa_i-1}_A[C_i](\sigma)B(\sigma)\Delta^{-1}(\sigma)$$

$$= e_i$$

and (41) is satisfied with $g_i(\sigma) = 1$. To address (42), write

$$L^{\kappa_i}_{A+BK}[C_i](t) = L_{A+BK}\left[L^{\kappa_i-1}_A[C_i](t)\right]$$

$$= L^{\kappa_i-1}_A[C_i](t)[A(t) + B(t)K(t)] + \frac{d}{dt}L^{\kappa_i-1}_A[C_i](t) \quad (44)$$

Choosing the gain

$$K(t) = -\Delta^{-1}(t)\left[\Omega(t)A(t) + \frac{d}{dt}\Omega(t)\right] \quad (45)$$

where

$$\Omega(t) = \begin{bmatrix} L_A^{\kappa_1-1}[C_1](t) \\ \vdots \\ L_A^{\kappa_m-1}[C_m](t) \end{bmatrix}$$

and substituting into (44) gives

$$L_{A+BK}^{\kappa_i}[C_i](t) = L_A^{\kappa_i-1}[C_i](t)A(t) - L_A^{\kappa_i-1}[C_i](t)B(t)\Delta^{-1}(t) \left[\Omega(t)A(t) + \frac{d}{dt}\Omega(t) \right]$$

$$+ \frac{d}{dt}L_A^{\kappa_i-1}[C_i](t)$$

$$= L_A^{\kappa_i-1}[C_i](t)A(t) - e_i\Omega(t)A(t) - e_i\frac{d}{dt}\Omega(t) + \frac{d}{dt}L_A^{\kappa_i-1}[C_i](t)$$

$$= 0$$

Therefore (42) is satisfied with $f_i(\sigma_{\kappa_i}, \sigma)$ identically zero. Since the feedback gains (43) and (45) are independent of the index i, noninteracting control is achieved for the corresponding closed-loop state equation.

To prove necessity of the invertibility condition on $\Delta(t)$, suppose $K(t)$ and $N(t)$ achieve noninteracting control, with $N(t)$ invertible at each t. Then (41) is satisfied, in particular. From the definition of relative degree and the invertibility of $N(\sigma)$, we have

$$g_i(\sigma) \neq 0, \quad \sigma \in [t_o, t_f]$$

This argument applies for $i = 1, \ldots, m$, and the collection of identities represented by (41) can be written as

$$\Delta(\sigma)N(\sigma) = \text{diagonal } \{ g_1(\sigma), \ldots, g_m(\sigma) \}$$

It follows that $\Delta(\sigma)$ is invertible at each $\sigma \in [t_o, t_f]$.
□□□

Specialization of Theorem 14.12 to the time-invariant case is almost immediate from the observability lineage of $L_A[C_i](t)$. The notion of constant relative degree deflates to existence of finite positive integers $\kappa_1, \ldots, \kappa_m$ such that

$$C_iA^jB = 0, \quad j = 0, \ldots, \kappa_i-2$$

$$C_iA^{\kappa_i-1}B \neq 0 \tag{46}$$

for $i = 1, \ldots, m$. It remains only to work out the specialized proof to verify that the time interval is immaterial, and that constant gains can be used (Exercise 14.13).

14.13 Corollary   Suppose the time-invariant linear state equation (22) with $p = m$ has relative degree $\kappa_1, \ldots, \kappa_m$. Then there exist constant feedback gains K and invertible N that achieve noninteracting control if and only if the $m \times m$ matrix

$$\Delta = \begin{bmatrix} C_1 A^{\kappa_1-1} B \\ \vdots \\ C_m A^{\kappa_m-1} B \end{bmatrix} \tag{47}$$

is invertible.

14.14 Example   For the plant

$$\dot{x}(t) = \begin{bmatrix} 0 & 1 & 0 & 0 \\ 0 & 0 & 1 & 0 \\ 0 & 0 & 0 & 1 \\ 1 & 1 & 0 & 1 \end{bmatrix} x(t) + \begin{bmatrix} 1 & 1 \\ b(t) & 0 \\ 0 & 0 \\ 1 & 1 \end{bmatrix} u(t)$$

$$y(t) = \begin{bmatrix} 0 & 0 & 1 & 0 \\ 0 & 1 & 0 & 0 \end{bmatrix} x(t)$$

simple calculations give

$$L_A^0[C_1](t)B(t) = \begin{bmatrix} 0 & 0 \end{bmatrix}$$
$$L_A[C_1](t)B(t) = \begin{bmatrix} 1 & 1 \end{bmatrix}$$
$$L_A^0[C_2](t)B(t) = \begin{bmatrix} b(t) & 0 \end{bmatrix}$$

If $[t_o, t_f]$ is an interval such that $b(t) \neq 0$ for $t \in [t_o, t_f]$, then the plant has constant relative degree $\kappa_1 = 2$, $\kappa_2 = 1$ on $[t_o, t_f]$. Furthermore

$$\Delta(t) = \begin{bmatrix} 1 & 1 \\ b(t) & 0 \end{bmatrix}$$

is invertible for $t \in [t_o, t_f]$. The gains in (43) and (45) yield the state feedback

$$u(t) = - \begin{bmatrix} 0 & 0 & 1/b(t) & 0 \\ 1 & 1 & -1/b(t) & 1 \end{bmatrix} x(t) + \begin{bmatrix} 0 & 1/b(t) \\ 1 & -1/b(t) \end{bmatrix} r(t) \tag{48}$$

and the resulting noninteracting closed-loop state equation is

$$\dot{x}(t) = \begin{bmatrix} 1 & 2 & 0 & 1 \\ 0 & 0 & 2 & 0 \\ 0 & 0 & 0 & 1 \\ 2 & 2 & 0 & 2 \end{bmatrix} x(t) + \begin{bmatrix} 1 & 0 \\ 0 & 1 \\ 0 & 0 \\ 1 & 0 \end{bmatrix} r(t)$$

$$y(t) = \begin{bmatrix} 0 & 0 & 1 & 0 \\ 0 & 1 & 0 & 0 \end{bmatrix} x(t)$$

EXERCISES

Exercise 14.1 Suppose the time-invariant linear state equation

$$\dot{x}(t) = Ax(t) + Bu(t)$$

is controllable, and the $n \times n$ matrix F has the characteristic polynomial $det\,(\lambda I - F) = p(\lambda)$. If the $m \times n$ matrix R and the invertible, $n \times n$ matrix Q are such that

$$AQ - QF = BR$$

show how to choose an $m \times n$ matrix K such that $A + BK$ has characteristic polynomial $p(\lambda)$.

Exercise 14.2 Establish the following version of Theorem 14.7. If the time-invariant linear state equation

$$\dot{x}(t) = Ax(t) + Bu(t)$$

is controllable, then for any $t_f > 0$ the time-invariant state feedback

$$u(t) = -B^T \left[\int_0^{t_f} e^{-A\tau} BB^T e^{-A^T\tau}\, d\tau \right]^{-1} x(t)$$

yields an exponentially stable closed-loop state equation. (*Hint*: Consider

$$(A + BK)Q + Q(A + BK)^T$$

where

$$Q = \int_0^{t_f} e^{-A\tau} BB^T e^{-A^T\tau}\, d\tau$$

and proceed as in Exercise 9.9.)

Exercise 14.3 Suppose that the time-invariant linear state equation

$$\dot{x}(t) = Ax(t) + Bu(t)$$

is controllable, and $A + A^T \leq 0$. Show that the state feedback

$$u(t) = -B^T x(t)$$

yields a closed-loop state equation that is exponentially stable. (*Hint*: One approach is to directly consider an arbitrary eigenvalue-eigenvector pair for $A - BB^T$.)

Exercise 14.4   Given the time-invariant linear state equation

$$\dot{x}(t) = Ax(t) + Bu(t)$$
$$y(t) = Cx(t)$$

with time-invariant state feedback

$$u(t) = Kx(t) + Nr(t)$$

show that the zero-state response of the resulting closed-loop state equation is the same as the zero-state response of the $2n$-dimensional state equation

$$\dot{x}(t) = Ax(t) + Bu(t)$$
$$\dot{z}(t) = [A + BK]z(t) + BNr(t)$$
$$u(t) = Kz(t) + Nr(t)$$
$$y(t) = Cx(t)$$

(This shows that the input-output behavior of the closed-loop state equation can be obtained by use of a *precompensator* instead of feedback.)

Exercise 14.5   Provide a proof of Theorem 14.8 via these steps:
(a) Consider the quadratic form $x^H Ax + x^H A^T x$ for x a unity-norm eigenvector of A, and show that $-(A^T + \alpha I)$ has negative-real-part eigenvalues.
(b) Use Theorem 7.10 to write the unique solution of (24), and show by contradiction that the controllability hypothesis implies $Q > 0$.
(c) For the linear state equation (26), substitute for BB^T from (24), and conclude (26) is exponentially stable.
(d) Apply Lemma 14.6 to complete the proof.

Exercise 14.6   Use Exercise 13.12 to give an alternate proof of Theorem 14.9.

Exercise 14.7   For a controllable, single-input linear state equation

$$\dot{x}(t) = Ax(t) + bu(t)$$

suppose a degree-n monic polynomial $p(\lambda)$ is given. Show that the state feedback gain

$$k = -[\,0 \quad \cdots \quad 0 \quad 1\,][\,b \quad Ab \quad \cdots \quad A^{n-1}b\,]^{-1}p(A)$$

is such that $det(\lambda I - A - bk) = p(\lambda)$. *Hint*: First show for the controller-form case (Example 10.12) that

$$k = -[\,1 \quad 0 \quad \cdots \quad 0\,]p(A)$$

and

$$[\,1 \quad 0 \quad \cdots \quad 0\,] = [\,0 \quad \cdots \quad 0 \quad 1\,][\,b \quad Ab \quad \cdots \quad A^{n-1}b\,]^{-1}$$

Exercise 14.8   For the time-invariant linear state equation

$$\dot{x}(t) = Ax(t) + Bu(t)$$

show that there exists a time-invariant state feedback

$$u(t) = Kx(t)$$

such that the closed-loop state equation is exponentially stable if and only if

$$\text{rank} \begin{bmatrix} \lambda I - A & B \end{bmatrix} = n$$

for each λ that is a nonnegative-real-part eigenvalue of A. (The property in question is called *stabilizability*.)

Exercise 14.9   Prove that the controllability indices and observability indices in Definition 13.5 and Definition 13.16, respectively, for the time-invariant linear state equation

$$\dot{x}(t) = [\, A + BLC \,]\, x(t) + Bu(t)$$

$$y(t) = Cx(t)$$

are independent of the choice of $m \times p$ output feedback gain L.

Exercise 14.10   Prove that the time-invariant linear state equation

$$\dot{x}(t) = Ax(t) + Bu(t)$$

$$y(t) = Cx(t)$$

cannot be made exponentially stable by output feedback

$$u(t) = Ly(t)$$

if $CB = 0$ and $\text{tr}\,[A] > 0$.

Exercise 14.11   Determine if the noninteracting control problem for the plant

$$\dot{x}(t) = \begin{bmatrix} 0 & 1 & 0 & 0 & 0 \\ 0 & 0 & 1 & 1 & 1 \\ 1 & 0 & 0 & 0 & e^t \\ 0 & 0 & 0 & 0 & 0 \\ 0 & 0 & 0 & 0 & 0 \end{bmatrix} x(t) + \begin{bmatrix} 0 & 0 \\ 0 & 0 \\ 0 & 0 \\ 1 & 0 \\ 0 & 1 \end{bmatrix} u(t)$$

$$y(t) = \begin{bmatrix} 1 & 0 & 0 & 0 & 0 \\ 0 & 0 & 1 & 0 & 0 \end{bmatrix} x(t)$$

can be solved on a suitable time interval. If so, compute a state feedback that solves the problem.

Exercise 14.12   Suppose a time-invariant linear state equation with $p = m$ is described by the transfer function $G(s)$. Interpret the relative degrees $\kappa_1, \ldots, \kappa_m$ in terms of simple features of $G(s)$.

Exercise 14.13   Write out a detailed proof of Corollary 14.13, including formulas for constant gains that achieve noninteracting control.

Exercise 14.14   Compute the transfer function of the closed-loop linear state equation resulting from the sufficiency proof of Theorem 14.12. (*Hint*: This is not an unreasonable request.)

Exercise 14.15   Changing notation from Definition 9.3, corresponding to the linear state equation

$$\dot{x}(t) = A(t)x(t) + B(t)u(t)$$

let

$$K_A[B](t) = -A(t)B(t) + \frac{d}{dt}B(t)$$

Show that the notion of constant relative degree in Definition 14.10 can be defined in terms of this linear operator. Then prove that Theorem 14.12 remains true if $\Delta(t)$ in (37) is replaced by

$$\begin{bmatrix} C_1(t)K_A^{\kappa_1-1}[B](t) \\ \vdots \\ C_m(t)K_A^{\kappa_m-1}[B](t) \end{bmatrix}$$

Hint: Show first that for $j, k \geq 0$,

$$L_A^j[C_i](t)K_A^k[B](t) = (-1)^k L_A^{j+k}[C_i](t)K_A^0[B](t) + \sum_{l=1}^{k}(-1)^{k+l}\frac{d}{dt}\left[L_A^{j+k-l}[C_i](t)K_A^{l-1}[B](t)\right]$$

NOTES

Note 14.1   The treatment of effects of feedback follows Section 19 of

R.W. Brockett, *Finite Dimensional Linear Systems,* John Wiley, New York, 1970

The representation of state feedback in terms of open-loop and closed-loop transfer functions is pursued further in Chapter 16 using the polynomial fraction description for transfer functions.

Note 14.2   Results on stabilization of time-varying linear state equations by state feedback using methods of optimal control are given in

R.E. Kalman, "Contributions to the theory of optimal control," *Boletin de la Sociedad Matematica Mexicana,* Vol. 5, pp. 102–119, 1960

See also

M. Ikeda, H. Maeda, S. Kodama, "Stabilization of linear systems," *SIAM Journal on Control and Optimization,* Vol. 10, No. 4, pp. 716–729, 1972

The proof of the stabilization result in Theorem 14.7 is based on

V.H.L. Cheng, "A direct way to stabilize continuous-time and discrete-time linear time-varying systems," *IEEE Transactions on Automatic Control,* Vol. 24, No. 4, pp. 641–643, 1979

For the time-invariant case Theorem 14.8 is attributed to R.W. Bass, and the result of Exercise 14.2 is due to D.L. Kleinman. Many additional aspects of stabilization are known, though only two are mentioned here. For slowly-time-varying linear state equations, stabilization results based on Theorem 8.7 are discussed in

E.W. Kamen, P.P. Khargonekar, A. Tannenbaum, "Control of slowly-varying linear systems," *IEEE Transactions on Automatic Control,* Vol. 34, No. 12, pp. 1283–1285, 1989

It is shown in

M.A. Rotea, P.P. Khargonekar, "Stabilizability of linear time-varying and uncertain linear systems," *IEEE Transactions on Automatic Control,* Vol. 33, No. 9, pp. 884–887, 1988

that if uniform exponential stability can be achieved by dynamic state feedback of the form

$$\dot{z}(t) = F(t)z(t) + G(t)x(t)$$

$$u(t) = H(t)z(t) + E(t)x(t)$$

then uniform exponential stability can be achieved by static state feedback of the form (2). However when other objectives are considered, for example noninteracting control with exponential stability in the time-invariant setting, dynamic state feedback offers more capability than static state feedback. See Note 19.4.

Note 14.3   Eigenvalue assignability for controllable, time-invariant, single-input linear state equations is clear from the single-input controller form, and has been understood since about 1960. The feedback gain formula in Exercise 14.7 is due to J. Ackermann, and other formulas are available. See Section 3.2 of

T. Kailath, *Linear Systems,* Prentice Hall, New York, 1980

For multi-input state equations the eigenvalue assignment result in Theorem 14.9 is proved in

W.M. Wonham, "On pole assignment in multi-input controllable linear systems," *IEEE Transactions on Automatic Control,* Vol. 12, No. 6, pp. 660–665, 1967

The approach suggested in Exercise 14.6 is due to M. Heymann.

Note 14.4   In contrast to the single-input case, a state-feedback gain K that assigns a specified set of eigenvalues for a multi-input plant is not unique. One way of using the resulting flexibility involves assigning closed-loop eigenvectors as well as eigenvalues. See

B.C. Moore, "On the flexibility offered by state feedback in multivariable systems beyond closed loop eigenvalue assignment," *IEEE Transactions on Automatic Control,* Vol. 21, No. 5, pp. 689–692, 1976

and

G. Klein, B.C. Moore, "Eigenvalue-generalized eigenvector assignment with state feedback," *IEEE Transactions on Automatic Control,* Vol. 22, No. 1, pp. 140–141, 1977

Another characterization of the flexibility involves the *invariant factors* of $A + BK$, and is due to H.H. Rosenbrock. See the treatment in

B.W. Dickinson, "On the fundamental theorem of linear state feedback," *IEEE Transactions on Automatic Control,* Vol. 19, No. 5, pp. 577–579, 1974

Note 14.5   Eigenvalue assignment capabilities of static output feedback is a famously difficult topic. Early contributions include

H. Kimura, "Pole assignment by gain output feedback," *IEEE Transactions on Automatic Control,* Vol. 20, No. 4, pp. 509–516, 1975

E.J. Davison, S.H. Wang, "On pole assignment in linear multivariable systems using output feedback," *IEEE Transactions on Automatic Control,* Vol. 20, No. 4, pp. 516–518, 1975

Recent studies that make use of the geometric theory in Chapter 18 are

C. Champetier, J.F. Magni, ''On eigenstructure assignment by gain output feedback,'' *SIAM Journal on Control and Optimization,* Vol. 29, No. 4, pp. 848–865, 1991

J.F. Magni, C. Champetier, ''A geometric framework for pole assignment algorithms,'' *IEEE Transactions on Automatic Control,* Vol. 36, No. 9, pp. 1105–1111, 1991

A survey paper focusing on methods of algebraic geometry is

C.I. Byrnes, ''Pole assignment by output feedback,'' in *Three Decades of Mathematical System Theory,* H. Nijmeijer, J.M. Schumacher, editors, Springer-Verlag Lecture Notes in Control and Information Sciences, No. 135, pp. 31–78, Berlin, 1989

Note 14.6   For a time-invariant linear state equation in controller form,

$$\dot{x}(t) = [\, A_o + B_o U P^{-1}\,] x(t) + B_o R u(t)$$

the linear state feedback

$$u(t) = -R^{-1} U P^{-1} x(t) + R^{-1} r(t)$$

gives a closed-loop state equation described by the integrator coefficient matrices,

$$\dot{x}(t) = A_o x(t) + B_o r(t)$$

In other words, for a controllable linear state equation there is a state variable change and state feedback yielding a closed-loop state equation with structure that depends only on the controllability indices. This is called *Brunovsky form* after

P. Brunovsky, ''A classification of linear controllable systems,'' *Kybernetika,* Vol. 6, pp. 173–188, 1970

If an output is specified, the additional operations of *output variable change* and *output injection* (See Exercise 15.6) permit simultaneous attainment of a special structure for C that has the form of B_o^T. A treatment using geometric tools of Chapters 18 and 19 can be found in

A.S. Morse, ''Structural invariants of linear multivariable systems,'' *SIAM Journal on Control and Optimization,* Vol. 11, No. 3, pp. 446–465, 1973

Note 14.7   The noninteracting control problem also is called the *decoupling problem.* For time-invariant linear state equations, the existence condition in Corollary 14.13 appears in

P.L. Falb, W.A. Wolovich, ''Decoupling in the design and synthesis of multivariable control systems,'' *IEEE Transactions on Automatic Control,* Vol. 12, No. 6, pp. 651–659, 1967

For time-varying linear state equations, the existence condition is discussed in

W.A. Porter, ''Decoupling of and inverses for time-varying linear systems,'' *IEEE Transactions on Automatic Control,* Vol. 14, No. 4, pp. 378–380, 1969

with additional work reported in

E. Freund, ''Design of time-variable multivariable systems by decoupling and by the inverse,'' *IEEE Transactions on Automatic Control,* Vol. 16, No. 2, pp. 183–185, 1971

W.J. Rugh, ''On the decoupling of linear time-variable systems,'' *Proceedings of the Fifth Conference on Information Sciences and Systems,* Princeton University, Princeton, New Jersey, pp. 490–494, 1971

Output controllability, used to impose nontrivial input-output behavior on each noninteracting closed-loop subsystem, is discussed in

E. Kriendler, P.E. Sarachik, "On the concepts of controllability and observability of linear systems," *IEEE Transactions on Automatic Control,* Vol. 9, pp. 129–136, 1964 (Correction: Vol. 10, No. 1, p. 118, 1965)

However, the definition used is slightly different from the definition in Exercise 9.10.

Details aside, we leave noninteracting control at an embryonic stage. Endearing magic occurs in the proof of Theorem 14.12 (see Exercise 14.14), yet many questions remain. For example characterizing the class of state feedback gains that yield noninteraction is crucial in assessing the possibility of achieving desirable input-output behavior — for example, stability if the time interval is infinite. Further developments are left to the literature of control theory, some of which is cited in Chapter 19 where a more general noninteracting control problem for time-invariant linear state equations is reconstituted in a geometric setting.

15

STATE OBSERVATION

An important application of the notion of feedback in linear system theory occurs in the theory of observers. In rough terms state observation involves using current and past values of the plant input and output signals to generate an estimate of the (assumed unknown) current state. Of course as the current time t gets larger there is more information available, and a better estimate is expected.

A more precise formulation is based on an idealized objective. Given a linear state equation

$$\dot{x}(t) = A(t)x(t) + B(t)u(t), \quad x(t_o) = x_o$$

$$y(t) = C(t)x(t) \tag{1}$$

with the initial state x_o unknown, the goal is to generate an $n \times 1$ vector function $\hat{x}(t)$ such that

$$\lim_{t \to \infty} \left[x(t) - \hat{x}(t) \right] = 0 \tag{2}$$

Again it is assumed that the procedure for producing $\hat{x}(t_a)$ at any $t_a \geq t_o$ can make use of the values of $u(t)$ and $y(t)$ for $t \in [t_o, t_a]$, as well as knowledge of the coefficient matrices in (1).

If (1) is observable on $[t_o, t_b]$, then an immediate suggestion for obtaining a state estimate is to first compute the initial state from knowledge of $u(t)$ and $y(t)$ for $t \in [t_o, t_b]$. Then solve (1) for $t \geq t_o$, yielding an estimate that is exact at any $t \geq t_o$, though not current. That is, the estimate is delayed because of the wait until t_b, the time required to compute x_o, then the time to compute the current state from this information. In any case observability plays an important role in the state observation problem. How feedback enters the problem is less clear, for it depends on another specific idea for obtaining a state estimate.

Observers

The standard approach to state observation, motivated partly on grounds of hindsight, is to generate an asymptotic estimate using another linear state equation that accepts as inputs the plant input and output signals, $u(t)$ and $y(t)$. As diagramed in Figure 15.1, consider the problem of choosing an n-dimensional linear state equation of the form

$$\dot{\hat{x}}(t) = F(t)\hat{x}(t) + G(t)u(t) + H(t)y(t) , \quad \hat{x}(t_o) = \hat{x}_o \tag{3}$$

with the property that (2) holds for any initial states x_o and $\hat{x}_o$. A natural requirement to impose is that if $\hat{x}_o = x_o$, then $\hat{x}(t) = x(t)$ for all $t \geq t_o$. Simple algebraic manipulation shows that this fidelity is attained if coefficients of (3) are chosen as

$$F(t) = A(t) - H(t)C(t)$$

$$G(t) = B(t)$$

Then (3) can be written in the form

$$\dot{\hat{x}}(t) = A(t)\hat{x}(t) + B(t)u(t) + H(t)\left[y(t) - \hat{y}(t) \right], \quad \hat{x}(t_o) = \hat{x}_o$$

$$\hat{y}(t) = C(t)\hat{x}(t) \tag{4}$$

where for convenience we have defined an output estimate $\hat{y}(t)$. The only remaining coefficient to specify is the $n \times p$ matrix function $H(t)$, and this final step is best motivated by considering the error in the state estimate.

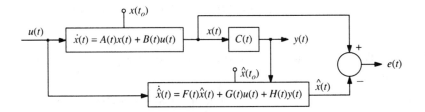

15.1 Figure Observer structure for generating a state estimate.

From (1) and (4) the estimate error

$$e(t) = x(t) - \hat{x}(t)$$

satisfies the linear state equation

$$\dot{e}(t) = \left[A(t) - H(t)C(t) \right]e(t) , \quad e(t_o) = x_o - \hat{x}_o \tag{5}$$

Therefore (2) is satisfied if $H(t)$ can be chosen so that (5) is uniformly exponentially stable. Such a selection of $H(t)$ completely specifies the linear state equation (4) that generates the estimate, and (4) then is called a *state observer* for the given plant. Of course uniform exponential stability of (5) is stronger than necessary for satisfaction of (2), but we choose to retain uniform exponential stability for reasons that will be clear when output-feedback stabilization is considered.

The problem of choosing an *observer gain* $H(t)$ to stabilize (5) obviously bears a resemblance to the problem of choosing a stabilizing state feedback gain $K(t)$ in Chapter 14. But the explicit connection is more elusive than might be expected. Recall that for the plant (1) the observability Gramian is given by

$$M(t_o, t_f) = \int_{t_o}^{t_f} \Phi^T(\tau, t_o)C^T(\tau)C(\tau)\Phi(\tau, t_o) \, d\tau$$

where $\Phi(t, \tau)$ is the transition matrix for $A(t)$. Mimicking the setup of Theorem 14.7 on state feedback stabilization, let

$$M_\alpha(t_o, t_f) = \int_{t_o}^{t_f} 2e^{4\alpha(\tau-t_f)}\Phi^T(\tau, t_o)C^T(\tau)C(\tau)\Phi(\tau, t_o) \, d\tau$$

15.2 Theorem   Suppose for the linear state equation (1) there exist positive constants δ, ε_1, and ε_2 such that

$$\varepsilon_1 I \le \Phi^T(t - \delta, t)M(t - \delta, t)\Phi(t - \delta, t) \le \varepsilon_2 I \tag{6}$$

for all t. Then given a positive constant α the observer gain

$$H(t) = \left[\Phi^T(t - \delta, t)M_\alpha(t - \delta, t)\Phi(t - \delta, t) \right]^{-1} C^T(t) \tag{7}$$

is such that the resulting observer-error state equation (5) is uniformly exponentially stable with rate α.

Proof   Given $\alpha > 0$, first note that from (6),

$$2\varepsilon_1 e^{-4\alpha\delta} I \le \Phi^T(t - \delta, t)M_\alpha(t - \delta, t)\Phi(t - \delta, t) \le 2\varepsilon_2 I$$

for all t, so that existence of the inverse in (7) is clear. To show that (7) yields an error state equation (5) that is uniformly exponentially stable with rate α, we will show that the gain

$$-H^T(-t) = -C(-t)\left[\Phi^T(-t - \delta, -t)M_\alpha(-t - \delta, -t)\Phi(-t - \delta, -t) \right]^{-1}$$

renders the linear state equation

$$\dot{f}(t) = \left[A^T(-t) + C^T(-t)[-H^T(-t)] \right]f(t) \tag{8}$$

uniformly exponentially stable with rate α. That this suffices follows easily from the relation between the transition matrices associated to (5) and (8), namely the identity $\Phi_e(t, \tau) = \Phi_f^T(-\tau, -t)$ established in Exercise 3.3. For if

$$\| \Phi_f(t, \tau) \| \le \gamma e^{-\alpha(t-\tau)}$$

for all t, τ with $t \ge \tau$, then

$$\| \Phi_e(t, \tau) \| = \| \Phi_f^T(-\tau, -t) \| = \| \Phi_f(-\tau, -t) \|$$

$$\le \gamma e^{-\alpha(t-\tau)}$$

for all t, τ with $t \ge \tau$. The beauty of this approach is that selection of $-H^T(-t)$ to render (8) uniformly exponentially stable with rate α is precisely the state-feedback stabilization problem solved in Theorem 14.7. All that remains is to complete the notation conversion so that (7) can be verified.

Writing $\tilde{A}(t) = A^T(-t)$ and $\tilde{B}(t) = C^T(-t)$ to minimize confusion, consider the linear state equation

$$\dot{z}(t) = \tilde{A}(t)z(t) + \tilde{B}(t)u(t) \tag{9}$$

Denoting the transition matrix for $\tilde{A}(t)$ by $\tilde{\Phi}(t, \tau)$, the controllability Gramian for (9) is given by

$$\tilde{W}(t_o, t_f) = \int_{t_o}^{t_f} \tilde{\Phi}(t_o, \sigma)\tilde{B}(\sigma)\tilde{B}^T(\sigma)\tilde{\Phi}^T(t_o, \sigma)\, d\sigma$$

$$= \int_{t_o}^{t_f} \Phi^T(-\sigma, -t_o)C^T(-\sigma)C(-\sigma)\Phi(-\sigma, -t_o)\, d\sigma$$

This expression can be used to evaluate $\tilde{W}(-t, -t+\delta)$, and then changing the integration variable to $\tau = -\sigma$ gives

$$\tilde{W}(-t, -t+\delta) = \int_{t-\delta}^{t} \Phi^T(\tau, t)C^T(\tau)C(\tau)\Phi(\tau, t)\, d\tau$$

$$= \Phi^T(t-\delta, t)M(t-\delta, t)\Phi(t-\delta, t)$$

Therefore (6) implies, since t can be replaced by $-t$ in that inequality,

$$\varepsilon_1 I \le \tilde{W}(t, t+\delta) \le \varepsilon_2 I$$

for all t. That is, the controllability Gramian for (9) satisfies the requisite condition for application of Theorem 14.7. Letting

$$\tilde{W}_\alpha(t_o, t_f) = \int_{t_o}^{t_f} 2e^{4\alpha(t_o-\sigma)}\tilde{\Phi}(t_o, \sigma)\tilde{B}(\sigma)\tilde{B}^T(\sigma)\tilde{\Phi}^T(t_o, \sigma)\, d\sigma \tag{10}$$

it remains only to check that

$$\tilde{W}_\alpha(t, t+\delta) = \Phi^T(-t-\delta, -t)M_\alpha(-t-\delta, -t)\Phi(-t-\delta, -t) \tag{11}$$

For then

$$-H^T(-t) = -\tilde{B}^T(t)\tilde{W}_\alpha^{-1}(t,\ t+\delta)$$

renders (9), and hence (8), uniformly exponentially stable with rate α, and this gain corresponds to $H(t)$ given in (7).

The verification of (11) proceeds as in our previous calculation of $\tilde{W}(t,\ t+\delta)$. From (10),

$$\tilde{W}_\alpha(t,\ t+\delta) = \int_t^{t+\delta} 2e^{4\alpha(t-\sigma)}\Phi^T(-\sigma,\ -t)C^T(-\sigma)C(-\sigma)\Phi(-\sigma,\ -t)\ d\sigma$$

$$= \Phi^T(-t-\delta,\ -t)\int_{-t-\delta}^{-t} 2e^{4\alpha(t+\tau)}\Phi^T(\tau,\ -t-\delta)C^T(\tau)C(\tau)$$

$$\cdot\ \Phi(\tau,\ -t-\delta)\ d\tau\ \Phi(-t-\delta,\ -t)$$

and this is readily recognized as (11).

Output Feedback Stabilization

An important application of state observation arises in the context of linear feedback when not all the state variables are available. This can be illustrated in terms of the stabilization problem for (1) using *linear dynamic output feedback*. Specifically to stabilize the plant (1) consider a state observer and linear feedback of the estimated state described by

$$\dot{\hat{x}}(t) - A(t)\hat{x}(t) + B(t)u(t) + H(t)\left[y(t) - C(t)\hat{x}(t)\right]$$

$$u(t) = K(t)\hat{x}(t) + N(t)r(t) \tag{12}$$

The overall closed-loop state equation, shown in Figure 15.3, can be written as a partitioned $2n$-dimension linear state equation,

$$\begin{bmatrix} \dot{x}(t) \\ \dot{\hat{x}}(t) \end{bmatrix} = \begin{bmatrix} A(t) & B(t)K(t) \\ H(t)C(t) & A(t) - H(t)C(t) + B(t)K(t) \end{bmatrix}\begin{bmatrix} x(t) \\ \hat{x}(t) \end{bmatrix} + \begin{bmatrix} B(t)N(t) \\ B(t)N(t) \end{bmatrix}r(t)$$

$$y(t) = \begin{bmatrix} C(t) & 0_{p\times n} \end{bmatrix}\begin{bmatrix} x(t) \\ \hat{x}(t) \end{bmatrix} \tag{13}$$

The problem is to choose the feedback gain $K(t)$, now applied to the state estimate, and the observer gain $H(t)$ to achieve uniform exponential stability of the zero-input response of (13). (Again the gain $N(t)$ plays no role in stabilization of the zero-input response.)

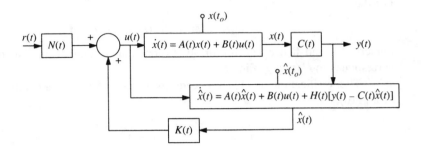

15.3 Figure   Observer-based dynamic output feedback.

15.4 Theorem   Suppose for the linear state equation (1) there exist positive constants δ, ε_1, ε_2, β_1, and β_2 such that

$$\varepsilon_1 I \leq W(t, \ t+\delta) \leq \varepsilon_2 I$$

$$\varepsilon_1 I \leq \Phi^T(t-\delta, \ t)M(t-\delta, \ t)\Phi(t-\delta, \ t) \leq \varepsilon_2 I$$

for all t, and

$$\int_{\tau}^{t} \|B(\sigma)\|^2 \, d\sigma \leq \beta_1 + \beta_2(t-\tau)$$

for all t, τ with $t \geq \tau$. Then given $\alpha > 0$, for any $\eta > 0$ the feedback and observer gains

$$K(t) = -B^T(t)W_{\alpha+\eta}^{-1}(t, \ t+\delta)$$

$$H(t) = \left[\Phi^T(t-\delta, \ t)M_{\alpha+\eta}(t-\delta, \ t)\Phi(t-\delta, \ t)\right]^{-1} C^T(t) \tag{14}$$

are such that the closed-loop state equation (13) is uniformly exponentially stable with rate α.

Proof   In considering uniform exponential stability for (13), $r(t)$ can be ignored. We first apply the state variable change (using suggestive notation)

$$\begin{bmatrix} x(t) \\ e(t) \end{bmatrix} = \begin{bmatrix} I_n & 0_n \\ I_n & -I_n \end{bmatrix} \begin{bmatrix} x(t) \\ \hat{x}(t) \end{bmatrix} \tag{15}$$

This is a Lyapunov transformation, and (13) is uniformly exponentially stable with rate α if and only if the state equation in the new state variables,

$$\begin{bmatrix} \dot{x}(t) \\ \dot{e}(t) \end{bmatrix} = \begin{bmatrix} A(t)+B(t)K(t) & -B(t)K(t) \\ 0_n & A(t)-H(t)C(t) \end{bmatrix} \begin{bmatrix} x(t) \\ e(t) \end{bmatrix} \tag{16}$$

is uniformly exponentially stable with rate α. Let $\Phi(t, \tau)$ denote the transition matrix corresponding to (16), and let $\Phi_x(t, \tau)$ and $\Phi_e(t, \tau)$ denote the $n \times n$ transition matrices for $A(t) + B(t)K(t)$ and $A(t) - H(t)C(t)$, respectively. Then invoking the result of Exercise 4.12 we have

$$\Phi(t,\ \tau) = \begin{bmatrix} \Phi_x(t,\ \tau) & -\int_\tau^t \Phi_x(t,\ \sigma)B(\sigma)K(\sigma)\Phi_e(\sigma,\ \tau)\ d\sigma \\[2ex] 0_n & \Phi_e(t,\ \tau) \end{bmatrix}$$

Writing $\Phi(t,\ \tau)$ as a sum of three matrices, each with one nonzero partition, the triangle inequality and Exercise 1.3 provide the inequality

$$\| \Phi(t,\ \tau) \| \le \| \Phi_x(t,\ \tau) \| + \| \Phi_e(t,\ \tau) \|$$

$$+ \| \int_\tau^t \Phi_x(t,\ \sigma)B(\sigma)K(\sigma)\Phi_e(\sigma,\ \tau)\ d\sigma \| \qquad (17)$$

Now given $\alpha > 0$ and any (presumably small) $\eta > 0$, the feedback and observer gains in (14) are such that there is a constant γ for which

$$\| \Phi_x(t,\ \tau) \|,\ \ \| \Phi_e(t,\ \tau) \| \le \gamma e^{-(\alpha+\eta)(t-\tau)}$$

for all t, τ with $t \ge \tau$. (Theorems 14.7 and 15.2.) Then

$$\| \int_\tau^t \Phi_x(t,\ \sigma)B(\sigma)K(\sigma)\Phi_e(\sigma,\ \tau)\ d\sigma \| \le \gamma^2 e^{-(\alpha+\eta)(t-\tau)} \int_\tau^t \| B(\sigma) \|\ \| K(\sigma) \|\ d\sigma$$

Using an inequality established in the proof of Theorem 14.7,

$$\| K(\sigma) \| \le \| B^T(\sigma) \|\ \| W_{\alpha+\eta}^{-1}(\sigma,\ \sigma+\delta) \| \le \frac{e^{4(\alpha+\eta)\delta}}{2\varepsilon_1} \| B(\sigma) \|$$

Thus for all t, τ with $t \ge \tau$,

$$\| \int_\tau^t \Phi_x(t,\ \sigma)B(\sigma)K(\sigma)\Phi_e(\sigma,\ \tau)\ d\sigma \| \le \frac{\gamma^2 e^{4(\alpha+\eta)\delta}}{2\varepsilon_1} e^{-(\alpha+\eta)(t-\tau)} \int_\tau^t \| B(\sigma) \|^2\ d\sigma$$

$$\le \frac{\gamma^2 e^{4(\alpha+\eta)\delta}}{2\varepsilon_1} e^{-(\alpha+\eta)(t-\tau)}[\ \beta_1 + \beta_2(t-\tau)] \quad (18)$$

Finally using the elementary bound (for $\eta > 0$)

$$te^{-\eta t} \le \frac{1}{\eta e},\quad t \ge 0$$

in (18) gives, for (17),

$$\| \Phi(t,\ \tau) \| \le \left[2\gamma + \frac{\gamma^2 e^{4(\alpha+\eta)\delta}}{2\varepsilon_1} \left(\beta_1 + \frac{\beta_2}{\eta e} \right) \right] e^{-\alpha(t-\tau)}$$

for all t, τ with $t \ge \tau$, and the proof is complete.

Time-Invariant Case

When specialized to the case of a time-invariant linear state equation,

$$\dot{x}(t) = Ax(t) + Bu(t) , \quad x(0) = x_o$$

$$y(t) = Cx(t) \tag{19}$$

the state observation problem can be connected to the state feedback stabilization problem in a much simpler fashion. The form of the observer is, from (4),

$$\dot{\hat{x}}(t) = A\hat{x}(t) + Bu(t) + H\left[y(t) - \hat{y}(t) \right] , \quad \hat{x}(0) = \hat{x}_o$$

$$\hat{y}(t) = C\hat{x}(t) \tag{20}$$

and the error state equation is, from (5),

$$\dot{e}(t) = \left[A - HC \right] e(t) , \quad e(0) = x_o - \hat{x}_o \tag{21}$$

Now the problems of choosing H so that (21) is exponentially stable with prescribed rate, or so that $A - HC$ has a prescribed characteristic polynomial, can be recast in a form familiar from Chapter 14. Let

$$\tilde{A} = A^T , \quad \tilde{B} = C^T , \quad \tilde{K} = -H^T$$

Then the characteristic polynomial of $A - HC$ is identical to the characteristic polynomial of

$$[A - HC]^T = \tilde{A} + \tilde{B}\tilde{K}$$

Also observability of (19) is equivalent to the controllability assumption needed to apply either Theorem 14.8 on stabilization or Theorem 14.9 on eigenvalue assignment. Alternatively observer form in Chapter 13 can be used to prove that if *rank C = p* and (19) is observable, then H can be chosen to obtain any desired characteristic polynomial for the observer error state equation in (21). (See Exercise 15.4.)

Specialization of Theorem 15.4 on output feedback stabilization to the time-invariant case can be described in terms of eigenvalue assignment. Time-invariant linear feedback of the estimated state yields a *2n*-dimension closed-loop state equation that follows directly from (13):

$$\begin{bmatrix} \dot{x}(t) \\ \dot{\hat{x}}(t) \end{bmatrix} = \begin{bmatrix} A & BK \\ HC & A-HC+BK \end{bmatrix} \begin{bmatrix} x(t) \\ \hat{x}(t) \end{bmatrix} + \begin{bmatrix} BN \\ BN \end{bmatrix} r(t)$$

$$y(t) = \begin{bmatrix} C & 0_{p \times n} \end{bmatrix} \begin{bmatrix} x(t) \\ \hat{x}(t) \end{bmatrix} \tag{22}$$

The state variable change (15) shows that the characteristic polynomial for (22) is precisely the same as the characteristic polynomial for the linear state equation

$$\begin{bmatrix} \dot{x}(t) \\ \dot{e}(t) \end{bmatrix} = \begin{bmatrix} A+BK & -BK \\ 0_n & A-HC \end{bmatrix} \begin{bmatrix} x(t) \\ e(t) \end{bmatrix} + \begin{bmatrix} BN \\ 0 \end{bmatrix} r(t)$$

$$y(t) = \begin{bmatrix} C & 0_{p \times n} \end{bmatrix} \begin{bmatrix} x(t) \\ e(t) \end{bmatrix} \qquad (23)$$

Taking advantage of block triangular structure, this characteristic polynomial is

$$\det \begin{bmatrix} \lambda I - A - BK \end{bmatrix} \det \begin{bmatrix} \lambda I - A + HC \end{bmatrix}$$

By this calculation we have uncovered a remarkable *eigenvalue separation property*. The $2n$ eigenvalues of the closed-loop state equation (22) are given by the n eigenvalues of the observer, and the n eigenvalues that would be obtained by linear state feedback (instead of linear estimated-state feedback). Of course if (19) is controllable and observable, then K and H can be chosen such that the characteristic polynomial for (22) is any specified monic, degree-$2n$ polynomial.

Another property of the closed-loop state equation that is equally remarkable concerns the input-output behavior. The transfer function for (22) is identical to the transfer function for (23), and a quick calculation, again making use of the block-triangular structure in (23), shows that this transfer function is

$$G(s) = C[sI - A - BK]^{-1}BN \qquad (24)$$

That is, linear estimated-state feedback leads to the same input-output behavior as does linear state feedback.

Reduced Dimension Observers

The discussion of state observers so far has ignored information about the state of the plant that is provided directly by the plant output signal. For example if plant output components are state components — that is, if each row of C has a single nonzero entry — why estimate what is available? We should be able to make use of this information, and construct an observer only for states that are not directly known from the output.

We will concentrate on the time-invariant case, leaving the time-varying case to Note 15.1. Assuming *rank* $C = p$, a state variable change can be employed that leads to the development of a *reduced-dimension observer* that has dimension $n-p$.

For the plant (19) define an $n \times n$ matrix P by

$$P^{-1} = \begin{bmatrix} C \\ P_b \end{bmatrix} \qquad (25)$$

where P_b is an $(n-p) \times n$ matrix that is arbitrary, subject to the requirement that P is invertible. Then letting $z(t) = P^{-1}x(t)$, the plant in the new state variables can be written in the partitioned form

$$\begin{bmatrix} \dot{z}_a(t) \\ \dot{z}_b(t) \end{bmatrix} = \begin{bmatrix} F_{11} & F_{12} \\ F_{21} & F_{22} \end{bmatrix} \begin{bmatrix} z_a(t) \\ z_b(t) \end{bmatrix} + \begin{bmatrix} G_1 \\ G_2 \end{bmatrix} u(t), \qquad \begin{bmatrix} z_a(0) \\ z_b(0) \end{bmatrix} = P^{-1} x_o$$

$$y(t) = \begin{bmatrix} I_p & 0_{p \times (n-p)} \end{bmatrix} \begin{bmatrix} z_a(t) \\ z_b(t) \end{bmatrix} \tag{26}$$

where F_{11} is $p \times p$, G_1 is $p \times m$, $z_a(t)$ is $p \times 1$, and the remaining partitions have corresponding dimensions. Obviously $z_a(t) = y(t)$, and the following argument shows that an asymptotic estimate of $z_b(t)$ is all that is needed to obtain an asymptotic estimate of $x(t)$.

Suppose for the moment that we have computed an $(n-p)$-dimensional observer for $z_b(t)$ that has the form

$$\dot{z}_c(t) = \tilde{F} z_c(t) + \tilde{G}_a u(t) + \tilde{G}_b z_a(t)$$

$$\hat{z}_b(t) = z_c(t) + H z_a(t) \tag{27}$$

That is, for known $u(t)$ and $z_a(t)$, but regardless of the initial values $z_b(0)$ and $z_c(0)$, the output of (27) satisfies

$$\lim_{t \to \infty} \left[z_b(t) - \hat{z}_b(t) \right] = 0$$

Then an asymptotic estimate for the state vector in (26), the first p components of which are in fact perfect estimates, can be written in the form

$$\begin{bmatrix} \hat{z}_a(t) \\ \hat{z}_b(t) \end{bmatrix} = \begin{bmatrix} I_p & 0_{p \times (n-p)} \\ H & I_{n-p} \end{bmatrix} \begin{bmatrix} z_a(t) \\ z_c(t) \end{bmatrix} \tag{28}$$

Correspondingly this provides an asymptotic estimate for $x(t)$, the original plant state, via

$$\hat{x}(t) = P \begin{bmatrix} I_p & 0_{p \times (n-p)} \\ H & I_{n-p} \end{bmatrix} \begin{bmatrix} y(t) \\ z_c(t) \end{bmatrix} \tag{29}$$

This variable-change argument shows that we need only consider the problem of computing an $(n-p)$-dimensional observer of the form (27) for an n-dimensional state equation in the special form (26). Of course the focus in this problem is on the $(n-p) \times 1$ error signal

$$e_b(t) = z_b(t) - \hat{z}_b(t)$$

that satisfies the error state equation

$$\dot{e}_b(t) = \dot{z}_b(t) - \dot{\hat{z}}_b(t)$$

$$= \dot{z}_b(t) - \dot{z}_c(t) - H \dot{z}_a(t)$$

$$= F_{21}z_a(t) + F_{22}z_b(t) + G_2u(t) - \widetilde{F}z_c(t)$$

$$- \widetilde{G}_au(t) - \widetilde{G}_bz_a(t) - HF_{11}z_a(t) - HF_{12}z_b(t) - HG_1u(t)$$

Using (27) to substitute for $z_c(t)$, and rearranging, gives

$$\dot{e}_b(t) = \widetilde{F}e_b(t) + \left[F_{22} - HF_{12} - \widetilde{F} \right]z_b(t) + \left[F_{21} + \widetilde{F}H - \widetilde{G}_b - HF_{11} \right]z_a(t)$$

$$+ \left[G_2 - \widetilde{G}_a - HG_1 \right]u(t) , \quad e_b(0) = z_b(0) - \hat{z}_b(0)$$

Again a reasonable requirement on the observer is that, regardless of $u(t)$, $z_a(0)$, and the resulting $z_a(t)$, the lucky occurrence $\hat{z}_b(0) = z_b(0)$ should yield $e_b(t) = 0$ for all $t \geq 0$. This objective is satisfied by the coefficient choices

$$\widetilde{F} = F_{22} - HF_{12}$$

$$\widetilde{G}_b = F_{21} + \widetilde{F}H - HF_{11}$$

$$\widetilde{G}_a = G_2 - HG_1 \tag{30}$$

with the resulting error state equation

$$\dot{e}_b(t) = \left[F_{22} - HF_{12} \right]e_b(t) , \quad e_b(0) = z_b(0) - \hat{z}_b(0) \tag{31}$$

To complete the specification of the reduced-dimension observer in (27), we need to show that the $(n-p) \times p$ gain matrix H can be chosen to yield any desired characteristic polynomial for (31). This is carried out in the proof of the following summary statement of the reduced-dimension observer development.

15.5 Theorem   Suppose the time-invariant linear state equation (19) is is observable and *rank C* $= p$. Given any degree-$(n-p)$ monic polynomial $q(\lambda)$ there is a gain H such that the reduced-dimension observer defined by (27), (29), and (30) has an error state equation (31) with characteristic polynomial $q(\lambda)$.

Proof   We need to show H can be chosen such that

$$\det (\lambda I - F_{22} + HF_{12}) = q(\lambda)$$

This can be done by proving that the observability hypothesis on (19) implies that the $(n-p)$-dimensional state equation

$$\dot{z}_d(t) = F_{22}z_d(t)$$

$$v(t) = F_{12}z_d(t) \tag{32}$$

is observable. Supposing the contrary, a contradiction is obtained as follows. If (32) is not observable, then by Theorem 13.14 there exists a nonzero $(n-p) \times 1$ vector l and a scalar η such that

$$F_{22}l = \eta l , \quad F_{12}l = 0$$

This implies, using the coefficients of (26),

$$
\begin{bmatrix} 0_{p \times 1} \\ l \end{bmatrix} \neq 0 \,, \quad
\begin{bmatrix} F_{11} & F_{12} \\ F_{21} & F_{22} \end{bmatrix}
\begin{bmatrix} 0_{p \times 1} \\ l \end{bmatrix} =
\begin{bmatrix} F_{12}l \\ F_{22}l \end{bmatrix} = \eta
\begin{bmatrix} 0_{p \times 1} \\ l \end{bmatrix}
$$

and, of course,

$$
\begin{bmatrix} I_p & 0 \end{bmatrix}
\begin{bmatrix} 0_{p \times 1} \\ l \end{bmatrix} = 0
$$

Therefore another application of Theorem 13.14 shows that the linear state equation (26) is not observable. But (26) is related to (19) by a state variable change, and thus a contradiction with the observability hypothesis for (19) is obtained.

A Servomechanism Problem

As another illustration of state observation and estimated-state feedback, we consider a plant affected by disturbances and pose multiple objectives for the closed-loop state equation. Specifically consider a plant described by a linear state equation of the form

$$
\dot{x}(t) = Ax(t) + Bu(t) + Ew(t) \,, \quad x(0) = x_o
$$
$$
y(t) = Cx(t) + Fw(t) \tag{33}
$$

We assume that $w(t)$ is a $q \times 1$ disturbance signal that is unavailable for use in feedback, and for simplicity we assume $p = m$. Using output feedback, the objectives for the closed-loop state equation are that the output signal should track any constant reference input with asymptotically-zero error in the face of unknown constant disturbance signals, and that the coefficients of the characteristic polynomial should be arbitrarily assignable. This type of problem often is called a *servomechanism problem*.

The basic idea in addressing this problem is to use an observer to generate asymptotic estimates of both the plant state and the constant disturbance. As in earlier observer constructions it is not apparent at the outset how to do this, but writing the plant (33) together with the constant disturbance $w(t)$ in the form of an 'augmented' plant provides the key. Namely we write the constant disturbance as the linear state equaton $\dot{w}(t) = 0$ so that

$$
\begin{bmatrix} \dot{x}(t) \\ \dot{w}(t) \end{bmatrix} =
\begin{bmatrix} A & E \\ 0 & 0 \end{bmatrix}
\begin{bmatrix} x(t) \\ w(t) \end{bmatrix} +
\begin{bmatrix} B \\ 0 \end{bmatrix} u(t)
$$
$$
y(t) = \begin{bmatrix} C & F \end{bmatrix}
\begin{bmatrix} x(t) \\ w(t) \end{bmatrix} \tag{34}
$$

and then adapt the observer structure suggested in (20) to this $(n+q)$-dimensional linear state equation. With the observer gain partitioned appropriately, this leads to the observer state equation

$$\begin{bmatrix} \dot{\hat{x}}(t) \\ \dot{\hat{w}}(t) \end{bmatrix} = \begin{bmatrix} A & E \\ 0 & 0 \end{bmatrix} \begin{bmatrix} \hat{x}(t) \\ \hat{w}(t) \end{bmatrix} + \begin{bmatrix} B \\ 0 \end{bmatrix} u(t) + \begin{bmatrix} H_1 \\ H_2 \end{bmatrix} \begin{bmatrix} y(t) - \hat{y}(t) \end{bmatrix}$$

$$\hat{y}(t) = \begin{bmatrix} C & F \end{bmatrix} \begin{bmatrix} \hat{x}(t) \\ \hat{w}(t) \end{bmatrix} \tag{35}$$

Since

$$\begin{bmatrix} A & E \\ 0 & 0 \end{bmatrix} - \begin{bmatrix} H_1 \\ H_2 \end{bmatrix} \begin{bmatrix} C & F \end{bmatrix} = \begin{bmatrix} A-H_1C & E-H_1F \\ -H_2C & -H_2F \end{bmatrix} \tag{36}$$

the error equation, in the obvious notation, satisfies

$$\begin{bmatrix} \dot{e}_x(t) \\ \dot{e}_w(t) \end{bmatrix} = \begin{bmatrix} A-H_1C & E-H_1F \\ -H_2C & -H_2F \end{bmatrix} \begin{bmatrix} e_x(t) \\ e_w(t) \end{bmatrix}$$

Now consider linear feedback of the form

$$u(t) = K_1\hat{x}(t) + K_2\hat{w}(t) + Nr(t) \tag{37}$$

The corresponding closed-loop state equation can be written as

$$\begin{bmatrix} \dot{x}(t) \\ \dot{\hat{x}}(t) \\ \dot{\hat{w}}(t) \end{bmatrix} = \begin{bmatrix} A & BK_1 & BK_2 \\ H_1C & A+BK_1-H_1C & E+BK_2-H_1F \\ H_2C & -H_2C & -H_2F \end{bmatrix} \begin{bmatrix} x(t) \\ \hat{x}(t) \\ \hat{w}(t) \end{bmatrix}$$

$$+ \begin{bmatrix} BN \\ BN \\ 0_{q \times m} \end{bmatrix} r(t) + \begin{bmatrix} E \\ H_1F \\ H_2F \end{bmatrix} w(t)$$

$$y(t) = \begin{bmatrix} C & 0 & 0 \end{bmatrix} \begin{bmatrix} x(t) \\ \hat{x}(t) \\ \hat{w}(t) \end{bmatrix} + Fw(t) \tag{38}$$

It is convenient to use the state-estimate error variable, and to change the sign of the disturbance estimate to simplify the analysis of this complicated linear state equation. With the state variable change

$$\begin{bmatrix} x(t) \\ e_x(t) \\ -\hat{w}(t) \end{bmatrix} = \begin{bmatrix} I_n & 0_n & 0_{n \times q} \\ I_n & -I_n & 0_{n \times q} \\ 0_{q \times n} & 0_{q \times n} & -I_q \end{bmatrix} \begin{bmatrix} x(t) \\ \hat{x}(t) \\ \hat{w}(t) \end{bmatrix}$$

the closed-loop state equation can be written as

$$
\begin{bmatrix} \dot{x}(t) \\ \dot{e}_x(t) \\ -\dot{\hat{w}}(t) \end{bmatrix} = \begin{bmatrix} A+BK_1 & -BK_1 & -BK_2 \\ 0 & A-H_1C & E-H_1F \\ 0 & -H_2C & -H_2F \end{bmatrix} \begin{bmatrix} x(t) \\ e_x(t) \\ -\hat{w}(t) \end{bmatrix}
$$

$$
+ \begin{bmatrix} BN \\ 0 \\ 0 \end{bmatrix} r(t) + \begin{bmatrix} E \\ E-H_1F \\ -H_2F \end{bmatrix} w(t)
$$

$$
y(t) = \begin{bmatrix} C & 0 & 0 \end{bmatrix} \begin{bmatrix} x(t) \\ e_x(t) \\ -\hat{w}(t) \end{bmatrix} + Fw(t) \tag{39}
$$

The characteristic polynomial of (39) is identical to the characteristic polynomial of (38). Because of the block-triangular structure of (39), it is clear that the closed-loop characteristic polynomial coefficients depend only on the choice of gains K_1, H_1, and H_2.

Assuming for the moment that (39) is exponentially stable, we can address the choice of gains N and K_2 to achieve the input-output objectives of asymptotic tracking and disturbance rejection. A careful partitioned multiplication verifies that

$$
\left(sI_{2n+q} - \begin{bmatrix} A+BK_1 & -BK_1 & -BK_2 \\ 0 & A-H_1C & E-H_1F \\ 0 & -H_2C & -H_2F \end{bmatrix} \right)^{-1} =
$$

$$
\begin{bmatrix} (sI-A-BK_1)^{-1} & -(sI-A-BK_1)^{-1}[BK_1 \ \ BK_2]\begin{bmatrix} sI-A+H_1C & -E+H_1F \\ H_2C & sI+H_2F \end{bmatrix}^{-1} \\ 0 & \begin{bmatrix} sI-A+H_1C & -E+H_1F \\ H_2C & sI+H_2F \end{bmatrix}^{-1} \end{bmatrix}
$$

and another gives

$$
Y(s) = C(sI-A-BK_1)^{-1}BN\mathsf{R}(s) + C(sI-A-BK_1)^{-1}E\mathsf{W}(s)
$$

$$
- \begin{bmatrix} C(sI-A-BK_1)^{-1}BK_1 & C(sI-A-BK_1)^{-1}BK_2 \end{bmatrix}
$$

$$
\cdot \begin{bmatrix} sI-A+H_1C & -E+H_1F \\ H_2C & sI+H_2F \end{bmatrix}^{-1} \begin{bmatrix} E-H_1F \\ -H_2F \end{bmatrix} \mathsf{W}(s) + F\mathsf{W}(s) \tag{40}
$$

Constant reference and disturbance inputs correspond to

$$
\mathsf{R}(s) = r_0 \frac{1}{s}, \quad \mathsf{W}(s) = w_0 \frac{1}{s}
$$

and the only terms in (40) that contribute to the asymptotic value of $y(t)$ are those partial-fraction-expansion terms for $\mathbf{Y}(s)$ corresponding to denominator roots at $s = 0$. Computing the coefficients of such terms using

$$\begin{bmatrix} -A+H_1C & -E+H_1F \\ H_2C & H_2F \end{bmatrix}^{-1} \begin{bmatrix} E-H_1F \\ -H_2F \end{bmatrix} = \begin{bmatrix} 0 \\ -I_q \end{bmatrix}$$

gives

$$\lim_{t \to \infty} y(t) = -C(A+BK_1)^{-1}BN r_o$$

$$+ \left[-C(A+BK_1)^{-1}E - C(A+BK_1)^{-1}BK_2 + F \right] w_o \qquad (41)$$

Alternatively the Final Value theorem for Laplace transforms can be used to obtain the same result.

At this point we are prepared to establish the eigenvalue assignment property using (38), and the tracking and disturbance rejection property using (41).

15.6 Theorem   Suppose the plant (33) is controllable for $E = 0$, the augmented plant (34) is observable, and the $(n+m) \times (n+m)$ matrix

$$\begin{bmatrix} A & B \\ C & 0 \end{bmatrix} \qquad (42)$$

is invertible. Then linear dynamic output feedback of the form (37), (35) has the following properties. The gains K_1, H_1, and H_2 can be chosen such that the closed-loop state equation (38) is exponentially stable with any desired characteristic polynomial coefficients. Furthermore the gains

$$N = -\left[C(A+BK_1)^{-1}B \right]^{-1}$$

$$K_2 = NC(A+BK_1)^{-1}E - NF \qquad (43)$$

are such that for any constant reference input $r(t) = r_o$ and constant disturbance $w(t) = w_o$, the response of the closed-loop state equation satisfies

$$\lim_{t \to \infty} y(t) = r_o \qquad (44)$$

Proof   By the observability assumption in conjunction with (36), and the controllability assumption in conjunction with $A + BK_1$, we know from previous results that K_1, H_1, and H_2 can be chosen to achieve any specified degree-$2n$ characteristic polynomial for (39), and thus for (38). Then Exercise 2.8 can be applied to conclude, under the invertibility condition on (42), that $C(A+BK_1)^{-1}B$ is invertible. Therefore the gains N and K_2 in (43) are well defined, and substituting (43) into (41) gives (44).

EXERCISES

Exercise 15.1   Suppose the time-invariant linear state equation

$$\dot{x}(t) = Ax(t) + Bu(t)$$

$$y(t) = Cx(t)$$

is controllable and observable, and *rank* $B = m$. Given an $(n-m) \times (n-m)$ matrix F and an $n \times p$ matrix H, consider dynamic output feedback

$$\dot{z}(t) = Fz(t) + Gv(t)$$

$$v(t) = y(t) + CLz(t)$$

$$u(t) = Mz(t) + Nv(t)$$

where the matrices G, L, M, and N satisfy

$$AL - BM = LF$$

$$LG + BN = -H$$

Show that the $2n-m$ eigenvalues of the closed-loop state equation are given by the eigenvalues of F and the eigenvalues of $A - HC$. *Hint*: Consider the coordinate change

$$\begin{bmatrix} w(t) \\ z(t) \end{bmatrix} = \begin{bmatrix} I & L \\ 0 & I \end{bmatrix} \begin{bmatrix} x(t) \\ z(t) \end{bmatrix}$$

Exercise 15.2   For the linear state equation

$$\dot{x}(t) = A(t)x(t)$$

$$y(t) = C(t)x(t)$$

show that if there exist positive constants γ, δ, ε_1, and ε_2 such that

$$\|A(t)\| \leq \gamma, \quad \varepsilon_1 I \leq M(t-\delta, t) \leq \varepsilon_2 I$$

for all t, then there exist positive constants ε_3 and ε_4 such that

$$\varepsilon_3 I \leq \Phi^T(t-\delta, t)M(t-\delta, t)\Phi(t-\delta, t) \leq \varepsilon_4 I$$

for all t. (*Hint*: See Exercise 6.2.)

Exercise 15.3   For the linear state equation

$$\dot{x}(t) = A(t)x(t) + B(t)u(t)$$

prove that if there exist positive constants γ, δ, ε_1, and ε_2 such that

$$\|A(t)\| \leq \gamma, \quad \varepsilon_1 I \leq W(t, t+\delta) \leq \varepsilon_2 I$$

for all t, then there exist positive constants β_1 and β_2 such that

$$\int_{\tau}^{t} \|B(\sigma)\|^2 \, d\sigma \leq \beta_1 + \beta_2(t-\tau)$$

for all t, τ with $t \geq \tau$. (*Hint*: Write

$$\int\limits_{\tau}^{\tau+\delta} \|B(\sigma)\|^2 \, d\sigma = \int\limits_{\tau}^{\tau+\delta} \|\Phi(\sigma, \tau)\Phi(\tau, \sigma)B(\sigma)B^T(\sigma)\Phi^T(\tau, \sigma)\Phi^T(\sigma, \tau)\|^2 \, d\sigma$$

bound this via Exercise 6.2, and Exercise 1.16, and add up the bounds over subintervals of $[\tau, t]$ of length δ.)

Exercise 15.4 Suppose the time-invariant linear state equation

$$\dot{x}(t) = Ax(t) + Bu(t)$$

$$y(t) = Cx(t)$$

is observable with *rank* $C = p$. Using a variable change to observer form, show how to compute an observer gain H such that characteristic polynomial $det\,(\lambda I - A + HC)$ has a specified set of coefficients.

Exercise 15.5 Suppose the time-invariant linear state equation

$$\dot{z}(t) = Az(t) + Bu(t)$$

$$y(t) = [\, I_p \quad 0_{p \times (n-p)} \,]z(t)$$

is controllable and observable. Consider an output feedback of the form

$$u(t) = K\hat{z}(t) + Nr(t)$$

where $\hat{z}(t)$ is an asymptotic state estimate generated via the reduced-dimension observer specified by (27), (28), and (30). Characterize the eigenvalues of the closed-loop state equation. What is the closed-loop transfer function?

Exercise 15.6 For the time-invariant linear state equation

$$\dot{x}(t) = Ax(t) + Bu(t)$$

$$y(t) = Cx(t)$$

show that there exists an $n \times p$ matrix H such that

$$\dot{x}(t) = (A + HC)x(t) + Bu(t)$$

$$y(t) = Cx(t)$$

is exponentially stable if and only if

$$\text{rank} \begin{bmatrix} C \\ \lambda I - A \end{bmatrix} = n$$

for each λ that is a nonnegative-real-part eigenvalue of A. (The property in question is called *detectability,* and the term *output injection* sometimes is used to describe how the second state equation is obtained from the first.)

NOTES

Note 15.1 Observer theory dates from the paper

D.G. Luenberger, "Observing the state of a linear system," *IEEE Transactions on Military Electronics,* Vol. 8, pp. 74–80, 1964 and an elementary review of early work is given in

D.G. Luenberger, "An introduction to observers," *IEEE Transactions on Automatic Control,* Vol. 16, No. 6, pp. 596–602, 1971

Observers for time-varying linear state equations, including reduced-dimension observers, are discussed in

Y.O. Yuksel, J.J. Bongiorno, "Observers for linear multivariable systems with applications," *IEEE Transactions on Automatic Control,* Vol. 16, No. 6, pp. 603–613, 1971

Note 15.2   Related to observability is the property of reconstructibility. Loosely speaking, an unforced linear state equation is *reconstructible on* $[t_o, t_f]$ if $x(t_f)$ can be determined from $y(t)$ for $t \in [t_o, t_f]$. This property is characterized by invertibility of the reconstructibility Gramian

$$N(t_o, t_f) = \int_{t_o}^{t_f} \Phi^T(\tau, t_f) C^T(\tau) C(\tau) \Phi(\tau, t_f)\, d\tau$$

The relation between this and the observability Gramian is

$$N(t_o, t_f) = \Phi^T(t_o, t_f) M(t_o, t_f) \Phi(t_o, t_f)$$

and thus the 'observability' hypotheses of Theorem 15.2 and Theorem 15.4 can be replaced by the more compact expression

$$\varepsilon_1 I \le N(t - \delta, t) \le \varepsilon_2 I$$

Reconstructibility is discussed in Chapter 2 of

R.E. Kalman, P.L. Falb, M.A. Arbib, *Topics in Mathematical System Theory,* McGraw-Hill, New York, 1969

and Chapter 1 of

J. O'Reilly, *Observers for Linear Systems,* Academic Press, London, 1983

Note 15.3   The proof of output feedback stabilization in Theorem 15.4 is from

M. Ikeda, H. Maeda, S. Kodama, "Estimation and feedback in linear time-varying systems: a deterministic theory," *SIAM Journal on Control and Optimization,* Vol. 13, No. 2, pp. 304–327, 1975

This paper contains an extensive taxonomy of concepts related to state estimation, stabilization, and even 'instabilization.' An approach to output feedback stabilization via linear optimal control theory is in the paper by Yuksel and Bongiorno cited in Note 15.1.

Note 15.4   The problem of state observation is closely related to the problem of statistical estimation of the state based on output signals corrupted by noise, and the well-known *Kalman filter*. A gentle introduction is given in

B.D.O. Anderson, J.B. Moore, *Optimal Control — Linear Quadratic Methods,* Prentice Hall, New York, 1990

On the other hand, the Kalman filtering problem is reinterpreted as a deterministic optimization problem in Section 7.7 of

E.D. Sontag, *Mathematical Control Theory,* Springer-Verlag, New York, 1990

Note 15.5 The design of a state observer for a linear system driven by *unknown* input signals is an interesting problem. For one approach, and references to earlier treatments, see

F. Yang, R.W. Wilde, "Observers for linear systems with unknown inputs," *IEEE Transactions on Automatic Control,* Vol. 33, No. 7, pp. 677–681, 1988

Note 15.6 The construction of an observer that provides asymptotically-zero error depends crucially on choosing observer coefficients in terms of plant coefficients. This is easily recognized in the process of deriving the observer error state equation (5). The behavior of the observer error when observer coefficients are mismatched with plant coefficients, and remedies for this situation, are subjects in *robust* observer theory. Consult

J.C. Doyle, G. Stein, "Robustness with observers," *IEEE Transactions on Automatic Control,* Vol. 24, No. 4, pp. 607–611, 1979

S.P. Bhattacharyya, "The structure of robust observers," *IEEE Transactions on Automatic Control,* Vol. 21, No. 4, pp. 581–588, 1976

K. Furuta, S. Hara, S. Mori, "A class of systems with the same observer," *IEEE Transactions on Automatic Control,* Vol. 21, No. 4, pp. 572–576, 1976

Note 15.7 The servomechanism problem treated in Theorem 15.6 is based on

H.W. Smith, E.J. Davison, "Design of industrial regulators: integral feedback and feedforward control," *Proceedings of the IEE,* Vol. 119, pp. 1210–1216, 1972

Significant extensions and generalizations of this problem — using many different approaches — can be found in the control literature.

16

POLYNOMIAL FRACTION DESCRIPTION

The polynomial fraction description is a mathematically efficacious representation for a matrix of rational functions. Applied to the transfer function of a multi-input, multi-output linear state equation, polynomial fraction descriptions can reveal structural features that, for example, permit natural generalization of minimal realization considerations noted for single-input, single-output state equations in Example 10.12. Additional applications include the formulation of appropriate notions of poles and zeros, and the representation of feedback in terms of open-loop and closed-loop transfer functions. These applications are considered in Chapter 17, following development of the basic properties of polynomial fraction descriptions here.

We assume that $G(s)$ is a $p \times m$ matrix of strictly-proper rational functions of s. Then, from Theorem 10.11, $G(s)$ is realizable by a time-invariant linear state equation with $D = 0$. (Helvetica-font for Laplace transforms is not used in the sequel, since no conflicting time-domain notation arises.)

Right Polynomial Fractions

Matrices of real-coefficient polynomials in s, equivalently polynomials in s with coefficients that are real matrices, provide the mathematical foundation for the new transfer function representation

16.1 Definition   A $p \times r$ *polynomial matrix* $P(s)$ is a matrix with entries that are real-coefficient polynomials in s. A square ($p = r$) polynomial matrix $P(s)$ is called *nonsingular* if $det\, P(s)$ is a nonzero polynomial, and *unimodular* if $det\, P(s)$ is a nonzero real number.

The determinant of a square polynomial matrix is a polynomial (a sum of products of the polynomial entries). Thus an alternative characterization is that a square polynomial matrix $P(s)$ is nonsingular if and only if $det\ P(s_o) \neq 0$ for all but a finite number of complex numbers s_o. It is unimodular if and only if $det\ P(s_o) \neq 0$ for all complex numbers s_o.

The adjugate-over-determinant formula shows that if $P(s)$ is square and non-singular, then $P^{-1}(s)$ exists and (each entry) is a rational function of s. An easy consequence is that $P^{-1}(s)$ is a polynomial matrix if $P(s)$ is square and unimodular. And by the reciprocal-determinant relationship between a matrix and its inverse, $P^{-1}(s)$ is unimodular if $P(s)$ is unimodular. Conversely if $P(s)$ and $P^{-1}(s)$ both are polynomial matrices, then both are unimodular.

16.2 Definition   A *right polynomial fraction* description for the $p \times m$ strictly-proper rational transfer function $G(s)$ is an expression of the form

$$G(s) = N(s)D^{-1}(s) \tag{1}$$

where $N(s)$ is a $p \times m$ polynomial matrix, and $D(s)$ is an $m \times m$ nonsingular polynomial matrix. A *left polynomial fraction* description for $G(s)$ is an expression of the form

$$G(s) = D_L^{-1}(s)N_L(s) \tag{2}$$

where $N_L(s)$ is a $p \times m$ polynomial matrix, and $D_L(s)$ is a $p \times p$ nonsingular polynomial matrix. The *degree* of a right polynomial fraction description is the degree of the polynomial $det\ D(s)$. A similar definition applies for the degree of a left polynomial fraction.

Of course this definition is familiar if $m = p = 1$. In the multi-input, multi-output case, a simple device can be used to exhibit polynomial fraction descriptions for a given $G(s)$. Suppose $d(s)$ is a least common multiple of the denominator polynomials of entries of $G(s)$. (In fact, any common multiple of the denominators can be used.) Then

$$N_d(s) = d(s)G(s)$$

is a $p \times m$ polynomial matrix, and we can write either a right or left polynomial fraction description:

$$G(s) = N_d(s)[d(s)I_m]^{-1} = [d(s)I_p]^{-1}N_d(s) \tag{3}$$

The degrees of the two descriptions are different in general, and it should not be surprising that lower-degree polynomial fraction descriptions typically can be found if some effort is invested.

In the single-input, single-output case, the issue of common factors in the numerator and denominator polynomials of $G(s)$ arises at this point. The utility of the polynomial fraction representation begins to emerge from the corresponding concept in the matrix case.

16.3 Definition   An $r \times r$ polynomial matrix $R(s)$ is called a *right divisor* of the $p \times r$ polynomial matrix $P(s)$ if there exists a $p \times r$ polynomial matrix $\tilde{P}(s)$ such that

$$P(s) = \tilde{P}(s)R(s)$$

If a right divisor $R(s)$ is nonsingular, then $P(s)R^{-1}(s)$ is a $p \times r$ polynomial matrix. Also if $P(s)$ is square and nonsingular, then every right divisor of $P(s)$ is non-singular.

To become accustomed to these notions it helps to reflect on the case of scalar polynomials. There a right divisor is simply a factor of the polynomial. For polynomial matrices the situation is roughly similar.

16.4 Example   For the polynomial matrix

$$P(s) = \begin{bmatrix} (s+1)^2(s+2) \\ (s+1)(s+2)(s+3) \end{bmatrix} \tag{4}$$

right divisors include the 1×1 polynomial matrices

$$R_a(s) = 1 , \quad R_b(s) = s+1 , \quad R_c(s) = s+2 , \quad R_d(s) = (s+1)(s+2)$$

In this simple case each right divisor is a common factor of the two scalar polynomials in $P(s)$, and $R_d(s)$ is a greatest-degree common factor of the scalar polynomials. For the slightly less-simple

$$P(s) = \begin{bmatrix} (s+1)^2(s+2) & (s+3)(s+5) \\ 0 & (s+4)(s+5) \end{bmatrix}$$

two right divisors are

$$\begin{bmatrix} (s+1) & 0 \\ 0 & s+5 \end{bmatrix}, \quad \begin{bmatrix} (s+1)^2 & 0 \\ 0 & s+5 \end{bmatrix}$$

□□□

Next we consider the matrix-polynomial extension of the concept of common factors of two scalar polynomials. Since one of the polynomial matrices always is square in our application to transfer function representation, attention is restricted to that situation.

16.5 Definition   Suppose $P(s)$ is a $p \times r$ polynomial matrix, and $Q(s)$ is a $r \times r$ polynomial matrix. If the $r \times r$ polynomial matrix $R(s)$ is a right divisor of both, then $R(s)$ is called a *common right divisor* of $P(s)$ and $Q(s)$. We call $R(s)$ a *greatest common right divisor* of $P(s)$ and $Q(s)$ if it is a common right divisor, and if any other common right divisor of $P(s)$ and $Q(s)$ is a right divisor of $R(s)$. If all common right divisors of $P(s)$ and $Q(s)$ are unimodular, then $P(s)$ and $Q(s)$ are called *right coprime*.

For polynomial fraction descriptions of a transfer function, one of the polynomial matrices always is nonsingular, so only nonsingular common right divisors occur. Suppose $G(s)$ is given by the right polynomial fraction description

$$G(s) = N(s)D^{-1}(s)$$

and that $R(s)$ is a common right divisor of $N(s)$ and $D(s)$. Then

$$\widetilde{N}(s) = N(s)R^{-1}(s) , \quad \widetilde{D}(s) = D(s)R^{-1}(s) \tag{5}$$

are polynomial matrices, and they provide another right polynomial fraction description for $G(s)$ since

$$\widetilde{N}(s)\widetilde{D}^{-1}(s) = N(s)R^{-1}(s) \, R(s)D^{-1}(s) = G(s)$$

The degree of this new polynomial fraction description is no greater than the degree of the original, since

$$\deg [\det D(s)] = \deg [\det \widetilde{D}(s)] + \deg [\det R(s)]$$

Of course the largest degree reduction occurs if $R(s)$ is a greatest common right divisor, and no reduction occurs if $N(s)$ and $D(s)$ are right coprime. This discussion indicates that extracting common right divisors of a right polynomial fraction is a generalization of the process of canceling common factors in a scalar transfer function.

The computation of greatest common right divisors can be based on capabilities of elementary row operations on a polynomial matrix — operations similar to elementary row operations on a matrix of real numbers. To set up this approach we present a preliminary result.

16.6 Theorem   Suppose $P(s)$ is a $p \times r$ polynomial matrix, and $Q(s)$ is a $r \times r$ polynomial matrix. If a unimodular $(p+r) \times (p+r)$ polynomial matrix $U(s)$ and a $r \times r$ polynomial matrix $R(s)$ are such that

$$U(s) \begin{bmatrix} Q(s) \\ P(s) \end{bmatrix} = \begin{bmatrix} R(s) \\ 0 \end{bmatrix} \tag{6}$$

then $R(s)$ is a greatest common right divisor of $P(s)$ and $Q(s)$.

Proof   Partition $U(s)$ in the form

$$U(s) = \begin{bmatrix} U_{11}(s) & U_{12}(s) \\ U_{21}(s) & U_{22}(s) \end{bmatrix} \tag{7}$$

where $U_{11}(s)$ is $r \times r$, and $U_{22}(s)$ is $p \times p$. Then the polynomial matrix $U^{-1}(s)$ can be partitioned similarly as

$$U^{-1}(s) = \begin{bmatrix} U_{\overline{11}}(s) & U_{\overline{12}}(s) \\ U_{\overline{21}}(s) & U_{\overline{22}}(s) \end{bmatrix}$$

Using this notation to rewrite (6) gives

$$\begin{bmatrix} Q(s) \\ P(s) \end{bmatrix} = \begin{bmatrix} U_{\overline{11}}(s) & U_{\overline{12}}(s) \\ U_{\overline{21}}(s) & U_{\overline{22}}(s) \end{bmatrix} \begin{bmatrix} R(s) \\ 0 \end{bmatrix}$$

That is,

$$Q(s) = U_{11}^{-}(s)R(s) , \quad P(s) = U_{21}^{-}(s)R(s)$$

Therefore $R(s)$ is a common right divisor of $P(s)$ and $Q(s)$. But, from (6) and (7),

$$R(s) = U_{11}(s)Q(s) + U_{12}(s)P(s) \tag{8}$$

so that if $R_a(s)$ is another common right divisor of $P(s)$ and $Q(s)$, say

$$Q(s) = \tilde{Q}_a(s)R_a(s) , \quad P(s) = \tilde{P}_a(s)R_a(s)$$

then (8) gives

$$R(s) = \left[U_{11}(s)\tilde{Q}_a(s) + U_{12}(s)\tilde{P}_a(s) \right] R_a(s)$$

That is, $R_a(s)$ also is a right divisor of $R(s)$, and thus $R(s)$ is a greatest common right divisor of $P(s)$ and $Q(s)$.
□□□

To calculate greatest common right divisors using Theorem 16.6, we consider three types of *elementary row operations* on a polynomial matrix. First is the interchange of two rows, and second is the multiplication of a row by a nonzero real number. The third is to add to any row a polynomial multiple of another row. Each of these elementary row operations can be represented by premultiplication by a unimodular matrix, as is easily seen by filling in the following argument.

Interchange of rows i and $j \neq i$ corresponds to premultiplying by a matrix E_a that has a very simple form. The diagonal entries are unity, except that $[E_a]_{ii} = [E_a]_{jj} = 0$, and the off-diagonal entries are zero, except that $[E_a]_{ij} = [E_a]_{ji} = 1$. Multiplication of the ith-row by a real number $\alpha \neq 0$ corresponds to premultiplication by a matrix E_b that is diagonal with all diagonal entries unity, except $[E_b]_{ii} = \alpha$. Finally adding to row i a polynomial $p(s)$ times row j, $j \neq i$, corresponds to premultiplication by a matrix $E_c(s)$ that has unity diagonal entries, with off-diagonal entries zero, except $[E_c]_{ij}(s) = p(s)$.

It is straightforward to show that the determinants of matrices of the form E_a, E_b, and $E_c(s)$ described above are nonzero real numbers, that is, the matrices are unimodular. Also it is easy to show that the inverse of any of these matrices corresponds to another elementary row operation. The diligent might prove that multiplication of a row by a polynomial is *not* an elementary row operation, thereby burying a frequent misconception.

At this point it is clear that a sequence of elementary row operations can be represented as premultiplication by a sequence of these elementary unimodular matrices, and thus as a single unimodular premultiplication. We also want to show the converse — that premultiplication by any unimodular matrix can be represented by a sequence of elementary row operations. Then Theorem 16.6 provides a method based on elementary row operations for computing a greatest common right divisor $R(s)$ via (6).

That any unimodular matrix can be written as a product of matrices of the form E_a, E_b, and $E_c(s)$ derives easily from a special form for polynomial matrices. We present this special form for the particular case where the polynomial matrix contains a nonsingular partition. This suffices for our application to polynomial fraction descriptions, and also avoids some fussy but trivial issues such as how to handle identical columns, or all-zero columns. Recall the terminology that a scalar polynomial is called *monic* if the coefficient of the highest power of s is unity, that the *degree* of a polynomial is the highest power of s with nonzero coefficient, and that the degree of the zero polynomial is, by convention, $-\infty$.

16.7 Theorem Suppose $P(s)$ is a $p \times r$ polynomial matrix, and $Q(s)$ is a $r \times r$, nonsingular polynomial matrix. Then elementary row operations can be used to transform

$$M(s) = \begin{bmatrix} Q(s) \\ P(s) \end{bmatrix} \tag{9}$$

into *row Hermite form* described as follows. For $k = 1, \ldots, r$, all entries of the kth-column below the k,k-entry are zero, and the k,k-entry is nonzero and monic with higher degree than every entry above it in column k. (If the k,k-entry is unity, then all entries above it are zero.)

Proof Row Hermite form can be obtained via a simple algorithm that is similar to the row reduction process for constant matrices.

Step *(i)*: In the first column of $M(s)$ use row interchange to bring to the first row a lowest-degree entry among nonzero first-column entries. (By nonsingularity of $Q(s)$, there is a nonzero first-column entry.)

Step *(ii)*: Multiply the first row by a real number so that the first column entry is monic.

Step *(iii)*: For each entry $m_{i1}(s)$ below the first row in the first column, use polynomial division to write

$$m_{i1}(s) = q_i(s)m_{11}(s) + r_{i1}(s), \quad i = 2, \ldots, p+r \tag{10}$$

where each remainder is such that $deg\ r_{i1}(s) < deg\ m_{11}(s)$. (If $m_{i1}(s) = 0$, that is $deg\ m_{i1}(s) = -\infty$, set $q_i(s) = r_{i1}(s) = 0$. If $deg\ m_{i1}(s) = 0$, then by Step *(i)* $deg\ m_{11}(s) = 0$. Therefore $deg\ q_i(s) = 0$ and $deg\ r_{i1} = -\infty$, that is, $r_{i1}(s) = 0$.)

Step *(iv)*: For $i = 2, \ldots, p+r$, add to the ith-row the product of $-q_i(s)$ and the first row. The resulting entries in the first column, below the first row, are $r_{21}(s), \ldots, r_{p+r,1}(s)$, all of which have degrees less than $deg\ m_{11}(s)$.

Step *(v)*: Repeat steps *(i)* through *(iv)* until all entries of the first column are zero except the first entry. Since the degrees of the entries below the first entry are lowered by at least one in each iteration, a finite number of operations is required.

Proceed to the second column of $M(s)$ and repeat the above steps while ignoring the first row. This results in a monic, nonzero entry $m_{22}(s)$, with all entries below it zero. If $m_{12}(s)$ does not have lower degree than $m_{22}(s)$, then polynomial division of $m_{12}(s)$ by $m_{22}(s)$ as in Step *(iii)*, and an elementary row operation as in Step *(iv)* replaces $m_{12}(s)$ by a polynomial of degree less than $\deg m_{22}(s)$. Next repeat the process for the third column of $M(s)$, while ignoring the first two rows. Continuing yields the claimed form on exhausting the columns of $M(s)$.
□□□

To complete the connection between unimodular matrices and elementary row operations, suppose in Theorem 16.7 that $p = 0$, and $Q(s)$ is unimodular. Of course the resulting row Hermite form is upper triangular. The diagonal entries must be unity, for a diagonal entry of positive degree would yield a determinant of positive degree, contradicting unimodularity. But then entries above the diagonal must have degree $-\infty$. Thus the row Hermite form for a unimodular matrix is the identity matrix. In other words for a unimodular polynomial matrix $U(s)$ there is a sequence of elementary row operations, say E_a, E_b, $E_c(s)$, $\ldots$, E_b, such that

$$\left[E_a\, E_b\, E_c(s)\ \cdots\ E_b \right] U(s) = I \tag{11}$$

But this obviously gives $U(s)$ as the sequence of elementary row operations on the identity specified by

$$U(s) = \left[E_b^{-1}\ \cdots\ E_c^{-1}(s)\, E_b^{-1}\, E_a^{-1} \right] I$$

and premultiplication of a matrix by $U(s)$ thus corresponds to application of a sequence of elementary row operations. Therefore Theorem 16.6 can be restated, for the case of nonsingular $Q(s)$, in terms of elementary row operations rather than premultiplication by a unimodular $U(s)$. If reduction to row Hermite form is used in implementing (6), then the greatest common right divisor $R(s)$ will be an upper-triangular polynomial matrix. Furthermore if $P(s)$ and $Q(s)$ are right coprime, then Theorem 16.7 shows that there is a unimodular $U(s)$ such that (6) is satisfied for $R(s) = I_r$.

16.8 Example   For

$$Q(s) = \begin{bmatrix} s^2+s+1 & s+1 \\ s^2-3 & 2s-2 \end{bmatrix}$$

$$P(s) = \begin{bmatrix} s+2 & 1 \end{bmatrix}$$

the calculation of a greatest common right divisor via Theorem 16.6 is a sequence of elementary row operations. (Each arrow represents one type of operation, and should be easy to decipher.)

$$M(s) = \begin{bmatrix} Q(s) \\ P(s) \end{bmatrix} = \begin{bmatrix} s^2+s+1 & s+1 \\ s^2-3 & 2s-2 \\ s+2 & 1 \end{bmatrix} \rightarrow \begin{bmatrix} s+2 & 1 \\ s^2-3 & 2s-2 \\ s^2+s+1 & s+1 \end{bmatrix}$$

$$\rightarrow \begin{bmatrix} s+2 & 1 \\ (s-2)(s+2)+1 & 2s-2 \\ (s-1)(s+2)+3 & s+1 \end{bmatrix} \rightarrow \begin{bmatrix} s+2 & 1 \\ 1 & s \\ 3 & 2 \end{bmatrix} \rightarrow \begin{bmatrix} 1 & s \\ s+2 & 1 \\ 3 & 2 \end{bmatrix}$$

$$\rightarrow \begin{bmatrix} 1 & s \\ 0 & -s^2-2s+1 \\ 0 & -3s+2 \end{bmatrix} \rightarrow \begin{bmatrix} 1 & s \\ 0 & -3s+2 \\ 0 & -s^2-2s+1 \end{bmatrix} \rightarrow \begin{bmatrix} 1 & s \\ 0 & s-2/3 \\ 0 & -s^2-2s+1 \end{bmatrix}$$

$$\rightarrow \begin{bmatrix} 1 & s \\ 0 & s-2/3 \\ 0 & -7/9 \end{bmatrix} \rightarrow \begin{bmatrix} 1 & s \\ 0 & 1 \\ 0 & s-2/3 \end{bmatrix} \rightarrow \begin{bmatrix} 1 & s \\ 0 & 1 \\ 0 & 0 \end{bmatrix} \rightarrow \begin{bmatrix} 1 & 0 \\ 0 & 1 \\ 0 & 0 \end{bmatrix}$$

This calculation shows that a greatest common right divisor is the identity, and $P(s)$ and $Q(s)$ are right coprime.
□□□

Two different characterizations of right coprimeness are used in the sequel.

16.9 Theorem   For a $p \times r$ polynomial matrix $P(s)$ and a nonsingular $r \times r$ polynomial matrix $Q(s)$, the following statements are equivalent.

(i) The polynomial matrices $P(s)$ and $Q(s)$ are right coprime.

(ii) There exist an $r \times p$ polynomial matrix $X(s)$ and a $r \times r$ polynomial matrix $Y(s)$ such that

$$X(s)P(s) + Y(s)Q(s) = I_r \tag{12}$$

(iii) For every complex number s_o,

$$\text{rank} \begin{bmatrix} Q(s_o) \\ P(s_o) \end{bmatrix} = r \tag{13}$$

Proof   Beginning a demonstration that each claim implies the next, first we show that *(i)* implies *(ii)*. If $P(s)$ and $Q(s)$ are right coprime, then reduction to row Hermite form as in (6) yields polynomial matrices $U_{11}(s)$ and $U_{12}(s)$ such that

$$U_{11}(s)Q(s) + U_{12}(s)P(s) = I_r$$

and this has the form of (12).

To prove that *(ii)* implies *(iii)*, write the condition (12) in the matrix form

$$\begin{bmatrix} Y(s) & X(s) \end{bmatrix} \begin{bmatrix} Q(s) \\ P(s) \end{bmatrix} = I_r$$

If s_o is a complex number for which

$$\text{rank} \begin{bmatrix} Q(s_o) \\ P(s_o) \end{bmatrix} < r$$

then we have a rank contradiction.

Finally to show *(iii)* implies *(i)*, suppose that (13) holds and $R(s)$ is a common right divisor of $P(s)$ and $Q(s)$. Then for some $p \times r$ polynomial matrix $\tilde{P}(s)$ and some $r \times r$ polynomial matrix $\tilde{Q}(s)$,

$$\begin{bmatrix} Q(s) \\ P(s) \end{bmatrix} = \begin{bmatrix} \tilde{Q}(s) \\ \tilde{P}(s) \end{bmatrix} R(s) \tag{14}$$

If $det\, R(s)$ is a polynomial of degree at least one, then with s_o a root of this polynomial we obtain the contradiction

$$\text{rank} \begin{bmatrix} Q(s_o) \\ P(s_o) \end{bmatrix} \leq \text{rank } R(s_o) < r$$

Therefore $det\, R(s)$ is a nonzero constant, that is, $R(s)$ is unimodular. This proves that $P(s)$ and $Q(s)$ are right coprime.
□□□

A right polynomial fraction description with $N(s)$ and $D(s)$ right coprime is called simply a *coprime right polynomial fraction description.* The next result shows that in an important sense all coprime right polynomial fraction descriptions of a given transfer function are equivalent. In particular they all have the same degree.

16.10 Theorem   Suppose we are given two coprime right polynomial fraction descriptions for a strictly-proper, rational transfer function:

$$G(s) = N(s)D^{-1}(s) = N_a(s)D_a^{-1}(s) \tag{15}$$

Then there exists a unimodular polynomial matrix $U(s)$ such that

$$N(s) = N_a(s)U(s) , \quad D(s) = D_a(s)U(s)$$

Proof   By Theorem 16.9 there exist polynomial matrices $X(s)$, $Y(s)$, $A(s)$, and $B(s)$ such that

$$X(s)N_a(s) + Y(s)D_a(s) = I_m \tag{16}$$

and

$$A(s)N(s) + B(s)D(s) = I_m$$

Since $N(s)D^{-1}(s) = N_a(s)D_a^{-1}(s)$, we have $N_a(s) = N(s)D^{-1}(s)D_a(s)$. Substituting this into (16) gives

$$X(s)N(s)D^{-1}(s)D_a(s) + Y(s)D_a(s) = I_m$$

or

$$X(s)N(s) + Y(s)D(s) = D_a^{-1}(s)D(s)$$

A similar calculation using $N(s) = N_a(s)D_a^{-1}(s)D(s)$ in (33) gives

$$A(s)N_a(s) + B(s)D_a(s) = D^{-1}(s)D_a(s)$$

Therefore both $D_a^{-1}(s)D(s)$ and $D^{-1}(s)D_a(s)$ are polynomial matrices, and since they are inverses of each other both must be unimodular. Let

$$U(s) = D_a^{-1}(s)D(s)$$

Then

$$N(s) = N_a(s)U(s) \ , \quad D(s) = D_a(s)D_a^{-1}(s)D(s) = D_a(s)U(s)$$

and the proof is complete.

Left Polynomial Fractions

Before going further we pause to consider left polynomial fraction descriptions and their relation to right polynomial fraction descriptions of the same transfer function. This means repeating much of the right-handed development, and proofs of the results are left as unlisted exercises.

16.11 Definition   A $q \times q$ polynomial matrix $L(s)$ is called a *left divisor* of the $q \times p$ polynomial matrix $P(s)$ if there exists a $q \times p$ polynomial matrix $\tilde{P}(s)$ such that

$$P(s) = L(s)\tilde{P}(s) \tag{17}$$

16.12 Definition   If $P(s)$ is a $q \times p$ polynomial matrix and $Q(s)$ is a $q \times q$ polynomial matrix, then a $q \times q$ polynomial matrix $L(s)$ is called a *common left divisor* of $P(s)$ and $Q(s)$ if $L(s)$ is a left divisor of both $P(s)$ and $Q(s)$. We call $L(s)$ a *greatest common left divisor* of $P(s)$ and $Q(s)$ if it is a common left divisor, and if any other common left divisor of $P(s)$ and $Q(s)$ is a left divisor of $L(s)$. If all common left divisors of $P(s)$ and $Q(s)$ are unimodular, then $P(s)$ and $Q(s)$ are called *left coprime*.

16.13 Example   Revisiting Example 16.4 from the other side exhibits the different look of right- and left-handed calculations. For

$$P(s) = \begin{bmatrix} (s+1)^2(s+2) \\ (s+1)(s+2)(s+3) \end{bmatrix} \tag{18}$$

one left divisor is

$$L(s) = \begin{bmatrix} (s+1)^2(s+2) & 0 \\ 0 & (s+1)(s+2)(s+3) \end{bmatrix}$$

where the corresponding 2×1 polynomial matrix $\widetilde{P}(s)$ has unity entries. In this simple case it should be clear how to write down many other left divisors.

16.14 Theorem   Suppose $P(s)$ is a $q \times p$ polynomial matrix, and $Q(s)$ is a $q \times q$ polynomial matrix. If a $(q+p) \times (q+p)$ unimodular polynomial matrix $U(s)$ and a $q \times q$ polynomial matrix $L(s)$ are such that

$$\begin{bmatrix} Q(s) & P(s) \end{bmatrix} U(s) = \begin{bmatrix} L(s) & 0 \end{bmatrix} \tag{19}$$

then $L(s)$ is a greatest common left divisor of $P(s)$ and $Q(s)$.

Three types of *elementary column operations* can be represented by post-multiplication by a unimodular matrix. The first is interchange of two columns, and the second is multiplication of any column by a nonzero real number. The third elementary column operation is addition to any column of a polynomial multiple of another column. It is easy to check that a sequence of these elementary column operations can be represented by post-multiplication by a unimodular matrix. That post-multiplication by any unimodular matrix can be represented by an appropriate sequence of elementary column operations is a consequence of another special form, introduced below for the class of polynomial matrices of interest.

16.15 Theorem   Suppose $P(s)$ is a $q \times p$ polynomial matrix, and $Q(s)$ is a $q \times q$ nonsingular polynomial matrix. Then elementary column operations can be used to transform

$$M(s) = \begin{bmatrix} Q(s) & P(s) \end{bmatrix}$$

into a *column Hermite form* described as follows. For $k = 1, \ldots, q$, all entries of the kth-row to the right of the k,k-entry are zero, and the k,k-entry is monic with higher degree than any entry to its left. (If the k,k-entry is unity, all entries to its left are zero.)

Theorem 16.14 and Theorem 16.15 together provide a method for computing greatest common left divisors using elementary column operations to obtain column Hermite form. The polynomial matrix $L(s)$ in (19) will be lower-triangular.

16.16 Theorem   For a $q \times p$ polynomial matrix $P(s)$ and a nonsingular $q \times q$ polynomial matrix $Q(s)$, the following statements are equivalent.

(i) The polynomial matrices $P(s)$ and $Q(s)$ are left coprime.

(ii) There exist a $p \times q$ polynomial matrix $X(s)$ and a $q \times q$ polynomial matrix $Y(s)$ such that

$$P(s)X(s) + Q(s)Y(s) = I_q \tag{20}$$

(iii) For every complex number s_o,

$$\text{rank} \begin{bmatrix} Q(s_o) & P(s_o) \end{bmatrix} = q \tag{21}$$

16.17 Theorem   Suppose we are given two coprime left polynomial fraction descriptions for a strictly-proper rational transfer function:

$$G(s) = D^{-1}(s)N(s) = D_a^{-1}(s)N_a(s)$$

Then there exists a unimodular polynomial matrix $U(s)$ such that

$$N(s) = U(s)N_a(s) , \quad D(s) = U(s)D_a(s)$$

Suppose that we begin with the elementary right polynomial fraction description and the elementary left polynomial fraction description in (3) for a given strictly-proper rational transfer function $G(s)$. Then appropriate greatest common divisors can be extracted to obtain a coprime right polynomial fraction description, and a coprime left polynomial fraction description for $G(s)$. We now show that these two coprime polynomial fraction descriptions have the same degree. An economical demonstration relies on a matrix-inversion fact.

16.18 Lemma   Suppose that $V_{11}(s)$ is a $m \times m$ nonsingular polynomial matrix, and

$$V(s) = \begin{bmatrix} V_{11}(s) & V_{12}(s) \\ V_{21}(s) & V_{22}(s) \end{bmatrix} \tag{22}$$

is a $(m+p) \times (m+p)$ nonsingular polynomial matrix. Then with $V_a(s) = V_{22}(s) - V_{21}(s)V_{11}^{-1}(s)V_{12}(s)$,

(i) $\det V(s) = \det [V_{11}(s)] \cdot \det [V_a(s)]$,

(ii) $\det V_a(s)$ is not identically zero,

(iii) the inverse of $V(s)$ is, dropping the argument s,

$$V^{-1} = \begin{bmatrix} V_{11}^{-1} + V_{11}^{-1}V_{12} V_a^{-1} V_{21}V_{11}^{-1} & -V_{11}^{-1}V_{12}V_a^{-1} \\ -V_a^{-1} V_{21}V_{11}^{-1} & V_a^{-1} \end{bmatrix}$$

Proof   A partitioned calculation verifies

$$\begin{bmatrix} I_m & 0_{m \times p} \\ -V_{21}(s)\, V_{11}^{-1}(s) & I_p \end{bmatrix} V(s) = \begin{bmatrix} V_{11}(s) & V_{12}(s) \\ 0 & V_a(s) \end{bmatrix} \tag{23}$$

Using the obvious determinental identity for block-triangular matrices, in particular

$$\det \begin{bmatrix} I_m & 0_{m \times p} \\ -V_{21}(s)\, V_{11}^{-1}(s) & I_p \end{bmatrix} = 1$$

gives

$$\det V(s) = \det V_{11}(s) \cdot \det V_a(s)$$

Since $V(s)$ and $V_{11}(s)$ are nonsingular, this proves that $\det V_a(s)$ is not identically zero, that is, $V_a^{-1}(s)$ exists. To establish *(iii)*, multiply (23) on the left by

$$\begin{bmatrix} V_{11}^{-1}(s) & 0 \\ 0 & V_a^{-1}(s) \end{bmatrix} \begin{bmatrix} I_m & -V_{12}(s)\, V_a^{-1}(s) \\ 0 & I_p \end{bmatrix}$$

to obtain

$$\begin{bmatrix} V_{11}^{-1} + V_{11}^{-1} V_{12}\, V_a^{-1}\, V_{21} V_{11}^{-1} & -V_{11}^{-1} V_{12} V_a^{-1} \\ -V_a^{-1}\, V_{21} V_{11}^{-1} & V_a^{-1} \end{bmatrix} V(s) = \begin{bmatrix} I_m & 0 \\ 0 & I_p \end{bmatrix}$$

and the proof is complete.

16.19 Theorem   Suppose that a strictly-proper rational transfer function is represented by a coprime right polynomial fraction, and a coprime left polynomial fraction,

$$G(s) = N(s)D^{-1}(s) = D_L^{-1}(s)N_L(s) \tag{24}$$

Then there exists a nonzero constant α such that $\det D(s) = \alpha \det D_L(s)$.

Proof   By right-coprimeness of $N(s)$ and $D(s)$ there exists an $(m+p) \times (m+p)$ unimodular polynomial matrix

$$U(s) = \begin{bmatrix} U_{11}(s) & U_{12}(s) \\ U_{21}(s) & U_{22}(s) \end{bmatrix}$$

such that

$$\begin{bmatrix} U_{11}(s) & U_{12}(s) \\ U_{21}(s) & U_{22}(s) \end{bmatrix} \begin{bmatrix} D(s) \\ N(s) \end{bmatrix} = \begin{bmatrix} I_m \\ 0 \end{bmatrix} \tag{25}$$

For notational convenience let

$$\begin{bmatrix} U_{11}(s) & U_{12}(s) \\ U_{21}(s) & U_{22}(s) \end{bmatrix}^{-1} = \begin{bmatrix} V_{11}(s) & V_{12}(s) \\ V_{21}(s) & V_{22}(s) \end{bmatrix}$$

Each $V_{ij}(s)$ is a polynomial matrix, and in particular (25) gives

$$V_{11}(s) = D(s), \quad V_{21}(s) = N(s)$$

Therefore $V_{11}(s)$ is nonsingular, and calling on Lemma 16.18 we have that

$$U_{22}(s) = \left[V_{22}(s) - V_{21}(s) \, V_{11}^{-1}(s) \, V_{12}(s) \right]^{-1}$$

which necessarily is a polynomial matrix, is nonsingular. Furthermore writing

$$\begin{bmatrix} U_{11}(s) & U_{12}(s) \\ U_{21}(s) & U_{22}(s) \end{bmatrix} \begin{bmatrix} V_{11}(s) & V_{12}(s) \\ V_{21}(s) & V_{22}(s) \end{bmatrix} = \begin{bmatrix} I_m & 0 \\ 0 & I_p \end{bmatrix}$$

gives, in the 2,2-block,

$$U_{21}(s)V_{12}(s) + U_{22}(s)V_{22}(s) = I_p$$

By Theorem 16.16 this implies that $U_{21}(s)$ and $U_{22}(s)$ are left coprime. Also, from the 2,1-block,

$$U_{21}(s)V_{11}(s) + U_{22}(s)V_{21}(s) = U_{21}(s)D(s) + U_{22}(s)N(s)$$

$$= 0 \tag{26}$$

Thus we can write, from (26),

$$G(s) = N(s)D^{-1}(s) = -U_{22}^{-1}(s)U_{21}(s) \tag{27}$$

This is a coprime left polynomial fraction description for $G(s)$. Again using Lemma 16.18, and the unimodularity of $V(s)$, there exists a nonzero constant α such that

$$\det \begin{bmatrix} V_{11}(s) & V_{12}(s) \\ V_{21}(s) & V_{22}(s) \end{bmatrix} = \det V_{11}(s) \cdot \det [\, V_{22}(s) - V_{21}(s)V_{11}^{-1}(s)V_{12}(s) \,]$$

$$= \det D(s) \cdot \det U_{22}^{-1}(s)$$

$$= \frac{\det D(s)}{\det U_{22}(s)} = \frac{1}{\alpha}$$

Therefore, for the coprime left polynomial fraction description in (27), we have $det\, U_{22}(s) = \alpha\, det\, D(s)$. Finally, using the unimodular relation between coprime left polynomial fractions in Theorem 16.17, such a determinant formula, with possibly a different nonzero constant, must hold for any coprime left polynomial fraction description for $G(s)$.

Column and Row Degrees

There is an additional technical consideration that complicates the representation of a strictly-proper rational transfer function by polynomial fraction descriptions. First we introduce terminology for matrix polynomials that is related to the notion of the degree of a scalar polynomial. Recall again conventions that the degree of a nonzero constant is zero, and the degree of the polynomial 0 is $-\infty$.

16.20 Definition   For a $p \times r$ polynomial matrix $P(s)$, the degree of the highest degree polynomial in the jth-column of $P(s)$, written $c_j[P]$, is called the jth-*column degree of $P(s)$*. The *column degree coefficient matrix* for $P(s)$, written P_{hc}, is the real $p \times r$ matrix with i,j-entry given by the coefficient of $s^{c_j[P]}$ in the i,j-entry of $P(s)$. If $P(s)$ is square and nonsingular, then it is called *column reduced* if

$$\deg [\det P(s)] = c_1[P] + \cdots + c_p[P] \tag{28}$$

If $P(s)$ is square, then the Laplace expansion of the determinant about columns shows that the degree of $det\ P(s)$ is never greater than $c_1[P] + \cdots + c_p[P]$. But it can be less.

The issue that requires attention involves the column degrees of $D(s)$ in a right polynomial fraction description for a strictly-proper rational transfer function. It is clear in the $m = p = 1$ case that this column degree plays an important role in realization considerations, for example. The same is true in the multi-input, multi-output case, and the complication is that some of the column degrees of $D(s)$ can be artificially high, and they can change in the process of post-multiplication by a unimodular matrix. Therefore two coprime right polynomial fraction descriptions for $G(s)$ — as in Theorem 16.10 — can be such that $D(s)$ and $D_a(s)$ have different column degrees, even though the degrees of the polynomials $det\ D(s)$ and $det\ D_a(s)$ are the same.

16.21 Example   The coprime right polynomial fraction description for

$$G(s) = \begin{bmatrix} \dfrac{2s-3}{s^2-1} & \dfrac{1}{s-1} \end{bmatrix} \tag{29}$$

specified by

$$N(s) = \begin{bmatrix} 1 & 2 \end{bmatrix}, \quad D(s) = \begin{bmatrix} 0 & s+1 \\ s-1 & 1 \end{bmatrix}$$

is such that $c_1[D] = 1$ and $c_2[D] = 1$. Choosing the unimodular matrix

$$U(s) = \begin{bmatrix} 1 & 0 \\ s^2-s+1 & 1 \end{bmatrix}$$

another coprime right polynomial fraction description for $G(s)$ is specified by

$$N_a(s) = N(s)U(s) = \begin{bmatrix} 2s^2-2s+3 & 2 \end{bmatrix}$$

$$D_a(s) = D(s)U(s) = \begin{bmatrix} s^3+1 & s+1 \\ s^2 & 1 \end{bmatrix}$$

with $c_1[D_a] = 3$ and $c_2[D_a] = 1$.
□□□

The first step in investigating this situation is to characterize column-reduced polynomial matrices in a way that does not involve computing a determinant. Using Definition 16.20 it is convenient to write a $p \times p$ polynomial matrix $P(s)$ in the form

$$
P(s) = P_{hc} \begin{bmatrix} s^{c_1[P]} & 0 & \cdots & 0 \\ 0 & s^{c_2[P]} & \cdots & 0 \\ \vdots & \vdots & \vdots & \vdots \\ 0 & 0 & \cdots & s^{c_p[P]} \end{bmatrix} + P_l(s) \tag{30}
$$

where $P_l(s)$ is a $p \times p$ polynomial matrix in which each entry of the jth-column has degree strictly less than $c_j[P]$. (We use this notation only when $P(s)$ is nonsingular, so that $c_1[P], \ldots, c_p[P] \geq 0$.)

16.22 Theorem   If $P(s)$ is a $p \times p$ nonsingular polynomial matrix, then $P(s)$ is column reduced if and only if P_{hc} is invertible.

Proof   We can write, using the representation (29),

$$
s^{-c_1[P]-\cdots-c_p[P]} \cdot \det P(s) = \det \left[P(s) \cdot \text{diagonal} \left\{ s^{-c_1[P]}, \ldots, s^{-c_p[P]} \right\} \right]
$$

$$
= \det \left[P_{hc} + P_l(s) \, \text{diagonal} \left\{ s^{-c_1[P]}, \ldots, s^{-c_p[P]} \right\} \right]
$$

$$
= \det \left[P_{hc} + \widetilde{P}(s^{-1}) \right] \tag{31}
$$

where $\widetilde{P}(s^{-1})$ is a matrix with entries that are polynomials in s^{-1} that have no constant terms, that is, no s^0 terms. The key facts in our argument are that letting the complex variable s grow large in magnitude (in any direction) yields $\widetilde{P}(s^{-1}) \to 0$, and that the determinant of a matrix is a continuous function of the matrix entries, so limit and determinant can be interchanged. In particular we can write

$$
\lim_{|s| \to \infty} \left[s^{-c_1[P]-\cdots-c_p[P]} \cdot \det P(s) \right] = \lim_{|s| \to \infty} \det \left[P_{hc} + \widetilde{P}(s^{-1}) \right]
$$

$$
= \det \left\{ \lim_{|s| \to \infty} \left[P_{hc} + \widetilde{P}(s^{-1}) \right] \right\}
$$

$$
= \det P_{hc} \tag{32}
$$

Using (28) the left side of (32) is a nonzero constant if and only if $P(s)$ is column reduced, and thus the proof is complete.
□□□

Consider a coprime right polynomial fraction description $N(s)D^{-1}(s)$, where $D(s)$ is not column reduced. We next show that elementary column operations on $D(s)$ (or post-multiplication by a unimodular matrix $U(s)$) can be used to reduce individual column degrees, and thus compute a new coprime right polynomial fraction description

$$\tilde{N}(s) = N(s)U(s), \quad \tilde{D}(s) = D(s)U(s) \tag{33}$$

where $\tilde{D}(s)$ is column reduced.

To describe the required sequence of calculations, suppose the column degrees of the $m \times m$ polynomial matrix $D(s)$ satisfy $c_1[D] \geq c_2[D], \ldots, c_m[D]$. This always can be achieved by a column interchange, a step that does no more than introduce a corresponding elementary unimodular matrix in the calculation. Using the notation

$$D(s) = D_{hc}\Delta(s) + D_l(s)$$

since $D(s)$ is not column reduced there exists a nonzero $m \times 1$ vector z such that $D_{hc}z = 0$. Define a corresponding polynomial vector by

$$z(s) = \begin{bmatrix} z_1 \\ z_2 s^{c_1[D]-c_2[D]} \\ \vdots \\ z_m s^{c_1[D]-c_m[D]} \end{bmatrix}$$

Then

$$D(s)z(s) = D_{hc}\Delta(s)z(s) + D_l(s)z(s)$$
$$= D_{hc}zs^{c_1[D]} + D_l(s)z(s)$$
$$= D_l(s)z(s)$$

and all entries of $D(s)z(s)$ have degree no greater than $c_1[D]-1$. Choosing the unimodular matrix

$$U(s) = \begin{bmatrix} z_1 & 0 & 0 & \cdots & 0 \\ z_2^{c_1[D]-c_2[D]} & 1 & 0 & \cdots & 0 \\ \vdots & & \vdots & \vdots & \vdots \\ z_m^{c_1[D]-c_m[D]} & 0 & 0 & \cdots & 1 \end{bmatrix}$$

gives that $\tilde{D}(s) = D(s)U(s)$ has column degrees that satisfy

$$c_1[\tilde{D}] < c_1[D]; \quad c_k[\tilde{D}] = c_k[D], \quad k = 2, \ldots, m$$

If $\tilde{D}(s)$ is not column reduced, then the process is repeated. A finite number of repetitions builds a unimodular $U(s)$ such that $\tilde{D}(s)$ in (33) is column reduced.

Another aspect of the column degree issue involves determining when a given $N(s)$ and $D(s)$ are such that $N(s)D^{-1}(s)$ is a strictly-proper rational transfer function. The relative column degrees of $N(s)$ and $D(s)$ play important roles, but not as simply as the single-input, single-output case suggests.

16.23 Example   Suppose a right polynomial fraction description is specified by

$$N(s) = \begin{bmatrix} s^2 & 1 \end{bmatrix}, \quad D(s) = \begin{bmatrix} s^3+1 & s+1 \\ s^2 & 1 \end{bmatrix}$$

Then

$$c_1[N] = 2, \quad c_2[N] = 0, \quad c_1[D] = 3, \quad c_2[D] = 1$$

and the column degrees of $N(s)$ are less than the respective column degrees of $D(s)$. But an easy calculation shows that $N(s)D^{-1}(s)$ is not a matrix of strictly-proper rational functions. This phenomenon is related again to the fact that

$$D_{hc} = \begin{bmatrix} 1 & 1 \\ 0 & 0 \end{bmatrix}$$

is not invertible.

16.24 Theorem   If the polynomial fraction description $N(s)D^{-1}(s)$ is a strictly-proper rational function, then $c_j[N] < c_j[D]$, $j = 1, \ldots, m$. If $D(s)$ is column reduced and $c_j[N] < c_j[D]$, $j = 1, \ldots, m$, then $N(s)D^{-1}(s)$ is a strictly-proper rational function.

Proof   Suppose $G(s) = N(s)D^{-1}(s)$ is strictly proper. Then $N(s) = G(s)D(s)$, and in particular

$$N_{ij}(s) = \sum_{k=1}^{m} G_{ik}(s)D_{kj}(s), \quad \begin{matrix} i = 1, \ldots, p \\ j = 1, \ldots, m \end{matrix} \tag{34}$$

Then for any fixed value of j,

$$N_{ij}(s) \, s^{-c_j[D]} = \sum_{k=1}^{m} G_{ik}(s)D_{kj}(s) \, s^{-c_j[D]}, \quad i = 1, \ldots, p$$

As $|s| \to \infty$, the strictly proper rational functions $G_{ik}(s)$ approach 0, and each $D_{kj}(s) \, s^{-c_j[D]}$ approaches a finite constant, possibly zero. In any case this gives

$$\lim_{|s| \to \infty} N_{ij}(s) \, s^{-c_j[D]} = 0, \quad i = 1, \ldots, p$$

Therefore $deg \, N_{ij}(s) < c_j[D]$, $i = 1, \ldots, p$, which implies $c_j[N] < c_j[D]$.

Now suppose that $D(s)$ is column reduced, and $c_j[N] < c_j[D]$, $j = 1, \ldots, m$. We can write

$$N(s)D^{-1}(s) = \left[N(s) \text{ diagonal} \left\{ s^{-c_1[D]}, \ldots, s^{-c_m[D]} \right\} \right]$$
$$\cdot \left[D(s) \text{ diagonal} \left\{ s^{-c_1[D]}, \ldots, s^{-c_m[D]} \right\} \right]^{-1} \qquad (35)$$

and since $c_j[N] < c_j[D]$, $j = 1, \ldots, m$,

$$\lim_{|s| \to \infty} \left[N(s) \text{ diagonal} \left\{ s^{-c_1[D]}, \ldots, s^{-c_m[D]} \right\} \right] = 0$$

From the adjugate-over-determinant formula it follows that each entry in the inverse of a matrix is a continuous function of the entries of the matrix. Thus limit can be interchanged with matrix inversion,

$$\lim_{|s| \to \infty} \left[D(s) \text{ diagonal} \left\{ s^{-c_1[D]}, \ldots, s^{-c_m[D]} \right\} \right]^{-1}$$
$$= \left[\lim_{|s| \to \infty} D(s) \text{ diagonal} \left\{ s^{-c_1[D]}, \ldots, s^{-c_m[D]} \right\} \right]^{-1}$$

and writing $D(s)$ in the form (29), the limit yields D_{hc}^{-1}. Finally, from (35),

$$\lim_{|s| \to \infty} N(s)D^{-1}(s) = 0 \cdot D_{hc}^{-1} = 0$$

which implies strict properness.
□□□

It remains to give the corresponding development for left polynomial fraction descriptions, though details are omitted.

16.25 Definition   For a $q \times p$ polynomial matrix $P(s)$, the degree of the highest degree polynomial in the ith- row of $P(s)$, written $r_i[P]$, is called the ith-*row degree of* $P(s)$. The *row degree coefficient matrix* of $P(s)$, written P_{hr}, is the real $q \times p$ matrix with i,j-entry given by the coefficient of $s^{r_i[P]}$ in $P_{ij}(s)$. If $P(s)$ is square and nonsingular, then it is called *row reduced* if

$$\deg [\det P(s)] = r_1[P] + \cdots + r_q[P] \qquad (36)$$

16.26 Theorem   If $P(s)$ is a $p \times p$ nonsingular polynomial matrix, then $P(s)$ is row reduced if and only if P_{hr} is invertible.

16.27 Theorem   If the polynomial fraction description $D^{-1}(s)N(s)$ is a strictly proper rational function, then $r_i[N] < r_i[D]$, $i = 1, \ldots, p$. If $D(s)$ is row reduced and $r_i[N] < r_i[D]$, $i = 1, \ldots, p$, then $D^{-1}(s)N(s)$ is a strictly-proper rational function.

Finally, if $G(s) = D^{-1}(s)N(s)$ is a polynomial fraction description and $D(s)$ is not row reduced, then a unimodular matrix $U(s)$ can be computed such that $D_b(s) = U(s)D(s)$ is row reduced. Letting $N_b(s) = U(s)N(s)$, the left polynomial fraction description

$$D_b^{-1}(s)N_b(s) = \left[U(s)D(s) \right]^{-1} U(s)N(s) = G(s) \tag{37}$$

has the same degree as the original.

Because of machinery developed in this chapter, polynomial fraction descriptions for a strictly-proper rational transfer function $G(s)$ can be assumed as either a coprime right polynomial fraction description with column-reduced $D(s)$, or a coprime left polynomial fraction with row-reduced $D_L(s)$. In either case the degree of the polynomial fraction description is the same, and is given by the sum of the column degrees, or, respectively, the sum of the row degrees.

EXERCISES

Exercise 16.1   Determine if the following pair of polynomial matrices is right coprime. If not, compute a greatest common right divisor.

$$P(s) = \begin{bmatrix} 0 & s^2 \\ -s & s^2 \end{bmatrix}, \quad Q(s) = \begin{bmatrix} 0 & (s + 1)^2(s + 3) \\ (s + 3)^2 & s + 3 \end{bmatrix}$$

Exercise 16.2   Determine if the following pair of polynomial matrices is right coprime. If not, compute a greatest common right divisor.

$$P(s) = \begin{bmatrix} s & s \\ 0 & s(s+1)^2-s \end{bmatrix}, \quad Q(s) = \begin{bmatrix} (s + 1)^2(s + 2)^2 & 0 \\ 0 & (s + 2)^2 \end{bmatrix}$$

Exercise 16.3   Show that the right polynomial fraction description

$$G(s) = N(s)D^{-1}(s)$$

is coprime if and only if there exist unimodular matrices $U(s)$ and $V(s)$ such that

$$U(s) \begin{bmatrix} D(s) \\ N(s) \end{bmatrix} V(s) = \begin{bmatrix} I \\ 0 \end{bmatrix}$$

If $N(s)D^{-1}(s)$ is right coprime, and $N_a(s)D_a^{-1}(s)$ is another right polynomial fraction description for $G(s)$, show that there is a polynomial matrix $R(s)$ such that

$$\begin{bmatrix} D_a(s) \\ N_a(s) \end{bmatrix} = \begin{bmatrix} D(s) \\ N(s) \end{bmatrix} R(s)$$

Exercise 16.4   Suppose that $D^{-1}(s)N(s)$ and $D_a^{-1}(s)N_a(s)$ are coprime left polynomial fraction descriptions for the same strictly-proper transfer function. Using Theorem 16.16, prove that $D(s)D_a^{-1}(s)$ is unimodular.

Exercise 16.5   Suppose $D_L^{-1}(s)N_L(s) = N(s)D^{-1}(s)$, and both are coprime polynomial fraction descriptions. Show that there exist $U_{11}(s)$ and $U_{12}(s)$ such that

$$\begin{bmatrix} U_{11}(s) & U_{12}(s) \\ N_L(s) & D_L(s) \end{bmatrix}$$

is unimodular, and

$$\begin{bmatrix} U_{11}(s) & U_{12}(s) \\ N_L(s) & D_L(s) \end{bmatrix} \begin{bmatrix} D(s) \\ -N(s) \end{bmatrix} = \begin{bmatrix} I \\ 0 \end{bmatrix}$$

Exercise 16.6   Suppose $D(s)$ is nonsingular, and in column Hermite form. Show that $D(s)$ is row reduced, but not necessarily column reduced. Explain how to compute a unimodular $U(s)$ such that $D(s)U(s)$ is column reduced.

Exercise 16.7   Suppose the inverse of the unimodular matrix

$$P(s) = P_\rho s^\rho + P_{\rho-1}s^{\rho-1} + \cdots + P_0$$

is written as

$$Q(s) = Q_\eta s^\eta + Q_{\eta-1}s^{\eta-1} + \cdots + Q_0$$

and that $\rho, \eta \geq 2$. Prove that if $P_{\rho-1}$ and $Q_{\eta-1}$ are invertible, then $P_\rho s + P_{\rho-1}$ is unimodular by exhibiting R_1 and R_0 such that

$$[\, P_\rho s + P_{\rho-1}\,]^{-1} = R_1 s + R_0$$

Exercise 16.8   Obtain a coprime, column-reduced right polynomial fraction description for

$$G(s) = \begin{bmatrix} s & s+2 \\ 1 & s+1 \end{bmatrix} \begin{bmatrix} s^2+2 & (s+1)^2 \\ s+1 & s \end{bmatrix}^{-1}$$

Exercise 16.9   Suppose $N(s)D^{-1}(s)$ and $\widetilde{N}(s)\widetilde{D}^{-1}(s)$ both are coprime right polynomial fraction descriptions for a strictly-proper, rational transfer function $G(s)$. Suppose also that $D(s)$ and $\widetilde{D}(s)$ both are column reduced, with column degrees that satisfy the ordering $c_1 \leq c_2 \leq \cdots \leq c_m$. Show that $c_j[D] = c_j[\widetilde{D}]$, $j = 1, \ldots, m$. (This shows that these column degrees are determined by the transfer function, not by a particular (coprime, column-reduced) right polynomial fraction description.) (*Hint*: Assume J is the least index for which $c_J[D] < c_J[\widetilde{D}]$, and express the unimodular relation between $D(s)$ and $\widetilde{D}(s)$ column-wise. Using linear independence of the columns of D_{hc}, and $\widetilde{D}_{hc}$, conclude that a submatrix of the unimodular matrix must be zero.)

NOTES

Note 16.1   A standard text and reference for polynomial fraction descriptions is

T. Kailath, *Linear Systems,* Prentice Hall, New York, 1980

At the beginning of Section 6.3 several references to the mathematical theory of polynomial matrices are provided. See also

S. Barnett, *Polynomials and Linear Control Systems,* Marcel Dekker, New York, 1983

A.I.G. Vardulakis, *Linear Multivariable Control,* John Wiley, Chichester, 1991

Note 16.2   The polynomial fraction description emerges from the time-domain description of input-output differential equations the form

$$L(p)y(t) = M(p)u(t)$$

This is an older notation where p represents the differential operator d/dt, and $L(p)$ and $M(p)$ are polynomial matrices in p. Early work based on this representation, much of it dealing with state-equation realization issues, includes

E. Polak, "An algorithm for reducing a linear, time-invariant differential system to state form," *IEEE Transactions on Automatic Control,* Vol. 11, No. 3, pp. 577–579, 1966

W.A. Wolovich, *Linear Multivariable Systems,* Applied Mathematical Sciences, Vol. 11, Springer-Verlag, New York, 1974

For more recent developments consult the book by Vardulakis cited in Note 16.1, and

H. Blomberg, R. Ylinen, *Algebraic Theory for Multivariable Linear Systems,* Mathematics in Science and Engineering, Vol. 166, Academic Press, London, 1983

Note 16.3   If $P(s)$ is a $p \times p$ polynomial matrix, it can be shown that there exist unimodular matrices $U(s)$ and $V(s)$ such that

$$U(s)P(s)V(s) = \text{diagonal } \{ \lambda_1(s), \ldots, \lambda_p(s) \}$$

where $\lambda_1(s), \ldots, \lambda_p(s)$ are monic polynomials with the property that $\lambda_k(s)$ divides $\lambda_{k+1}(s)$. A similar result holds in the nonsquare case, with the polynomials $\lambda_k(s)$ on the quasi-diagonal. This is called the *Smith form* for polynomial matrices. The polynomial fraction description can be developed using this form, and the related *Smith-McMillan form* for rational matrices, instead of Hermite forms. See Section 22 of

D.F. Delchamps, *State Space and Input-Output Linear Systems,* Springer-Verlag, New York, 1988

Note 16.4   For an approach to polynomial fraction descriptions for time-varying linear systems, see

A. Ilchmann, I. Nurnberger, W. Schmale, "Time-varying polynomial matrix systems," *International Journal of Control,* Vol. 40, No. 2, pp. 329–362, 1984

17

POLYNOMIAL FRACTION
APPLICATIONS

In this chapter we apply polynomial fraction descriptions for a transfer function in three ways. First computation of a minimal realization from a polynomial fraction description is considered, as well as the reverse computation of a polynomial fraction description for given linear state equation. Then we define notions of poles and zeros of a transfer function in terms of polynomial fraction descriptions, and characterize these concepts in terms of response properties. Finally linear state feedback is considered from the viewpoint of polynomial fraction descriptions for the open-loop and closed-loop transfer functions.

Minimal Realization

We assume that a $p \times m$ strictly-proper rational transfer function is specified by a coprime right polynomial fraction description

$$G(s) = N(s)D^{-1}(s) \tag{1}$$

with $D(s)$ column reduced. Then the column degrees of $N(s)$ and $D(s)$ satisfy $c_j[N] < c_j[D]$, $j = 1, \ldots, m$. Some simplification occurs if one uninteresting case is ruled out. If $c_j[D] = 0$ for some j, then by Theorem 16.24 $G(s)$ is strictly proper if and only if all entries of the jth-column of $N(s)$ are zero, that is, $c_j[N] = -\infty$. Therefore we assume throughout this chapter that (1) also is such that $c_1[D], \ldots, c_m[D] \geq 1$. Recall that the degree of the polynomial fraction description (1) is $c_1[D] + \cdots + c_m[D]$, since $D(s)$ is column reduced.

From Chapter 10 we know there exists a minimal realization for $G(s)$,

$$\dot{x}(t) = Ax(t) + Bu(t)$$

$$y(t) = Cx(t) \tag{2}$$

In exploring the connection between a transfer function and its minimal realizations, an additional bit of terminology is convenient.

17.1 Definition   Suppose $N(s)D^{-1}(s)$ is a coprime right polynomial fraction description for the $p \times m$, strictly-proper, rational transfer function $G(s)$. Then the degree of this polynomial fraction description is called the *McMillan degree* of $G(s)$.

The first objective is to show that the McMillan degree of $G(s)$ is precisely the dimension of minimal realizations of $G(s)$. Our roundabout strategy is to prove that minimal realizations cannot have dimension less than the McMillan degree, and then compute a realization of dimension equal to the McMillan degree. This forces the conclusion that the computed realization is a minimal realization.

17.2 Lemma   The dimension of any realization of a strictly-proper rational transfer function $G(s)$ is at least the McMillan degree of $G(s)$.

Proof   Suppose that the linear state equation (2) is a dimension-n minimal realization for the $p \times m$ transfer function $G(s)$. Then (2) is both controllable and observable, and

$$G(s) = C(sI - A)^{-1}B$$

Define a $n \times m$ strictly-proper transfer function $H(s)$ by the left polynomial fraction description

$$H(s) = D_L^{-1}(s)N_L(s) = (sI - A)^{-1}B \tag{3}$$

Clearly this left polynomial fraction description has degree n. Since the state equation (2) is controllable, Theorem 13.4 gives

$$\text{rank}\begin{bmatrix} D_L(s_o) & N_L(s_o) \end{bmatrix} = \text{rank}\begin{bmatrix} (s_oI - A) & B \end{bmatrix}$$

$$= n$$

for every complex s_o. Thus by Theorem 16.16 the left polynomial fraction description (3) is coprime. Now suppose $N_a(s)D_a^{-1}(s)$ is a coprime right polynomial fraction description for $H(s)$. Then this right polynomial fraction description also has degree n, and

$$G(s) = [C \, N_a(s)] \, D_a^{-1}(s)$$

is a degree-n right polynomial fraction description for $G(s)$, though not necessarily coprime. Therefore the McMillan degree of $G(s)$ is no greater than n, the dimension of a minimal realization of $G(s)$.
□□□

For notational assistance in the construction of a minimal realization, recall the integrator coefficient matrices corresponding to a set of k positive integers, $\alpha_1, \ldots, \alpha_k$, with $\alpha_1 + \cdots + \alpha_k = n$. From Definition 13.7 these matrices are

$$A_o = \text{block diagonal} \left\{ \begin{bmatrix} 0 & 1 & 0 & \cdots & 0 \\ 0 & 0 & 1 & \cdots & 0 \\ \vdots & \vdots & \vdots & \vdots & \vdots \\ 0 & 0 & 0 & \cdots & 1 \\ 0 & 0 & 0 & \cdots & 0 \end{bmatrix}_{(\alpha_i \times \alpha_i)} , \quad i = 1, \ldots, k \right\}$$

$$B_o = \text{block diagonal} \left\{ \begin{bmatrix} 0 \\ \vdots \\ 0 \\ 1 \end{bmatrix}_{(\alpha_i \times 1)} , \quad i = 1, \ldots, k \right\}$$

Define the corresponding *integrator polynomial matrices* by

$$\Psi(s) = \text{block diagonal} \left\{ \begin{bmatrix} 1 \\ s \\ \vdots \\ s^{\alpha_i - 1} \end{bmatrix} , \quad i = 1, \ldots, k \right\}$$

$$\Delta(s) = \text{diagonal} \left\{ s^{\alpha_1}, \ldots, s^{\alpha_k} \right\} \tag{4}$$

The terminology couldn't be more appropriate, as we now demonstrate.

17.3 Lemma   The integrator polynomial matrices provide a right polynomial fraction description for the corresponding integrator state equation. That is,

$$(sI - A_o)^{-1} B_o = \Psi(s) \Delta^{-1}(s) \tag{5}$$

Proof   To verify (5), first multiply on the left by $(sI - A_o)$ and on the right by $\Delta(s)$ to obtain

$$B_o \Delta(s) = s \Psi(s) - A_o \Psi(s) \tag{6}$$

This expression is easy to check in a column-by-column fashion using the structure of the various matrices. For example the first column of (6) is the obvious

$$\begin{bmatrix} 0 \\ \vdots \\ 0 \\ s^{\alpha_1} \\ 0 \\ \vdots \\ 0 \end{bmatrix} = \begin{bmatrix} s \\ \vdots \\ s^{\alpha_1-1} \\ s^{\alpha_1} \\ 0 \\ \vdots \\ 0 \end{bmatrix} - \begin{bmatrix} s \\ \vdots \\ s^{\alpha_1-1} \\ 0 \\ 0 \\ \vdots \\ 0 \end{bmatrix}$$

Proceeding similarly through the remaining columns in (6) yields the proof.
□□□

Completing our minimal realization strategy now reduces to comparing a special representation for the polynomial fraction description and a special structure for a dimension-n state equation.

17.4 Theorem   Suppose that a strictly-proper transfer function is described by a coprime right polynomial fraction description (1), where $D(s)$ is column reduced with column degrees $c_1[D], \ldots, c_m[D] \geq 1$. Then the McMillan degree of $G(s)$ is given by $n = c_1[D] + \cdots + c_m[D]$, and minimal realizations of $G(s)$ have dimension n. Furthermore, writing

$$N(s) = N_l \Psi(s)$$

$$D(s) = D_{hc} \Delta(s) + D_l \Psi(s) \tag{7}$$

where $\Psi(s)$ and $\Delta(s)$ are the integrator polynomial matrices corresponding to $c_1[D], \ldots, c_m[D]$, a minimal realization for $G(s)$ is

$$\dot{x}(t) = \left[A_o - B_o D_{hc}^{-1} D_l \right] x(t) + B_o D_{hc}^{-1} u(t)$$

$$y(t) = N_l x(t) \tag{8}$$

where A_o and B_o are the integrator coefficient matrices corresponding to $c_1[D], \ldots, c_m[D]$.

Proof   First we verify that (8) is a realization for $G(s)$. It is straightforward to write down the representation in (7), where N_l and D_l are constant matrices that select for appropriate polynomial entries of $N(s)$, and $D_l(s)$. Then solving for $\Delta(s)$ in (7) and substituting into (6) gives

$$B_o D_{hc}^{-1} D(s) = s\Psi(s) - A_o \Psi(s) + B_o D_{hc}^{-1} D_l \Psi(s)$$

$$= \left[sI - A_o + B_o D_{hc}^{-1} D_l \right] \Psi(s)$$

This implies

$$\left[sI - A_o + B_o D_{hc}^{-1} D_l \right]^{-1} B_o D_{hc}^{-1} = \Psi(s) D^{-1}(s) \tag{9}$$

from which the transfer function for (8) is

$$N_l \left[sI - A_o + B_o D_{hc}^{-1} D_l \right]^{-1} B_o D_{hc}^{-1} = N(s) D^{-1}(s)$$

Invoking Lemma 17.2 we conclude that the McMillan degree of $G(s)$ is the dimension of minimal realizations of $G(s)$.
□□□

In the minimal realization (8), note that if D_{hc} is upper triangular with unity diagonal entries, then the realization is in the controller form discussed in Chapter 13. (Upper triangular structure for D_{hc} can be obtained by elementary column operations on the original polynomial fraction description.) If (8) is in controller form, then the controllability indices are precisely $\rho_1 = c_1[D], \ldots, \rho_m = c_m[D]$. Summoning Theorem 10.10 and Exercise 13.10, we conclude that all minimal realizations of $N(s)D^{-1}(s)$ have the same controllability indices, up to reordering. Then Exercise 16.10 shows that all minimal realizations of a strictly-proper rational transfer function $G(s)$ have the same controllability indices, up to reordering.

Calculations similar to those in the proof of Theorem 17.4 can be used to display a right polynomial fraction description for a given linear state equation.

17.5 Theorem   Suppose the linear state equation (2) is controllable with controllability indices $\rho_1, \ldots, \rho_m \geq 1$. Then the transfer function for (2) is given by the right polynomial fraction description

$$C(sI - A)^{-1} B = N(s) D^{-1}(s)$$

where

$$N(s) = CP^{-1}\Psi(s)$$

$$D(s) = R^{-1}\Delta(s) - R^{-1}UP^{-1}\Psi(s) \tag{10}$$

and $D(s)$ is column reduced. Here $\Psi(s)$ and $\Delta(s)$ are the integrator polynomial matrices corresponding to $\rho_1, \ldots, \rho_m$, P is the controller-form variable change, and U and R are the coefficient matrices defined in Theorem 13.9. If the state equation (2) also is observable, then $N(s)D^{-1}(s)$ is coprime with degree n.

Proof   By Theorem 13.9 we can write

$$PAP^{-1} = A_o + B_o UP^{-1}, \quad PB = B_o R$$

where A_o and B_o are the integrator coefficient matrices corresponding to $\rho_1, \ldots, \rho_m$. Let $\Delta(s)$ and $\Psi(s)$ be the corresponding integrator polynomial matrices. Using (10) to substitute for $\Delta(s)$ in (6) gives

$$B_o RD(s) + B_o UP^{-1}\Psi(s) = s\Psi(s) - A_o\Psi(s)$$

Rearranging this expression yields

$$\Psi(s)D^{-1}(s) = \left[sI - A_o - B_o U P^{-1} \right]^{-1} B_o R \tag{11}$$

and therefore

$$N(s)D^{-1}(s) = CP^{-1} \left[sI - A_o - B_o U P^{-1} \right]^{-1} B_o R$$

$$= CP^{-1} \left[sI - PAP^{-1} \right]^{-1} PB$$

$$= C(sI - A)^{-1} B$$

This calculation verifies that the polynomial fraction description defined by (10) represents the transfer function of the linear state equation (2). Also, $D(s)$ in (10) is column reduced because $D_{hc} = R^{-1}$. Since the degree of the polynomial fraction description is n, if the state equation also is observable, hence a minimal realization of its transfer function, then n is the McMillan degree of the polynomial fraction description (10).
□□□

For left polynomial fraction descriptions, the strategy for right fraction descriptions applies since the McMillan degree of $G(s)$ also is the degree of any coprime left polynomial fraction description for $G(s)$. The only details that remain in proving a left-handed version of Theorem 17.4 involve construction of a minimal realization. But this construction is not difficult to deduce from a summary statement.

17.6 Theorem   Suppose that a strictly-proper transfer function is described by a coprime left polynomial fraction description $D^{-1}(s)N(s)$, where $D(s)$ is row reduced with row degrees $r_1[D], \ldots, r_p[D] \geq 1$. Then the McMillan degree of $G(s)$ is given by $n = r_1[D] + \cdots + r_p[D]$, and minimal realizations of $G(s)$ have dimension n. Furthermore, writing

$$N(s) = \Psi^T(s)N_l$$

$$D(s) = \Delta(s)D_{hr} + \Psi^T(s)D_l \tag{12}$$

where $\Psi(s)$ and $\Delta(s)$ are the integrator polynomial matrices corresponding to $r_1[D], \ldots, r_p[D]$, a minimal realization for $G(s)$ is

$$\dot{x}(t) = \left[A_o^T - D_l D_{hr}^{-1} B_o^T \right] x(t) + N_l u(t)$$

$$y(t) = D_{hr}^{-1} B_o^T x(t)$$

where A_o and B_o are the integrator coefficient matrices corresponding to $r_1[D], \ldots, r_p[D]$.

Analogous to the discussion following Theorem 17.4, in the setting of Theorem 17.6 the observability indices of minimal realizations of $D^{-1}(s)N(s)$ are the same, up to reordering, as the row degrees of $D(s)$.

For the record we state a left-handed version of Theorem 17.5, leaving the proof to Exercise 17.3.

17.7 Theorem   Suppose the linear state equation (2) is observable with observability indices $\eta_1, \ldots, \eta_p \geq 1$. Then the transfer function for (2) is given by the left polynomial fraction description

$$C(sI - A)^{-1}B = D^{-1}(s)N(s)$$

where

$$N(s) = \Psi^T(s)Q^{-1}B$$

$$D(s) = \Delta(s)S^{-1} - \Psi^T(s)Q^{-1}VS^{-1} \tag{13}$$

and $D(s)$ is row reduced. Here $\Psi(s)$ and $\Delta(s)$ are the integrator polynomial matrices corresponding to $\eta_1, \ldots, \eta_p$, Q is the observer-form variable change, and V and S are the coefficient matrices defined in Theorem 13.17. If the state equation (2) also is controllable, then $D^{-1}(s)N(s)$ is coprime with degree n.

Poles and Zeros

The connections between a coprime polynomial fraction description for a strictly-proper rational transfer function $G(s)$ and minimal realizations of $G(s)$ can be used to define notions of poles and zeros of $G(s)$ that generalize the familiar notions for scalar transfer functions. In addition we characterize these concepts in terms of response properties of a minimal realization of $G(s)$.

Given coprime polynomial fraction descriptions

$$G(s) = N(s)D^{-1}(s) = D_L^{-1}(s)N_L(s) \tag{14}$$

it follows from Theorem 16.19 that the polynomials $det\, D(s)$ and $det\, D_L(s)$ have the same roots. Furthermore from Theorem 16.10 it is clear that these roots are the same for every coprime polynomial description. This permits introduction of terminology in terms of either a right or left polynomial fraction description, though we adhere to a societal bias and use right.

17.8 Definition   Suppose $G(s)$ is a strictly-proper rational transfer function. A complex number s_o is called a *pole* of $G(s)$ if $det\, D(s_o) = 0$, where $N(s)D^{-1}(s)$ is a coprime right polynomial fraction description for $G(s)$. The *multiplicity* of a pole s_o is the multiplicity of s_o as a root of the polynomial $det\, D(s)$.

This terminology is compatible with customary usage in the $m = p = 1$ case, and it agrees with the definition used in Chapter 12. Specifically, if s_o is a pole of $G(s)$, then some entry $G_{ij}(s)$ is such that $|G_{ij}(s_o)| = \infty$. Conversely if some entry of $G(s)$ has infinite magnitude when evaluated at the complex number s_o, then s_o is a pole of $G(s)$. (Detailed reasoning that substantiates these claims is left to Exercise 17.9.) Also Theorem 12.9 stands in this terminology: A linear state equation with transfer function

$G(s)$ is uniformly bounded-input, bounded-output stable if and only if all poles of $G(s)$ have negative real parts, that is all roots of $det\, D\,(s)$ have negative real parts.

The relation between eigenvalues of A in the linear state equation (2) and poles of the corresponding transfer function

$$G(s) = C(sI - A)^{-1}B$$

is a crucial feature in some of our arguments. Writing $G(s)$ in terms of a coprime right polynomial fraction description gives

$$\frac{N(s)\, \mathrm{adj}\, D(s)}{\det D(s)} = \frac{C\, \mathrm{adj}\,(sI - A)B}{\det(sI - A)} \tag{15}$$

Using Lemma 17.2, (15) reveals that if s_o is a pole of $G(s)$ with multiplicity σ_o, then s_o is an eigenvalue of A with multiplicity at least σ_o. But simple single-input, single-output examples confirm that multiplicities can be different, and in particular that an eigenvalue of A might not be a pole of $G(s)$. The remedy for this displeasing situation is to assume (2) is controllable and observable. Then (15) shows that, since the denominator polynomials are identical up to a constant multiplier, the set of poles of $G(s)$ is identical to the set of eigenvalues of a minimal realization of $G(s)$.

This discussion leads to an interpretation of a pole of a transfer function in terms of zero-input response properties of a minimal realization of the transfer function.

17.9 Theorem   Suppose the linear state equation (2) is controllable and observable. Then the complex number s_o is a pole of

$$G(s) = C(sI - A)^{-1}B$$

if and only if there exists a complex $n \times 1$ vector x_o and a complex $p \times 1$ vector $y_o \neq 0$ such that

$$Ce^{At}x_o = y_o e^{s_o t}, \quad t \geq 0 \tag{16}$$

Proof   If s_o is a pole of $G(s)$, then s_o is an eigenvalue of A. With x_o an eigenvector of A corresponding to the eigenvalue s_o, we have

$$e^{At}x_o = e^{s_o t}x_o$$

This easily gives (16), where $y_o = Cx_o$ is nonzero by the observability of (2) and the corresponding eigenvector criterion in Theorem 13.14.

On the other hand if (16) holds, then taking Laplace transforms gives

$$C(sI - A)^{-1}x_o = y_o(s - s_o)^{-1}$$

or,

$$(s - s_o)\, C\, [\, \mathrm{adj}\,(sI - A)\,]x_o = y_o \det(sI - A) \tag{17}$$

Evaluating this at $s = s_o$ shows that, since $y_o \neq 0$, $det\,(s_o I - A) = 0$. Therefore s_o is an eigenvalue of A and, by minimality of the state equation, a pole of $G(s)$.
□□□

Of course if s_o is a real pole of $G(s)$, then (16) directly gives a corresponding zero-input response property of minimal realizations of $G(s)$. If s_o is complex, then the real initial state $x_o + \overline{x}_o$ gives an easily-computed real response that can be written as a product of an exponential with exponent $(Re\ [s_o])t$ and a sinusoid with frequency $Im\ [s_o]$.

The concept of a zero of a transfer function is more delicate. In part this is because we want a characterization in terms of identically-zero response of a minimal realization to a particular initial state and particular input signal. The complication is that, given $G(s)$ with $m \geq 2$, there can exist a nonzero $m \times 1$ vector $U(s)$ of proper rational functions such that $G(s)U(s) = 0$. In this situation multiplying all the denominators in $U(s)$ by the same nonzero polynomial generates whole families of inputs for which the response is identically zero. This inconvenience always occurs when $p < m$.

To define the concept of a zero, the underlying assumption we make is that $rank\ G(s) = min\ [m, p]$ for almost all complex values of s. (By 'almost all' we mean 'all but a finite number.') In particular at poles of $G(s)$ at least one entry of $G(s)$ is ill-defined, and so poles are among those values of s we ignore when checking rank. (Another phrasing of this assumption is that $G(s)$ is assumed to have rank $min\ [m, p]$ over the field of rational functions, a more sophisticated terminology that we do not further employ.) Now consider coprime polynomial fraction descriptions

$$G(s) = N(s)D^{-1}(s) = D_L^{-1}(s)N_L(s) \tag{18}$$

for $G(s)$. Since both $D(s)$ and $D_L(s)$ are nonsingular polynomial matrices, assuming $rank\ G(s) = min\ [m, p]$ for almost all complex values of s is equivalent to assuming $rank\ N(s) = min\ [m, p]$ for almost all complex values of s, and also equivalent to assuming $rank\ N_L(s) = min\ [m, p]$ for almost all complex values of s. The agreeable feature of polynomial fraction descriptions is that $N(s)$ and $N_L(s)$ are well-defined for all values of s. Either right or left polynomial fractions can be adopted as the basis for defining transfer-function zeros.

17.10 Definition   Suppose $G(s)$ is a strictly-proper rational transfer function with $rank\ G(s) = min\ [m, p]$ for almost all complex numbers s. A complex number s_o is called a *transmission zero* of $G(s)$ if $rank\ N(s_o) < min\ [m, p]$, where $N(s)D^{-1}(s)$ is any coprime right polynomial fraction description for $G(s)$.

This reduces to the customary definition in the single-input, single-output case, where a zero is a root of the numerator polynomial. But a concrete look at multi-input, multi-output examples reveals subtleties in the concept of transmission zero.

17.11 Example   Consider the transfer function with coprime right polynomial fraction description

$$G(s) = \begin{bmatrix} \dfrac{s+2}{(s+1)^2} & 0 \\ 0 & \dfrac{s+1}{(s+2)^2} \end{bmatrix} = \begin{bmatrix} s+2 & 0 \\ 0 & s+1 \end{bmatrix} \begin{bmatrix} (s+1)^2 & 0 \\ 0 & (s+2)^2 \end{bmatrix}^{-1} \tag{19}$$

This transfer function has multiplicity-two poles at $s = -1$ and $s = -2$, and transmission zeros at $s = -1$ and $s = -2$. Thus a multi-input, multi-output transfer function can have coincident poles and transmission zeros — something that cannot happen in the $m = p = 1$ case according to a careful reading of Definition 17.10.

17.12 Example   The transfer function with coprime left polynomial fraction description

$$
G(s) = \begin{bmatrix} \dfrac{s+1}{(s+3)^2} & 0 \\[2mm] 0 & \dfrac{s+2}{(s+4)^2} \\[2mm] \dfrac{s+2}{(s+5)^2} & \dfrac{s+1}{(s+5)^2} \end{bmatrix} = \begin{bmatrix} (s+3)^2 & 0 & 0 \\ 0 & (s+4)^2 & 0 \\ 0 & 0 & (s+5)^2 \end{bmatrix}^{-1} \begin{bmatrix} s+1 & 0 \\ 0 & s+2 \\ s+2 & s+1 \end{bmatrix} \quad (20)
$$

has no transmission zeros, even though various entries of $G(s)$, viewed as single-input, single-output transfer functions, have transmission zeros at $s = -1$ or $s = -2$.
□□□

Now we turn to the promised characterization of transmission zeros in terms of response properties, adding an assumption that guarantees there are at least as many output components as input components. (See Exercise 17.5 for the case $p < m$.) The basic idea is to devise an input $U(s)$ such that the zero-state response component contains exponential terms due solely to poles of the transfer function, and such that these exponential terms can be canceled by terms in the zero-input response component.

17.13 Theorem   Suppose the linear state equation (2) is controllable and observable, and

$$
G(s) = C(sI - A)^{-1}B \quad (21)
$$

has rank m for almost all complex numbers s. If the complex number s_o is not a pole of $G(s)$, then it is a transmission zero of $G(s)$ if and only if there is a nonzero, complex $m \times 1$ vector u_o and a complex $n \times 1$ vector x_o such that

$$
Ce^{At}x_o + \int_0^t Ce^{A(t-\sigma)}Bu_o e^{s_o\sigma}\, d\sigma = 0, \quad t \geq 0 \quad (22)
$$

Proof   Suppose $N(s)D^{-1}(s)$ is a coprime right polynomial fraction description for (21). If s_o is not a pole of $G(s)$ then $D(s_o)$ is invertible, and s_o is not an eigenvalue of A. If x_o and $u_o \neq 0$ are such that (22) holds, then the Laplace transform of (22) gives

$$
C(sI - A)^{-1}x_o + N(s)D^{-1}(s)u_o(s-s_o)^{-1} = 0
$$

or

$$
(s-s_o)C(sI - A)^{-1}x_o + N(s)D^{-1}(s)u_o = 0
$$

Evaluating this expression at $s = s_o$ gives

$$N(s_o)D^{-1}(s_o)u_o = 0$$

and this implies that *rank* $N(s_o) \leq m$. That is, s_o is a transmission zero of $G(s)$.

On the other hand suppose s_o is not a pole of $G(s)$. Using the easily-verified identity

$$(s_oI - A)^{-1}(s-s_o)^{-1} = (sI - A)^{-1}(s_oI - A)^{-1} + (sI - A)^{-1}(s-s_o)^{-1} \qquad (23)$$

we can write, for any $m \times 1$ complex vector u_o and corresponding $n \times 1$ complex vector $x_o = (s_oI - A)^{-1}Bu_o$, the Laplace transform expression

$$\mathbf{L}\left[Ce^{At}x_o + \int_0^t Ce^{A(t-\sigma)}Bu_o e^{s_o\sigma} \, d\sigma \right]$$

$$= C(sI - A)^{-1}x_o + C(sI - A)^{-1}Bu_o(s-s_o)^{-1}$$

$$= C\left[(sI - A)^{-1}(s_oI - A)^{-1} + (sI - A)^{-1}(s-s_o)^{-1} \right]Bu_o$$

$$= G(s_o)u_o(s-s_o)^{-1}$$

$$= N(s_o)D^{-1}(s_o)u_o(s-s_o)^{-1}$$

Taking the inverse Laplace transform gives, for the particular choice of x_o above,

$$Ce^{At}x_o + \int_0^t Ce^{A(t-\sigma)}Bu_o e^{s_o\sigma} \, d\sigma = N(s_o)D^{-1}(s_o)u_o e^{s_o t}, \quad t \geq 0 \qquad (24)$$

Clearly the $m \times 1$ vector u_o can be chosen so that this expression is zero for $t \geq 0$ if *rank* $N(s_o) < m$, that is, if s_o is a transmission zero of $G(s)$.
□□□

Of course if a transmission zero s_o is real, and not a pole, then we can take u_o real and the corresponding $x_o = (s_oI - A)^{-1}Bu_o$ is real. Then (22) shows that the complete response for $x(0) = x_o$ and $u(t) = u_o e^{s_o t}$ is identically zero. If s_o is a complex transmission zero, then specification of a real input and real initial state that provides identically-zero response is left as a mild exercise.

State Feedback

Properties of linear state feedback

$$u(t) = Kx(t) + Mr(t)$$

applied to a linear state equation (2) are discussed in Chapter 14 (in a slightly different notation). For some purposes a transfer function formulation is more convenient, and in any case a second viewpoint often enlightens. Polynomial fraction descriptions and an adroit formulation both are essential because, as noted following Theorem 14.3, a direct approach to relating the closed-loop and plant transfer functions is unpromising in the case of state feedback.

We assume that a strictly-proper rational transfer function for the plant is given as a coprime right polynomial fraction $G(s) = N(s)D^{-1}(s)$, with $D(s)$ column reduced. To represent linear state feedback it is convenient to write the input-output description

$$Y(s) = N(s)D^{-1}(s)U(s) \tag{25}$$

as a pair of equations with polynomial matrix coefficients,

$$D(s)\xi(s) = U(s)$$

$$Y(s) = N(s)\xi(s) \tag{26}$$

The $m \times 1$ vector $\xi(s)$ is called the *pseudo-state* of the plant. This terminology can be motivated by considering a minimal realization of the form (8) for $G(s)$. From (9) we write

$$\Psi(s)\xi(s) = \Psi(s)D^{-1}(s)U(s)$$

$$= \left[sI - A_o + B_o D_{hc}^{-1} D_l \right]^{-1} B_o D_{hc}^{-1} U(s)$$

or,

$$s\Psi(s)\xi(s) = (A_o - B_o D_{hc}^{-1} D_l)\Psi(s)\xi(s) + B_o D_{hc}^{-1} U(s) \tag{27}$$

Defining the $n \times 1$ vector $x(t)$ as the inverse Laplace transform

$$x(t) = \mathbf{L}^{-1} \left[\Psi(s)\xi(s) \right]$$

we see that (27) is the Laplace transform representation of the linear state equation (8) with zero initial state. Beyond motivation for terminology, this development shows that linear state feedback for a linear state equation corresponds directly to feedback of $\Psi(s)\xi(s)$ in the associated pseudo-state representation (26).

Now, as shown in Figure 17.14, consider linear state feedback for (26) represented by

$$U(s) = K\Psi(s)\xi(s) + MR(s) \tag{28}$$

where K and M are constant matrices of dimensions $m \times n$ and $m \times m$, respectively. We assume that M is invertible. To develop a polynomial fraction description for the resulting closed-loop transfer function, substitute (28) into (26) to obtain

$$\left[D(s) - K\Psi(s) \right]\xi(s) = MR(s)$$

$$Y(s) = N(s)\xi(s)$$

Nonsingularity of the polynomial matrix $D(s) - K\Psi(s)$ is assured, since its column degree coefficient matrix is the same as the assumed-invertible column degree

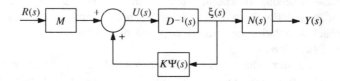

17.14 Figure   Laplace transform diagram for state feedback.

coefficient matrix for $D(s)$. Therefore we can write

$$\xi(s) = \Big[D(s) - K\Psi(s) \Big]^{-1} MR(s)$$
$$Y(s) = N(s)\xi(s) \tag{29}$$

Since M is invertible this gives a right polynomial fraction description for the closed-loop transfer function:

$$N(s)\hat{D}^{-1}(s) = N(s) \Big[M^{-1}D(s) - M^{-1}K\Psi(s) \Big]^{-1} \tag{30}$$

This description is not necessarily coprime, though $\hat{D}(s)$ is column reduced.

Calm reflection on (30) reveals that choices of K and invertible M provide complete freedom to specify the coefficients of $\hat{D}(s)$. In detail, suppose

$$D(s) = D_{hc}\Delta(s) + D_l\Psi(s)$$

and suppose the desired $\hat{D}(s)$ is

$$\hat{D}(s) = \hat{D}_{hc}\Delta(s) + \hat{D}_l\Psi(s)$$

Then the feedback gains

$$M = D_{hc}\hat{D}_{hc}^{-1}, \quad K = -M\hat{D}_l + D_l$$

accomplish the task. Although the choices of K and M do not directly affect $N(s)$, there is an indirect effect in that (30) might not be coprime. This occurs in a more obvious fashion in the single-input, single-output case when linear state feedback places a root of the denominator polynomial coincident with a root of the numerator polynomial.

EXERCISES

Exercise 17.1   If $G(s) = D^{-1}(s)N(s)$ is coprime and $D(s)$ is row reduced, show how to use the right polynomial fraction description

$$G^T(s) = N^T(s)[D^T(s)]^{-1}$$

and controller form to compute a minimal realization for $G(s)$.

Exercise 17.2   Suppose the linear state equation

$$\dot{x}(t) = Ax(t) + Bu(t)$$

$$y(t) = Cx(t)$$

is controllable and observable, and

$$C(sI - A)^{-1}B = N(s)D^{-1}(s)$$

is a coprime polynomial fraction description with $D(s)$ column reduced. Given any $p \times n$ matrix C_a, show that there exists a polynomial matrix $N_a(s)$ such that

$$C_a(sI - A)^{-1}B = N_a(s)D^{-1}(s)$$

Conversely show that if $N_a(s)$ is a $p \times m$ polynomial matrix such that $N_a(s)D^{-1}(s)$ is strictly proper, then there exists a C_a such that this relation holds.

Exercise 17.3   Write out a detailed proof of Theorem 17.7.

Exercise 17.4   Suppose the linear state equation

$$\dot{x}(t) = Ax(t) + Bu(t)$$

$$y(t) = Cx(t)$$

is controllable and observable, with $m = p$. Use the product

$$\begin{bmatrix} I_n & 0 \\ C(sI - A)^{-1} & I_m \end{bmatrix} \begin{bmatrix} sI - A & B \\ -C & 0 \end{bmatrix}$$

to give a characterization of transmission zeros of $C(sI - A)^{-1}B$ that are not also poles in terms of the matrix

$$\begin{bmatrix} sI - A & B \\ C & 0 \end{bmatrix}$$

Exercise 17.5   Suppose the linear state equation

$$\dot{x}(t) = Ax(t) + Bu(t)$$

$$y(t) = Cx(t)$$

with $p < m$ is controllable and observable, and

$$G(s) = C(sI - A)^{-1}B$$

has rank p for almost all complex values of s. Suppose the complex number s_o is not a pole of $G(s)$. Prove that s_o is a transmission zero of $G(s)$ if and only if there is a nonzero complex $1 \times p$ vector h with the property that for any complex $m \times 1$ vector u_o there is a complex $n \times 1$ vector x_o such that

$$hCe^{At}x_o + \int_0^t hCe^{A(t-\sigma)}Bu_o e^{s_o\sigma}\, d\sigma = 0, \quad t \geq 0$$

Phrase this result as a characterization of transmission zeros in terms of a complete-response property, and contrast the result with Theorem 17.13.

Exercise 17.6   Given a strictly-proper transfer function $G(s)$, let $n(s)$ be the greatest common divisor of the numerators of all the entries of $G(s)$. The roots of the polynomial $n(s)$ are called the *blocking zeros* of $G(s)$. Show that every blocking zero of $G(s)$ is a transmission zero. Show that the converse holds if either $m = 1$ or $p = 1$, but not otherwise.

Exercise 17.7   What are the transmission zeros of the transfer function

$$G(s) = \begin{bmatrix} s-1 & s+1 \\ s+1 & s \end{bmatrix} \begin{bmatrix} (s+\lambda)^2 & 0 \\ 0 & (s+4)^2 \end{bmatrix}^{-1}$$

if $\lambda = 3$? If $\lambda = 1$?

Exercise 17.8   Consider a linear state equation

$$\dot{x}(t) = Ax(t) + Bu(t)$$

$$y(t) = Cx(t)$$

where both B and C are square and invertible. What are the poles and transmission zeros of

$$G(s) = C(sI - A)^{-1}B$$

Exercise 17.9   Prove in detail that s_o is a pole of $G(s)$ in the sense of Definition 17.8 if and only if some entry of $G(s)$ satisfies $|G_{ij}(s_o)| = \infty$.

Exercise 17.10   For a plant described by the right polynomial fraction

$$Y(s) = N(s)D^{-1}(s)U(s)$$

with dynamic output feedback described by the left polynomial fraction

$$U(s) = D_c^{-1}(s)N_c(s)Y(s) + MR(s)$$

show that the closed-loop transfer function can be written as

$$Y(s) = N(s)\left[D_c(s)D(s) - N_c(s)N(s) \right]^{-1} D_c(s)MR(s)$$

What natural assumption on the plant and feedback guarantees nonsingularity of the polynomial matrix $D_c(s)D(s) - N_c(s)N(s)$?

NOTES

Note 17.1   Constructions for various forms of minimal realizations from polynomial fraction descriptions are given in Chapter 6 of

T. Kailath, *Linear System Theory*, Prentice Hall, New York, 1980

Also discussed are special forms for the polynomial fraction description that imply additional properties of particular minimal realizations. A method for computing coprime left and right polynomial fraction descriptions for a given linear state equation is presented in

C.H. Fang, ''A new approach for calculating doubly-coprime matrix fraction descriptions,'' *IEEE Transactions on Automatic Control*, Vol. 37, No. 1, pp. 138–141, 1992

Note 17.2   Transmission zeros of a linear state equation can be characterized in terms of rank properties of the *system matrix*

$$\begin{bmatrix} sI - A & B \\ -C & 0 \end{bmatrix}$$

thereby avoiding the transfer function. Another alternative is to characterize transmission zeros in terms of the *Smith-McMillan* form for the transfer function. Original sources for various approaches include

H.H. Rosenbrock, *State Space and Multivariable Theory*, Wiley Interscience, New York, 1970

C.A. Desoer, J.D. Schulman, "Zeros and poles of matrix transfer functions and their dynamical interpretation," *IEEE Transactions on Circuits and Systems*, Vol. 21, No. 1, pp. 3–8, 1974

See also the survey

C.B. Schrader, M.K. Sain, "Research in system zeros: A survey," *International Journal of Control*, Vol. 50, No. 4, pp. 1407–1433, 1989

Note 17.3   Various efforts have been made to extend the concepts of poles and zeros to the time-varying case. This requires more sophisticated algebraic constructs, as indicated by the reference

E.W. Kamen, "Poles and zeros of linear time-varying systems," *Linear Algebra and Its Applications*, Vol. 98, pp. 263–289, 1988

Note 17.4   The standard observer, estimated-state-feedback approach to output feedback is treated in terms of polynomial fractions in

B.D.O. Anderson, V.V. Kucera, "Matrix fraction construction of linear compensators," *IEEE Transactions on Automatic Control*, Vol. 30, No. 11, pp. 1112–1114, 1985

Further material regarding applications of polynomial fractions in linear control theory can be found in the books by Wolovich and Vardulakis cited in Note 16.2, and

F.M. Callier, C.A. Desoer, *Multivariable Feedback Systems*, Springer-Verlag, New York, 1982

C.T. Chen, *Linear System Theory and Design*, Holt, Rinehart, and Winston, New York, 1984

18

GEOMETRIC THEORY

We begin the study of subspace constructions that can be used to characterize the fine structure of a time-invariant linear state equation. After a brief review of relevant linear-algebraic notions, subspaces related to the concepts of controllability, observability, and stability are introduced. Then we consider corresponding definitions that facilitate extension of these concepts to a closed-loop state equation resulting from state feedback.

Definitions of the subspaces of interest are offered in a coordinate-free manner, that is, the definitions do not presuppose any choice of basis for the ambient vector space. However implications of the definitions are most clearly exhibited in terms of particular basis choices. Therefore the significance of various constructions often will be interpreted in terms of the structure of a linear state equation after a state-variable change corresponding to a particular change in basis. Additional subspace properties and related computational algorithms are developed in Chapter 19 while addressing sample problems in linear control theory.

Subspaces

The geometric theory rests on fundamentals of vector spaces, rather than the matrix algebra emphasized in earlier chapters. Therefore a review of the axioms for finite-dimensional linear vector spaces, and the properties of such spaces, is recommended. Basic notions such as the *span* of a set of vectors and a *basis* for a vector space are used freely, though we pause to recapitulate concepts related to subspaces of a vector space.

The vector spaces of interest can be viewed as R^k, for appropriate dimension k, though a more abstract notation is convenient and traditional. Suppose $\mathcal{V}$ and $\mathcal{W}$ are vector subspaces of a vector space X over the real field R. In this chapter the symbol '=' often means subspace equality, for example $\mathcal{V} = \mathcal{W}$. The symbol '$\subset$' denotes subspace inclusion, for example $\mathcal{V} \subset \mathcal{W}$, where this is not interpreted as strict inclusion. Thus $\mathcal{V} = \mathcal{W}$ is equivalent to the pair of inclusions $\mathcal{V} \subset \mathcal{W}$ and $\mathcal{W} \subset \mathcal{V}$. The usual method for proving that subspaces are identical is to show both inclusions. Also the symbol '0' means the zero vector, zero scalar, or the subspace 0, as indicated by context.

Various other subspaces of X arise from subspaces $\mathcal{V}$ and $\mathcal{W}$. The *intersection* of $\mathcal{V}$ and $\mathcal{W}$ is defined by

$$\mathcal{V} \cap \mathcal{W} = \left\{ v \mid v \in \mathcal{V}; \ v \in \mathcal{W} \right\}$$

and the *sum* of subspaces is

$$\mathcal{V} + \mathcal{W} = \left\{ v + w \mid v \in \mathcal{V}; \ w \in \mathcal{W} \right\} \tag{1}$$

It is not difficult to verify that these indeed are subspaces. If $\mathcal{V} + \mathcal{W} = X$ and $\mathcal{V} \cap \mathcal{W} = 0$, then we write the *direct sum* $X = \mathcal{V} \oplus \mathcal{W}$. These basic operations extend to any finite number of subspaces in a natural way.

Linear maps on vector spaces evoke additional subspaces. If $\mathcal{Y}$ is another vector space over R and A is a linear map, $A : X \to \mathcal{Y}$, then the *kernel* or *null space* of A is

$$Ker\,[A] = \left\{ x \mid x \in X; \ Ax = 0 \right\}$$

and the *image* or *range space* of A is

$$Im\,[A] = \left\{ Ax \mid x \in X \right\}$$

Confirmation that these are subspaces is straightforward, though it should be emphasized that $Ker\,[A] \subset X$, while $Im\,[A] \subset \mathcal{Y}$. Finally, if $\mathcal{V} \subset X$ and $\mathcal{Z} \subset \mathcal{Y}$, then the *image* of $\mathcal{V}$ under A is the subspace of $\mathcal{Y}$ given by

$$A\mathcal{V} = \left\{ Av \mid v \in \mathcal{V} \right\}$$

Of course $Im\,[A]$ is the same subspace as the image of X under A. The *inverse image* of $\mathcal{Z}$ with respect to A is the subspace of X

$$A^{-1}\mathcal{Z} = \left\{ x \mid x \in X; \ Ax \in \mathcal{Z} \right\}$$

These notations should be used carefully. Although $A(\mathcal{V} + \mathcal{W}) = A\mathcal{V} + A\mathcal{W}$, note that $(A_1 + A_2)\mathcal{V}$ typically is not the same subspace as $A_1\mathcal{V} + A_2\mathcal{V}$. However

$$(A_1 + A_2)\mathcal{V} \subset A_1\mathcal{V} + A_2\mathcal{V}$$

and

$$A_1\mathcal{V} + (A_1 + A_2)\mathcal{V} = A_1\mathcal{V} + A_2\mathcal{V} \tag{2}$$

Also the notation $A^{-1}z$ does not mean that A^{-1} is applied to anything, or even that A is an invertible linear map. On choosing bases for X and $\mathcal{Y}$ the map A is represented by a real matrix that also is denoted by A, with confidence that the chance of confusion is slight.

Invariant Subspaces

Throughout this chapter we deal with concepts associated to the m-input, p-output, n-dimensional, time-invariant linear state equation

$$\dot{x}(t) = Ax(t) + Bu(t), \quad x(0) = x_o$$

$$y(t) = Cx(t) \tag{3}$$

The coefficient matrices presume bases choices for the state, input, and output spaces, namely R^n, R^m, and R^p. However, adhering to tradition in the geometric theory, we adopt a more abstract view and write the state space R^n as X, the input space R^m as $\mathcal{U}$, and the output space R^p as $\mathcal{Y}$. Then the coefficient matrices in (3) are viewed as representing linear maps according to

$$A:X \to X, \quad B:\mathcal{U} \to X, \quad C:X \to \mathcal{Y}$$

State variable changes in (3) yielding $P^{-1}AP$, $P^{-1}B$, and CP usually are discussed in the language of basis changes in the state space X. The subspace $Im[B] \subset X$ occurs frequently, and is given the special symbol $\mathcal{B} = Im[B]$. Various additional subspaces are generated, but the dependence on (3) is suppressed routinely to simplify the notation and language.

The foundation upon which the development is erected should be familiar from basic linear algebra.

18.1 Definition   A subspace $\mathcal{V} \subset X$ is called an *invariant subspace* for $A:X \to X$ if $A\mathcal{V} \subset \mathcal{V}$.

18.2 Example   The subspaces 0, X, $Ker[A]$, and $Im[A]$ of X all are invariant subspaces for A. If $\mathcal{V}$ is an invariant subspace for A, then so is $A^k\mathcal{V}$ for any nonnegative integer k. Other subspaces associated with (3) such as $\mathcal{B}$ and $Ker[C]$ are not invariant subspaces for A, in general.
□□□

An important reason invariant subspaces are of interest for linear state equations can be explained in terms of the zero-input solution for (3). Suppose $\mathcal{V}$ is an invariant subspace for A. Then recalling the representation for the matrix exponential in Property 5.8,

$$e^{At}\mathcal{V} = \left(\sum_{k=0}^{n-1} \alpha_k(t)A^k \right) \mathcal{V} \subset \sum_{k=0}^{n-1} \alpha_k(t)A^k\mathcal{V}$$

$$\subset \mathcal{V} \tag{4}$$

for any value of $t \geq 0$. Therefore if $x_o \in \mathcal{V}$, then the zero-input solution of (3) satisfies $x(t) \in \mathcal{V}$ for all $t \geq 0$. (Notice that the calculation in (4) involves sums of matrices in the first term on the right side, then sums of subspaces in the second. This kind of mixing occurs frequently, though usually without comment.) Conversely a simple contradiction argument shows that if a subspace $\mathcal{V}$ is endowed with the property that $x_o \in \mathcal{V}$ implies the zero input solution of (3) satisfies $x(t) \in \mathcal{V}$ for all $t \geq 0$, then $\mathcal{V}$ is an invariant subspace for A.

Bringing the input signal into play, we consider first a special subspace and associated standard notation and terminology.

18.3 Definition   The subspace of X given by

$$<A \,|\, \mathcal{B}> = \mathcal{B} + A\mathcal{B} + \cdots + A^{n-1}\mathcal{B} \tag{5}$$

is called the *controllable subspace* for the linear state equation (3)

The Cayley-Hamilton theorem immediately implies that $<A \,|\, \mathcal{B}>$ is an invariant subspace for A. Also it is easy to show that $<A \,|\, \mathcal{B}>$ is the smallest subspace of X that contains $\mathcal{B}$ and is invariant under A. That is, every subspace that contains $\mathcal{B}$ and is invariant under A contains $<A \,|\, \mathcal{B}>$. Finally we note that the computation of $<A \,|\, \mathcal{B}>$, more specifically the computation of a basis for the subspace, involves selecting linearly independent columns from the set of matrices $B, AB, \ldots, A^{n-1}B$.

An important property of $<A \,|\, \mathcal{B}>$ relates to the solution of (3) with nonzero input signal. By invariance, $x_o \in <A \,|\, \mathcal{B}>$ implies

$$e^{At}x_o \in <A \,|\, \mathcal{B}>, \quad t \geq 0$$

If $u(t)$ is a continuous input signal (for consistency with our default assumptions), then

$$\int_0^t e^{A(t-\sigma)}Bu(\sigma)d\sigma = \sum_{k=0}^{n-1} A^k B \int_0^t \alpha_k(t-\sigma)u(\sigma)\,d\sigma$$

$$\in <A \,|\, \mathcal{B}>, \quad t \geq 0$$

The integral term on the right side provides, for each $t \geq 0$, an $m \times 1$ vector that describes the kth-summand as a linear combination of columns of $A^k B$. The immediate conclusion is that if $x_o \in <A \,|\, \mathcal{B}>$, then for any continuous input signal the corresponding solution of (3) satisfies $x(t) \in <A \,|\, \mathcal{B}>$ for all $t \geq 0$. But to justify the terminology in Definition 18.3, we need to refine the notion of controllability introduced in Chapter 9.

18.4 Definition   A vector $x_o \in X$ is called a *controllable state* for (3) if for $x(0) = x_o$ there is a finite time $t_a > 0$ and a continuous input signal $u_a(t)$, defined for $t \in [0, t_a]$, such that the corresponding solution of (3) satisfies $x(t_a) = 0$.

Recalling the controllability Gramian, in the present context written as

$$W(0, t) = \int_0^t e^{-A\sigma} BB^T e^{-A^T\sigma} \, d\sigma \tag{6}$$

we first establish a preliminary result.

18.5 Lemma   For any $t_a > 0$,

$$<A \mid \mathcal{B}> = Im \, [W(0, t_a)]$$

Proof   Fixing $t_a > 0$, for any $n \times 1$ vector x_o,

$$W(0, t_a)x_o = \int_0^{t_a} e^{-A\sigma} BB^T e^{-A^T\sigma} x_o \, d\sigma$$

$$= \sum_{k=0}^{n-1} A^k B \int_0^{t_a} \alpha_k(-\sigma) B^T e^{-A^T\sigma} x_o \, d\sigma$$

Since each column of $A^k B$ is in $A^k\mathcal{B}$, and the kth-summand above is a linear combination of columns of $A^k B$,

$$W(0, t_a)x_o \in \mathcal{B} + A\mathcal{B} + \cdots + A^{n-1}\mathcal{B}$$

This gives

$$Im \, [W(0, t_a)] \subset <A \mid \mathcal{B}>$$

To establish the reverse containment we use the proof of Theorem 13.1 to define a convenient basis. Clearly $<A \mid \mathcal{B}>$ is the range space of the controllability matrix

$$\begin{bmatrix} B & AB & \cdots & A^{n-1}B \end{bmatrix} \tag{7}$$

for the linear state equation (3). Define an invertible $n \times n$ matrix P column-wise by choosing a basis for $<A \mid \mathcal{B}>$ and extending to a basis for X. Then changing state variables according to $z(t) = P^{-1}x(t)$ leads to a new linear state equation in $z(t)$ with the coefficient matrices

$$P^{-1}AP = \begin{bmatrix} \hat{A}_{11} & \hat{A}_{12} \\ 0 & \hat{A}_{22} \end{bmatrix}, \quad P^{-1}B = \begin{bmatrix} \hat{B}_{11} \\ 0 \end{bmatrix}$$

These expressions can be used to write $W(0, t_a)$ in (6) as

$$W(0, t_a) = P \int_0^{t_a} \exp\left(\begin{bmatrix} -\hat{A}_{11} & -\hat{A}_{12} \\ 0 & -\hat{A}_{22} \end{bmatrix}\sigma\right) \begin{bmatrix} \hat{B}_{11} \\ 0 \end{bmatrix} \begin{bmatrix} \hat{B}_{11}^T & 0 \end{bmatrix} \exp\left(\begin{bmatrix} -\hat{A}_{11}^T & 0 \\ -\hat{A}_{12}^T & -\hat{A}_{22}^T \end{bmatrix}\sigma\right) d\sigma \, P^T$$

$$= P \begin{bmatrix} \hat{W}_1(0, t_a) & 0 \\ 0 & 0 \end{bmatrix} P^T$$

where

$$\hat{W}_1(0, t_a) = \int_0^{t_a} e^{-\hat{A}_{11}\sigma} \hat{B}_{11} \hat{B}_{11}^T e^{-\hat{A}_{11}^T \sigma} \, d\sigma$$

is an invertible matrix. This representation shows that $Im\,[W(0, t_a)]$ contains any vector of the form

$$P \begin{bmatrix} z \\ 0 \end{bmatrix} \tag{8}$$

for setting

$$x = [\,P^T\,]^{-1} \begin{bmatrix} \hat{W}_1^{-1}(0, t_a)z \\ 0 \end{bmatrix}$$

we obtain

$$W(0, t_1)x = P \begin{bmatrix} z \\ 0 \end{bmatrix}$$

Since

$$A^k B = P \begin{bmatrix} \hat{A}_{11}^k \hat{B}_{11} \\ 0 \end{bmatrix}, \quad k = 0, 1, \cdots$$

has the form (8), it follows that $<A\,|\,\mathcal{B}> \subset Im\,[W(0, t_a)]$.
□□□

    Lemma 18.5 provides the tool needed to show that $<A\,|\,\mathcal{B}>$ is exactly the set of controllable states.

18.6 Theorem   A vector $x_o \in X$ is a controllable state for the linear state equation (3) if and only if $x_o \in\ <A\,|\,\mathcal{B}>$.

    Proof   Fix $t_a > 0$. If $x_o \in\ <A\,|\,\mathcal{B}>$ then Lemma 18.5 implies that there exists a vector $z \in X$ such that $x_o = W(0, t_a)z$. Setting

$$u(t) = -B^T e^{-A^T t} z \tag{9}$$

the solution of (3) with $x(0) = x_o$ is, when evaluated at $t = t_a$,

$$x(t_a) = e^{At_a} x_o - \int_0^{t_a} e^{A(t_a - \sigma)} BB^T e^{-A^T \sigma} z \, d\sigma$$

$$= e^{At_a} [\, x_o - W(0, t_a)z \,]$$

$$= 0$$

Conversely if x_o is a controllable state then there is a finite time $t_a > 0$ and a continuous input $u_a(t)$ such that

$$0 = e^{At_a} x_o + \int_0^{t_a} e^{A(t_a - \sigma)} B u_a(\sigma)\, d\sigma \tag{10}$$

Therefore

$$x_o = -\int_0^{t_a} e^{-A\sigma} B u_a(\sigma)\, d\sigma$$

$$= \sum_{k=0}^{n-1} A^k B \int_0^{t_a} -\alpha_k(-\sigma) u_a(\sigma)\, d\sigma \tag{11}$$

and this implies $x_o \in\ <A\,|\,\mathcal{B}>$.
□□□

The proof of Theorem 18.6 shows that a linear state equation is controllable in the sense of Definition 9.1 if and only if every state is a controllable state. (The fact that t_a can be fixed independent of the initial state is crucial — the diligent should supply reasoning.) Of course this can be stated in geometric language.

18.7 Corollary   The linear state equation (3) is controllable if and only if $<A\,|\,\mathcal{B}> = X$.

It can be shown that $<A\,|\,\mathcal{B}>$ also is precisely the set of states that can be reached from the zero initial state in finite time using a continuous input signal. Such a characterization of $<A\,|\,\mathcal{B}>$ as the set of *reachable states* is pursued in Exercise 18.8.

Using the state variable change in the proof of Lemma 18.5, (3) can be written in terms of $z(t) = P^{-1} x(t)$ as a partitioned linear state equation

$$\begin{bmatrix} \dot{z}_c(t) \\ \dot{z}_{nc}(t) \end{bmatrix} = \begin{bmatrix} \hat{A}_{11} & \hat{A}_{12} \\ 0 & \hat{A}_{22} \end{bmatrix} \begin{bmatrix} z_c(t) \\ z_{nc}(t) \end{bmatrix} + \begin{bmatrix} \hat{B}_{11} \\ 0 \end{bmatrix} u(t)$$

$$y(t) = CPz(t) \tag{12}$$

Assuming $dim\ <A\,|\,\mathcal{B}> = q < n$, the submatrix $\hat{A}_{11}$ is $q \times q$, while $\hat{B}_{11}$ is $q \times m$. The component of the state equation (12) that describes $z_c(t)$,

$$\dot{z}_c(t) = \hat{A}_{11} z_c(t) + \hat{A}_{12} z_{nc}(t) + \hat{B}_{11} u(t)$$

is controllable. That is,

$$\mathrm{rank}\begin{bmatrix} \hat{B}_{11} & \hat{A}_{11}\hat{B}_{11} & \cdots & \hat{A}_{11}^{q-1}\hat{B}_{11} \end{bmatrix} = q$$

(The extra term $\hat{A}_{12} z_{nc}(t)$, known from $z_{nc}(0)$, does not change the ability to drive an initial state $z_c(0)$ to the origin in finite time.) Obviously the component of the state equation (12) describing $z_{nc}(t)$, namely

$$\dot{z}_{nc}(t) = \hat{A}_{22}z_{nc}(t)$$

is not controllable. The structure of (12) is exhibited in Figure 18.8.

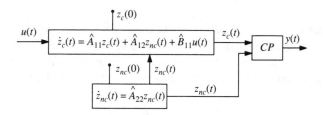

18.8 Figure   Decomposition of the state equation (12).

Coordinate changes of this type are used to display the structure of linear state equations relative to other invariant subspaces, and formal terminology is convenient.

18.9 Definition   Suppose $\mathcal{V} \subset X$ is an invariant subspace for $A : X \to X$. Then a basis $p_1, \ldots, p_n$ for X such that $p_1, \ldots, p_v$ span $\mathcal{V}$ is said to be *adapted* to the subspace $\mathcal{V}$.

In general, for the linear state equation (3), suppose $\mathcal{V}$ is a dimension-v invariant subspace for A, not necessarily containing $\mathcal{B}$. Suppose also that columns of the $n \times n$ matrix P form a basis for X adapted to $\mathcal{V}$. Then the state variable change $z(t) = P^{-1}x(t)$ yields

$$\begin{bmatrix} \dot{z}_a(t) \\ \dot{z}_b(t) \end{bmatrix} = \begin{bmatrix} \hat{A}_{11} & \hat{A}_{12} \\ 0 & \hat{A}_{22} \end{bmatrix} \begin{bmatrix} z_a(t) \\ z_b(t) \end{bmatrix} + \begin{bmatrix} \hat{B}_{11} \\ \hat{B}_{21} \end{bmatrix} u(t)$$

$$y(t) = CPz(t) \tag{13}$$

In terms of the basis $p_1, \ldots, p_n$ for X, an $n \times 1$ vector $z \in X$ satisfies $z \in \mathcal{V}$ if and only if it has the form

$$z = \begin{bmatrix} z_a \\ 0_{(n-v) \times 1} \end{bmatrix}$$

The action of A on $\mathcal{V}$ is described in the new basis by the partition $\hat{A}_{11}$ since

$$\begin{bmatrix} \hat{A}_{11} & \hat{A}_{12} \\ 0 & \hat{A}_{22} \end{bmatrix} \begin{bmatrix} z_a \\ 0 \end{bmatrix} = \begin{bmatrix} \hat{A}_{11}z_a \\ 0 \end{bmatrix}$$

Clearly $\hat{A}_{11}$ inherits features from A, for example eigenvalues. These features can be interpreted as properties of the partitioned linear state equation (13) as follows.

The linear state equation (13) can be written as two component state equations

$$\dot{z}_a(t) = \hat{A}_{11}z_a(t) + \hat{A}_{12}z_b(t) + \hat{B}_{11}u(t)$$
$$\dot{z}_b(t) = \hat{A}_{22}z_b(t) + \hat{B}_{21}u(t) \tag{14}$$

the first of which we specifically call the *component state equation corresponding to* $\mathcal{V}$. Exponential stability of (13) (equivalent to exponential stability of (3)) is equivalent to exponential stability of both state equations in (14). Also an easy exercise shows that controllability of (13) (equivalent to controllability of (3)) implies

$$\text{rank} \left[\hat{B}_{21} \; \hat{A}_{22}\hat{B}_{21} \; \cdots \; \hat{A}_{22}^{n-\nu-1}\hat{B}_{21} \right] = n - \nu$$

However simple examples show that controllability of (13) does not imply that

$$\left[\hat{B}_{11} \; \hat{A}_{11}\hat{B}_{11} \; \cdots \; \hat{A}_{11}^{\nu-1}\hat{B}_{11} \right]$$

has rank ν. In case this is puzzling in relation to the special case where $\mathcal{V} = \langle A | \mathcal{B} \rangle$ in (12), note that if (12) is controllable then $z_{nc}(t)$ is vacuous.

Often geometric features of a linear state equation are discussed in a way that leaves understood the variable change. As with subspaces, the various properties we consider — controllability, observability, stability, and eigenvalue assignment — are uninfluenced by state variable change. At times it is convenient to address these proper- ties in a particular set of coordinates, but other times it is convenient to leave the vari- able change unmentioned.

The geometric treatment of observability for the linear state equation (3) will not be pursued in such detail. The basic definition starts from a converse notion, and just as in Chapter 9 we consider only the zero-input response.

18.10 Definition   The subspace $\mathcal{N} \subset X$ given by

$$\mathcal{N} = \bigcap_{k=0}^{n-1} Ker[CA^k]$$

is called the *unobservable subspace* for (3).

Another way of writing the unobservable subspace for (3) involves a slight exten- sion of our inverse-image notation:

$$\mathcal{N} = Ker[C] \cap A^{-1}Ker[C] \cap \cdots \cap A^{-(n-1)}Ker[C]$$

It is easy to verify that $\mathcal{N}$ is an invariant subspace for A, and it is the largest subspace contained in $Ker[C]$ that is invariant under A. Also $\mathcal{N}$ is the null space of the observa- bility matrix

$$
\begin{bmatrix}
C \\
CA \\
\vdots \\
CA^{n-1}
\end{bmatrix}
\tag{15}
$$

By showing that, for any $t_a > 0$,

$$
\mathcal{N} = Ker\,[M\,(0,\,t_a)]
$$

where

$$
M\,(0,\,t) = \int_0^t e^{A^T\sigma} C^T C e^{A\sigma}\,d\sigma
\tag{16}
$$

is the observability Gramian for (3), the following results derive from an omitted linear-algebra argument.

18.11 Theorem   Suppose the linear state equation (3) with zero input and unknown initial state x_o yields the output signal $y(t)$. Then for any $t_a > 0$, x_o can be determined up to an additive $n \times 1$ vector in $\mathcal{N}$ from knowledge of $y(t)$ for $t \in [0,\,t_a]$.

18.12 Corollary   The linear state equation (3) is observable if and only if $\mathcal{N} = 0$.

Additional invariant subspaces of importance are related to the internal stability properties of (3). Suppose that the characteristic polynomial of A is factored into a product of polynomials

$$
det\,(\lambda I - A) = p^-(\lambda) p^+(\lambda)
$$

where all roots of $p^-(\lambda)$ have negative real parts, and all roots of $p^+(\lambda)$ have nonnegative real parts. Each polynomial has real coefficients, and we denote the respective polynomial degrees by n^- and n^+.

18.13 Definition   The subspace of X given by

$$
X^- = Ker\,[p^-(A)]
$$

is called the *stable subspace* for the linear state equation (3), and

$$
X^+ = Ker\,[p^+(A)]
$$

is called the *unstable subspace* for (3).

Obviously X^- and X^+ are subspaces of X. Also both are invariant subspaces for A; the key to proving this is that $Ap(A) = p(A)A$ for any polynomial $p(\lambda)$. The stability terminology is justified by a fundamental decomposition property.

18.14 Theorem   The stable and unstable subspaces for the linear state equation (3) provide the direct sum decomposition

$$X = X^- \oplus X^+ \qquad (17)$$

Furthermore in a basis adapted to X^- and X^+ the component state equation corresponding to X^- is exponentially stable, while all eigenvalues of the component state equation corresponding to X^+ have nonnegative real parts.

Proof   Since the polynomials $p^-(\lambda)$ and $p^+(\lambda)$ are coprime (have no roots in common), there exist polynomials $q_1(\lambda)$ and $q_2(\lambda)$ such that

$$p^-(\lambda)q_1(\lambda) + p^+(\lambda)q_2(\lambda) = 1$$

(This standard result from algebra is a special case of Theorem 16.9. The polynomials $q_1(\lambda)$ and $q_2(\lambda)$ can be computed by elementary row operations as described in Theorem 16.6.) Then

$$p^-(A)q_1(A) + p^+(A)q_2(A) = I \qquad (18)$$

For any vector $z \in X$, multiplying (18) on the right by z shows that we can write

$$z = z^+ + z^-$$

where

$$z^+ = p^-(A)q_1(A)z$$
$$z^- = p^+(A)q_2(A)z$$

The superscript notation z^- and z^+ is suggestive, and indeed the Cayley-Hamilton theorem gives

$$p^-(A)z^- = p^-(A)p^+(A)q_1(A)z = 0$$
$$p^+(A)z^+ = p^+(A)p^-(A)q_2(A)z = 0$$

That is,

$$z^- \in X^-, \quad z^+ \in X^+ \qquad (19)$$

and thus $X = X^- + X^+$. To show that $X^- \cap X^+ = 0$, we note that if $z \in X^- \cap X^+$, then

$$p^-(A)z = p^+(A)z = 0$$

Using (18), and commutativity of polynomials in A, gives

$$z = p^-(A)q_1(A)z + p^+(A)q_2(A)z$$
$$= 0$$

Therefore (17) is verified.

Now suppose the columns of P form a basis for X adapted to X^-. Then the first n^- columns of P form a basis for X^-, the remaining n^+ columns form a basis for X^+, and the state variable change $z(t) = P^{-1}x(t)$ yields the partitioned linear state equation

$$\begin{bmatrix} \dot{z}_a(t) \\ \dot{z}_b(t) \end{bmatrix} = \begin{bmatrix} \hat{A}_{11} & 0 \\ 0 & \hat{A}_{22} \end{bmatrix} \begin{bmatrix} z_a(t) \\ z_b(t) \end{bmatrix} + \begin{bmatrix} \hat{B}_{11} \\ \hat{B}_{21} \end{bmatrix} u(t)$$

$$y(t) = CPz(t) \tag{20}$$

Since the characteristic polynomials of the component state equations corresponding to X^- and X^+ are, respectively,

$$\det(\lambda I - \hat{A}_{11}) = p^-(\lambda) , \quad \det(\lambda I - \hat{A}_{22}) = p^+(\lambda)$$

the eigenvalue claims are obvious.

18.15 Example As usual, a diagonal-form state equation provides a helpful sanity check. Let $X = R^4$ with the standard basis $e_1, \ldots, e_4$, and consider the state equation

$$\dot{x}(t) = \begin{bmatrix} 1 & 0 & 0 & 0 \\ 0 & 2 & 0 & 0 \\ 0 & 0 & -3 & 0 \\ 0 & 0 & 0 & -4 \end{bmatrix} x(t) + \begin{bmatrix} 1 \\ 0 \\ 0 \\ 1 \end{bmatrix} u(t) \tag{21}$$

$$y(t) = \begin{bmatrix} 0 & 1 & 1 & 1 \end{bmatrix} x(t)$$

Then the controllable subspace $<A|\mathcal{B}>$ is spanned by e_1, e_4, the unobservable subspace $\mathcal{N}$ is spanned by e_1, the stable subspace X^- is spanned by e_3, e_4, and the unstable subspace X^+ is spanned by e_1, e_2. Checking these answers both from basic intuition and from definitions of the subspaces is highly recommended.

Controlled Invariant Subspaces

Linear state feedback can be used to modify some of the invariant subspaces for a given linear state equation. (Not all can be modified: X is an example of an invariant subspace that remains invariant for all state feedback gains.) This leads to the formulation of feedback control problems in terms of specified invariant subspaces for the closed-loop state equation. However we begin by showing that the controllable subspace for (3) cannot be modified by state feedback. Then the effect of feedback on other types of invariant subspaces is considered.

In a departure from the notation of Chapter 14, but consonant with the geometric literature, we write linear state feedback as

$$u(t) = Fx(t) + Gv(t) \tag{22}$$

where F is $m \times n$, G is $m \times m$, and $v(t)$ represents the $m \times 1$ reference input. The resulting closed-loop state equation is

$$\dot{x}(t) = (A + BF)x(t) + BGv(t)$$

$$y(t) = Cx(t) \qquad\qquad (23)$$

In Exercise 13.11 the objective is to show that for $G = I$ the closed-loop state equation is controllable if the open-loop state equation is controllable, regardless of F. We generalize this by showing that the set of controllable states does not change under such state feedback. The result holds also for any G that is invertible, since invertibility of G guarantees $\mathcal{B} = Im\,[BG]$.

18.16 Theorem   For any F,

$$<A+BF\,|\,\mathcal{B}> \; = \; <A\,|\,\mathcal{B}> \qquad\qquad (24)$$

Proof   For any F and any subspace $\mathcal{W}$ we can write (similar to (2))

$$\mathcal{B} + (A + BF)\mathcal{W} = \mathcal{B} + A\mathcal{W}$$

This immediately provides the first step of an induction proof:

$$\mathcal{B} + (A + BF)\mathcal{B} = \mathcal{B} + A\mathcal{B}$$

Assume K is a positive integer and

$$\mathcal{B} + (A + BF)\mathcal{B} + \;\cdots\; + (A + BF)^{K}\mathcal{B} = \mathcal{B} + A\mathcal{B} + \;\cdots\; + A^{K}\mathcal{B}$$

Then

$$\mathcal{B} + (A + BF)\mathcal{B} + \;\cdots\; + (A + BF)^{K+1}\mathcal{B} = \mathcal{B} + (A + BF)[\;\mathcal{B} + \;\cdots\; + (A + BF)^{K}\mathcal{B}\,]$$

$$= \mathcal{B} + (A + BF)[\;\mathcal{B} + \;\cdots\; + A^{K}\mathcal{B}\,]$$

$$= \mathcal{B} + A[\;\mathcal{B} + \;\cdots\; + A^{K}\mathcal{B}\,]$$

$$= \mathcal{B} + A\mathcal{B} + \;\cdots\; + A^{K+1}\mathcal{B}$$

This induction argument proves (24)
□□□

Consider again the linear state equation (3) written, after state variable change, in the form (12). Applying the partitioned state feedback

$$u(t) = \begin{bmatrix} F_{11} & F_{12} \end{bmatrix} \begin{bmatrix} z_c(t) \\ z_{nc}(t) \end{bmatrix} + v(t)$$

to (12) yields the closed-loop state equation

$$\begin{bmatrix} \dot{z}_c(t) \\ \dot{z}_{nc}(t) \end{bmatrix} = \begin{bmatrix} \hat{A}_{11}+\hat{B}_{11}F_{11} & \hat{A}_{12}+\hat{B}_{11}F_{12} \\ 0 & \hat{A}_{22} \end{bmatrix} \begin{bmatrix} z_c(t) \\ z_{nc}(t) \end{bmatrix} + \begin{bmatrix} \hat{B}_{11} \\ 0 \end{bmatrix} v(t)$$

$$y(t) = CPz(t) \qquad\qquad (25)$$

From the discussion following (12) it is clear that F_{11} can be chosen so that $\hat{A}_{11} + \hat{B}_{11}F_{11}$ has any desired eigenvalues. It is also important to note that regardless of F the eigenvalues of $\hat{A}_{22}$ in (25) remain fixed. That is, there is a factor of the characteristic polynomial for (25) that cannot be changed by state feedback.

Basic terminology used to discuss additional invariant subspaces for the closed-loop state equation is introduced next.

18.17 Definition   A subspace $\mathcal{V} \subset X$ is called a *controlled invariant subspace* for the linear state equation (3) if there exists an $m \times n$ matrix F such that $\mathcal{V}$ is an invariant subspace for $(A + BF)$. Such an F is called a *friend* of $\mathcal{V}$.

The subspaces 0, $<A \mid \mathcal{B}>$, and X all are controlled invariant subspaces for (3), and typically there are many more. Motivation for considering such subspaces can be provided by again considering properties achievable by state feedback.

18.18 Example   Suppose $\mathcal{V}$ is a controlled invariant subspace for (3), with $\mathcal{V} \subset Ker[C]$. Using a friend F of $\mathcal{V}$ to define the linear state feedback

$$u(t) = Fx(t)$$

yields

$$\dot{x}(t) = (A + BF)x(t) , \quad x(0) = x_o$$

$$y(t) = Cx(t)$$

This closed-loop state equation has the property that $x_o \in \mathcal{V}$ implies $y(t) = 0$ for all $t \geq 0$. Therefore the state feedback is such that $\mathcal{V}$ is contained in the unobservable subspace $\mathcal{N}$ for the closed-loop state equation.
□□□

We next provide a fundamental characterization of controlled invariant subspaces that removes explicit involvement of F.

18.19 Theorem   A subspace $\mathcal{V} \subset X$ is a controlled invariant subspace for (3) if and only if

$$A\mathcal{V} \subset \mathcal{V} + \mathcal{B} \tag{26}$$

Proof   If $\mathcal{V}$ is a controlled invariant subspace for (3), then there is a friend F of $\mathcal{V}$ such that $(A + BF)\mathcal{V} \subset \mathcal{V}$. Thus

$$A\mathcal{V} = (A + BF - BF)\mathcal{V}$$

$$\subset (A + BF)\mathcal{V} + BF\mathcal{V}$$

$$\subset \mathcal{V} + \mathcal{B}$$

Now suppose $\mathcal{V} \subset X$, and (26) holds. The following procedure constructs a friend of $\mathcal{V}$ to demonstrate that $\mathcal{V}$ is a controlled invariant subspace. With v denoting the dimension of $\mathcal{V}$, let $n \times 1$ vectors $v_1, \ldots, v_n$ be a basis for X adapted to $\mathcal{V}$. By hypothesis there exist $n \times 1$ vectors $w_1, \ldots, w_\nu \in \mathcal{V}$ and $m \times 1$ vectors $u_1, \ldots, u_\nu \in \mathcal{U}$ such that

$$Av_k = w_k - Bu_k, \quad k = 1, \ldots, \nu$$

Now let $u_{\nu+1}, \ldots, u_n$ be arbitrary $m \times 1$ vectors, all zero if simplicity is desired, and let

$$F = \begin{bmatrix} u_1 & \cdots & u_n \end{bmatrix} \begin{bmatrix} v_1 & \cdots & v_n \end{bmatrix}^{-1} \tag{27}$$

Then for $k = 1, \ldots, \nu$, with e_k the kth-column of I_n,

$$(A + BF)v_k = Av_k + BFv_k$$

$$= Av_k + B \begin{bmatrix} u_1 & \cdots & u_n \end{bmatrix} e_k$$

$$= Av_k + Bu_k$$

$$= w_k \in \mathcal{V}$$

Since any $v \in \mathcal{V}$ can be expressed as a linear combination of $v_1, \ldots, v_\nu$, we have that $\mathcal{V}$ is an invariant subspace for $(A + BF)$.
□□□

If $\mathcal{V}$ is a controlled invariant subspace, then by definition there exists at least one friend of $\mathcal{V}$. More generally it is useful to characterize all friends of $\mathcal{V}$.

18.20 Theorem   Suppose the $m \times n$ matrix F^a is a friend of $\mathcal{V}$. Then the $m \times n$ matrix F^b is a friend of $\mathcal{V}$ if and only if

$$(F^a - F^b)\mathcal{V} \subset B^{-1}\mathcal{V} \tag{28}$$

Proof   If F^a and F^b both are friends of $\mathcal{V}$, then for any $v \in \mathcal{V}$ there exist $v_a, v_b \in \mathcal{V}$ such that

$$(A + BF^a)v = v_a$$

$$(A + BF^b)v = v_b$$

Subtracting the second expression from the first gives

$$B(F^a - F^b)v = v_a - v_b$$

and since $v_a - v_b \in \mathcal{V}$ this calculation shows that (28) holds.

On the other hand if F^a is a friend of $\mathcal{V}$ and (28) holds, then given any $v_a \in \mathcal{V}$ there is a $v_b \in \mathcal{V}$ such that

$$B(F^a - F^b)v_a = (BF^a - BF^b)v_a = v_b$$

Therefore

$$(A + BF^a)v_a - (A + BF^b)v_a = v_b$$

Since F^a is a friend of $\mathcal{V}$ there exists a $v_c \in \mathcal{V}$ such that $(A + BF^a)v_a = v_c$. This gives

$$(A + BF^b)v_a = v_c - v_b \in \mathcal{V} \tag{29}$$

which shows that F^b also is a friend of $\mathcal{V}$.
□□□

Notice that this proof is carried out in terms of arbitrary vectors in $\mathcal{V}$ rather than in terms of the subspace $\mathcal{V}$ as a whole. One reason is that $(F^a - F^b)\mathcal{V}$ does not obey seductive algebraic manipulations. Namely $(F^a - F^b)\mathcal{V}$ is not necessarily the same subspace as $F^a\mathcal{V} - F^b\mathcal{V}$, nor is it the same as $(F^a + F^b)\mathcal{V}$.

Controllability Subspaces

In considering capabilities of linear state feedback with regard to stability or eigenvalue assignment, it is a displeasing fact that some controlled invariant subspaces are too large. Of course $<A\,|\,\mathcal{B}>$ is a controlled invariant subspace for (3), and eigenvalue assignability for the component of the closed-loop state equation corresponding to $<A\,|\,\mathcal{B}>$ is guaranteed. But the whole state space X also is a controlled invariant subspace for (3), and if (3) is not controllable, then eigenvalue assignment for the closed-loop state equation on X is not possible. We begin by defining a special type of controlled invariant subspace of X, and then work toward showing how it is related to the eigenvalue-assignment issue.

18.21 Definition   A subspace $\mathcal{R} \subset X$ is called a *controllability subspace* for the linear state equation (3) if there exists an $m \times n$ matrix F and an $m \times m$ matrix G such that

$$\mathcal{R} = <A+BF\,|\,Im\,[BG\,]> \tag{30}$$

The differences in terminology are subtle: A controllability subspace for (3) is the controllable subspace for a corresponding closed-loop state equation

$$\dot{x}(t) = (A + BF)x(t) + BGv(t)$$

for some choice of F and G. It should be clear that a controllability subspace for (3) is a controlled invariant subspace for (3). Also, since $Im\,[BG\,] \subset \mathcal{B}$ for any choice of G,

$$<A+BF\,|\,Im\,[BG\,]> \subset <A+BF\,|\,\mathcal{B}> = <A\,|\,\mathcal{B}>$$

for any G. That is, every controllability subspace for (3) is a subspace of the controllable subspace for (3). In the single-input case the only controllability subspaces are 0 and the controllable subspace $<A\,|\,\mathcal{B}>$, depending on whether the scalar G is nonzero. However for multi-input state equations controllability subspaces are richer geometric concepts. As a simple example, in addition to the role of F, the gain G is not necessarily invertible and can be used to isolate components of the input signal.

18.22 Example   For the linear state equation

$$\dot{x}(t) = \begin{bmatrix} 1 & 2 & 0 \\ 0 & 3 & 0 \\ 0 & 4 & 5 \end{bmatrix} x(t) + \begin{bmatrix} 0 & 1 \\ 2 & 0 \\ 3 & 0 \end{bmatrix} u(t)$$

a quick calculation shows that the controllable subspace is $X = R^3$. To show that

$$\text{span } \{e_1\} = \text{span } \left\{ \begin{bmatrix} 1 \\ 0 \\ 0 \end{bmatrix} \right\}$$

is a controllability subspace, let

$$G = \begin{bmatrix} 0 & 0 \\ 1 & 0 \end{bmatrix}, \quad F = \begin{bmatrix} 0 & -4/3 & 0 \\ 0 & -2 & 0 \end{bmatrix}$$

Then the closed-loop state equation is

$$\dot{x}(t) = \begin{bmatrix} 1 & 0 & 0 \\ 0 & 1/3 & 0 \\ 0 & 0 & 5 \end{bmatrix} x(t) + \begin{bmatrix} 1 & 0 \\ 0 & 0 \\ 0 & 0 \end{bmatrix} v(t)$$

Since $Im[BG] = \text{span } \{e_1\}$ and $A + BF$ is diagonal, it is easy to verify that $\mathcal{R} = \text{span } \{e_1\}$ satisfies (30).
□□□

Often it is convenient for theoretical purposes to remove explicit involvement of the matrix G in the definition of controllability subspaces. However this does leave an implicit characterization that must be unraveled when computing related state feedback gains.

18.23 Theorem   A subspace $\mathcal{R} \subset X$ is a controllability subspace for (3) if and only if there exists an $m \times n$ matrix F such that

$$\mathcal{R} = \, <A + BF \,| \, \mathcal{B} \cap \mathcal{R}> \tag{31}$$

Proof   Suppose F is such that (31) holds. Let the $n \times 1$ vectors $p_1, \ldots, p_q$, $q \le m$, be a basis for $\mathcal{B} \cap \mathcal{R} \subset X$. Then for some linearly independent set of $m \times 1$ vectors $u_1, \ldots, u_q \in \mathcal{U}$, we can write $p_1 = Bu_1, \ldots, p_q = Bu_q$. Complete this set to a basis $u_1, \ldots, u_m$ for $\mathcal{U}$, and let

$$G = \begin{bmatrix} u_1 & \cdots & u_q & 0_{m \times (m-q)} \end{bmatrix} \begin{bmatrix} u_1 & \cdots & u_m \end{bmatrix}^{-1}$$

Then it is easy to verify that

$$BGu_k = \begin{cases} p_k, & k = 1, \ldots, q \\ 0, & k = q+1, \ldots, m \end{cases}$$

Therefore $Im[BG] = \mathcal{B} \cap \mathcal{R}$, that is

$$\mathcal{R} = <A+BF \mid Im[BG]> \tag{32}$$

and $\mathcal{R}$ is a controllability subspace for (3).

Conversely if $\mathcal{R}$ is a controllability subspace for (3), then there exist matrices F and G such that (32) holds. From the basic definitions,

$$Im[BG] \subset \mathcal{B}, \quad Im[BG] \subset \mathcal{R}$$

and so $Im[BG] \subset \mathcal{B} \cap \mathcal{R}$. Therefore $\mathcal{R} \subset <A+BF \mid \mathcal{B} \cap \mathcal{R}>$. Also $\mathcal{R}$ is an invariant subspace for $(A + BF)$, from which $(A + BF)(\mathcal{B} \cap \mathcal{R}) \subset \mathcal{R}$. Thus $<A+BF \mid \mathcal{B} \cap \mathcal{R}> \subset \mathcal{R}$, and we have established (31).
□□□

As mentioned earlier a controllability subspace $\mathcal{R}$ for (3) also is a controlled invariant subspace for (3), and thus must have friends. We next show that any such friend can be used to characterize $\mathcal{R}$ as a controllability subspace.

18.24 Theorem   Suppose $\mathcal{R} \subset X$ is a controllability subspace for (3). If F is such that $(A + BF)\mathcal{R} \subset \mathcal{R}$, then

$$\mathcal{R} = <A+BF \mid \mathcal{B} \cap \mathcal{R}> \tag{33}$$

Proof   If $\mathcal{R}$ is a controllability subspace, then there exists an $m \times n$ matrix F^a such that

$$\mathcal{R} = <A+BF^a \mid \mathcal{B} \cap \mathcal{R}>$$

Now suppose F^b is a friend of $\mathcal{R}$, that is, $(A + BF^b)\mathcal{R} \subset \mathcal{R}$, and let

$$\mathcal{R}_b = <A+BF^b \mid \mathcal{B} \cap \mathcal{R}>$$

Clearly $\mathcal{R}_b \subset \mathcal{R}$, and we now want to show the reverse containment.

To set up an induction argument, first note that

$$(A + BF^a)^0(\mathcal{B} \cap \mathcal{R}) = \mathcal{B} \cap \mathcal{R} \subset \mathcal{R}_b$$

Assuming that for a positive integer K,

$$(A + BF^a)^K(\mathcal{B} \cap \mathcal{R}) \subset \mathcal{R}_b$$

we can write

$$(A + BF^a)^{K+1}(\mathcal{B} \cap \mathcal{R}) = (A + BF^a)\left[(A + BF^a)^K(\mathcal{B} \cap \mathcal{R})\right]$$

$$\subset (A + BF^a)\mathcal{R}_b$$

$$= [A + BF^b + B(F^a - F^b)]\mathcal{R}_b$$

$$\subset (A + BF^b)\mathcal{R}_b + [B(F^a - F^b)]\mathcal{R}_b \qquad (34)$$

By definition

$$(A + BF^b)\mathcal{R}_b \subset \mathcal{R}_b$$

Also $[B(F^a - F^b)]\mathcal{R}_b \subset \mathcal{B}$, and since $\mathcal{R}_b \subset \mathcal{R}$,

$$[B(F^a - F^b)]\mathcal{R}_b \subset [B(F^a - F^b)]\mathcal{R}$$

By Theorem 18.20, $[B(F^a - F^b)]\mathcal{R} \subset \mathcal{R}$. Therefore

$$[B(F^a - F^b)]\mathcal{R}_b \subset \mathcal{B} \cap \mathcal{R} \subset \mathcal{R}_b$$

and the right side of (34) is contained in $\mathcal{R}_b$. This completes an induction proof for

$$(A + BF^a)^k(\mathcal{B} \cap \mathcal{R}) \subset \mathcal{R}_b , \quad k = 0, 1, \cdots$$

and thus

$$\mathcal{R} = <A + BF^a \mid \mathcal{B} \cap \mathcal{R}> \subset \mathcal{R}_b$$

□□□

The last two results provide a method for checking if a controlled invariant subspace $\mathcal{V}$ is a controllability subspace. Pick any friend F of the controlled invariant subspace $\mathcal{V}$, and confront the condition

$$\mathcal{V} = <A + BF \mid \mathcal{B} \cap \mathcal{V}> \qquad (35)$$

If this holds, then $\mathcal{V}$ is a controllability subspace for (3) by Theorem 18.23. If the condition (35) fails, then Theorem 18.24 implies that $\mathcal{V}$ is not a controllability subspace.

18.25 Example   Suppose $\mathcal{R}$ is a controllability subspace for (3), and that F is any friend of $\mathcal{R}$. Then (33) holds, and we can choose a basis for X as follows. Select G such that

$$Im[BG] = \mathcal{B} \cap \mathcal{R} \qquad (36)$$

Then let $p_1, \ldots, p_q, q \le m$ be a basis for $\mathcal{B} \cap \mathcal{R}$. First extend to a basis $p_1, \ldots, p_\rho$, $q \le \rho \le n$, for $\mathcal{R}$, and then further extend to a basis $p_1, \ldots, p_n$ for X. The corresponding state variable change $z(t) = P^{-1}x(t)$ applied to the closed-loop state equation

$$\dot{x}(t) = (A + BF)x(t) + BGv(t)$$

gives

$$\begin{bmatrix} \dot{z}_r(t) \\ \dot{z}_{nr}(t) \end{bmatrix} = \begin{bmatrix} \hat{A}_{11} & \hat{A}_{12} \\ 0 & \hat{A}_{22} \end{bmatrix} \begin{bmatrix} z_r(t) \\ z_{nr}(t) \end{bmatrix} + \begin{bmatrix} \hat{B}_{11} \\ 0 \end{bmatrix} v(t) \tag{37}$$

where the $\rho \times m$ matrix $\hat{B}_{11}$ has the further structure

$$\hat{B}_{11} = \begin{bmatrix} \tilde{B}_{11} \\ 0 \end{bmatrix}$$

with $\tilde{B}_{11}$ of dimension $q \times m$.
□□□

    Finally, returning to the original motivation, we show the relation of controllability subspaces to the eigenvalue assignment issue.

18.26 Theorem   Suppose $\mathcal{R} \subset X$ is a controllability subspace for (3) of dimension $\rho \geq 1$. Then given any degree-ρ, real-coefficient polynomial $p(\lambda)$ there exists a state feedback

$$u(t) = Fx(t) + Gv(t)$$

with F a friend of $\mathcal{R}$ such that in a basis adapted to $\mathcal{R}$ the component of the closed-loop state equation corresponding to $\mathcal{R}$ has characteristic polynomial $p(\lambda)$.

    Proof   To construct a feedback with the desired property, first select G such that

$$Im[BG] = \mathcal{B} \cap \mathcal{R}$$

by following the construction in the proof of Theorem 18.23. The choice of F is more complicated, and begins with selection of a friend F^a of $\mathcal{R}$ so that

$$\mathcal{R} = <A+BF^a \,|\, \mathcal{B} \cap \mathcal{R}> = <A+BF^a \,|\, Im[BG]>$$

Choosing a basis adapted to $\mathcal{R}$, the corresponding variable change $z(t) = P^{-1}x(t)$ is such that the state equation

$$\dot{x}(t) = (A + BF^a)x(t) + BGv(t)$$

can be rewritten in partitioned form as

$$\begin{bmatrix} \dot{z}_r(t) \\ \dot{z}_{nr}(t) \end{bmatrix} = \begin{bmatrix} \hat{A}_{11} & \hat{A}_{12} \\ 0 & \hat{A}_{22} \end{bmatrix} \begin{bmatrix} z_r(t) \\ z_{nr}(t) \end{bmatrix} + \begin{bmatrix} \hat{B}_{11} \\ 0 \end{bmatrix} v(t)$$

The component of this state equation corresponding to $\mathcal{R}$, namely

$$\dot{z}_r(t) = \hat{A}_{11}z_r(t) + \hat{A}_{12}z_{nr}(t) + \hat{B}_{11}v(t)$$

is controllable, and thus there is a matrix F_{11}^b such that

$$\det(\lambda I - \hat{A}_{11} - \hat{B}_{11}F_{11}^b) = p(\lambda) \tag{38}$$

Now we verify that

$$F = F^a + G \begin{bmatrix} F^b_{11} & 0 \end{bmatrix} P^{-1}$$

is a friend of $\mathcal{R}$ that provides the desired characteristic polynomial for the component of the closed-loop state equation corresponding to $\mathcal{R}$. We have that $x \in \mathcal{R}$ if and only if x has the form

$$x = P \begin{bmatrix} z_r \\ 0 \end{bmatrix}$$

Since F^a is a friend of $\mathcal{R}$, and

$$F - F^a = G \begin{bmatrix} F^b_{11} & 0 \end{bmatrix} P^{-1}$$

we can write, for any $x \in \mathcal{R}$,

$$\begin{aligned} B(F - F^a)x = BG \begin{bmatrix} F^b_{11} & 0 \end{bmatrix} P^{-1} x \\ = P \begin{bmatrix} \hat{B}_{11} \\ 0 \end{bmatrix} \begin{bmatrix} F^b_{11} & 0 \end{bmatrix} \begin{bmatrix} z_r \\ 0 \end{bmatrix} \\ = P \begin{bmatrix} \hat{B}_{11} F^b_{11} z_r \\ 0 \end{bmatrix} \end{aligned} \tag{39}$$

Therefore $B(F - F^a)\mathcal{R} \subset \mathcal{R}$, that is,

$$(F - F^a)\mathcal{R} \subset B^{-1}\mathcal{R}$$

and F is a friend of $\mathcal{R}$ by Theorem 18.20. To complete the proof compute

$$\begin{aligned} P^{-1}(A + BF)P = P^{-1}\left(A + BF^a + BG \begin{bmatrix} F^b_{11} & 0 \end{bmatrix} P^{-1} \right) Pz \\ = P^{-1}(A + BF^a)Pz + P^{-1}BG \begin{bmatrix} F^b_{11} & 0 \end{bmatrix} \\ = \begin{bmatrix} \hat{A}_{11} + \hat{B}_{11}F^b_{11} & \hat{A}_{12} \\ 0 & \hat{A}_{22} \end{bmatrix} \end{aligned}$$

and from (38) the characteristic polynomial of the component corresponding to $\mathcal{R}$ is $p(\lambda)$.

□□□

Our main application of this result is in addressing eigenvalue assignability while preserving invariance of a specified subspace for the closed-loop state equation. To motivate, we note the following refinement of the discussion below Definition 18.9. If (13) results from a state variable change adapted to a controllability subspace, $\mathcal{V} = \mathcal{R}$, then controllability of (13) implies controllability of both component state equations in (14). More generally suppose for an uncontrollable state equation that $\mathcal{V}$ is a controlled invariant subspace, and $\mathcal{R}$ is a controllability subspace contained in $\mathcal{V}$. Then eigenvalues can be assigned for the component of the closed-loop state equation corresponding to $\mathcal{R}$ using a friend of $\mathcal{V}$. This is treated in detail in Chapter 19.

Stabilizability and Detectability

Stability properties of a closed-loop state equation also are of fundamental importance, and the geometric approach to this issue involves the stable and unstable subspaces of the open-loop state equation, and a concept briefly introduced in Exercise 14.8.

18.27 Definition   The linear state equation (3) is called *stabilizable* if there exists a state feedback gain F such that the closed-loop state equation

$$\dot{x}(t) = (A + BF)x(t) \tag{40}$$

is exponentially stable.

18.28 Theorem   The linear state equation (3) is stabilizable if and only if

$$X^+ \subset <A \mid \mathcal{B}> \tag{41}$$

Proof   Changing state variables using a basis adapted to $<A \mid \mathcal{B}>$ yields

$$\begin{bmatrix} \dot{z}_c(t) \\ \dot{z}_{nc}(t) \end{bmatrix} = \begin{bmatrix} \hat{A}_{11} & \hat{A}_{12} \\ 0 & \hat{A}_{22} \end{bmatrix} \begin{bmatrix} z_c(t) \\ z_{nc}(t) \end{bmatrix} + \begin{bmatrix} \hat{B}_{11} \\ 0 \end{bmatrix} u(t)$$

In terms of this basis, if $X^+ \subset <A \mid \mathcal{B}>$, then all eigenvalues of $\hat{A}_{22}$ have negative real parts. Therefore (3) is stabilizable since the component state equation corresponding to $<A \mid \mathcal{B}>$ is controllable.

On the other hand suppose that (3) is not stabilizable. Then $\hat{A}_{22}$ has at least one eigenvalue with nonnegative real part, and thus X^+ is not contained in $<A \mid \mathcal{B}>$.
□□□

An alternate statement of Theorem 18.28 sometimes is more convenient.

18.29 Corollary   The linear state equation (3) is stabilizable if and only if

$$X^- + <A \mid \mathcal{B}> = X \tag{42}$$

Stabilizability obviously is a weaker property than controllability, though stabilizability has intuitive interpretations as 'controllability on the infinite interval $0 \le t < \infty$,' or 'stability of uncontrollable states.' Further geometric treatment of issues involving stabilization is based on another special type of controlled invariant subspace called a *stabilizability subspace*. This is not pursued further here, except to suggest references in Note 18.5.

There is a similar weakening of the concept of observability that is of interest. Motivation stems from the observer theory in Chapter 15, with eigenvalue assignment in the error state equation replaced by exponential stability of the error state equation.

18.30 Definition   The linear state equation (3) is called *detectable* if there exists an $n \times p$ matrix H such that

$$\dot{x}(t) = (A + HC)x(t)$$

is exponentially stable.

The issue here is one of 'stability of unobservable states.' Proof of the following detectability criterion is left as an exercise, though Exercise 15.6 supplies an underlying calculation.

18.31 Theorem   The linear state equation (3) is detectable if and only if

$$\chi^+ \cap \mathcal{N} = 0$$

EXERCISES

Exercise 18.1   Suppose X is a vector space, $\mathcal{V}$, $\mathcal{W} \subset X$ are subspaces, and $A : X \to X$. Give proofs or counterexamples for the following claims.
(a) $\mathcal{V} \subset \mathcal{W}$ implies $A\mathcal{V} \subset A\mathcal{W}$
(b) $A^{-1}\mathcal{V} \subset \mathcal{W}$ implies $\mathcal{V} \subset A\mathcal{W}$
(c) $\mathcal{V} \subset \mathcal{W}$ implies $A^{-1}\mathcal{V} \subset A^{-1}\mathcal{W}$
(d) $\mathcal{V} \subset A\mathcal{W}$ implies $A^{-1}\mathcal{V} \subset \mathcal{W}$

Exercise 18.2   Suppose X is a vector space, $\mathcal{V}$, $\mathcal{W} \subset X$ are subspaces, and $A : X \to X$. Show that
(a) $A(A^{-1}\mathcal{V}) = \mathcal{V} \cap Im[A]$
(b) $A^{-1}(A\mathcal{V}) = \mathcal{V} + Ker[A]$
(c) $A\mathcal{V} \subset \mathcal{W}$ if and only if $\mathcal{V} \subset A^{-1}\mathcal{W}$

Exercise 18.3   If $\mathcal{V}$, $\mathcal{W} \subset X$ are subspaces that are invariant for $A : X \to X$, give proofs or counterexamples to the following claims.
(a) $\mathcal{V} \cap \mathcal{W}$ is an invariant subspace for A
(b) $A^{-1}(\mathcal{V} \cap \mathcal{W})$ is an invariant subspace for A
(c) $\mathcal{V} + \mathcal{W}$ is an invariant subspace for A
(d) $\mathcal{V} \cup \mathcal{W}$ is an invariant subspace for A (*Hint*: Don't be tricked.)

Exercise 18.4   If $\mathcal{V}$, $\mathcal{W}_a$, $\mathcal{W}_b \subset X$ are subspaces, show that

$$\mathcal{W}_a \cap \mathcal{V} + \mathcal{W}_b \cap \mathcal{V} \subset (\mathcal{W}_a + \mathcal{W}_b) \cap \mathcal{V}$$

If $\mathcal{W}_a \subset \mathcal{V}$, show that

$$(\mathcal{W}_a + \mathcal{W}_b) \cap \mathcal{V} = \mathcal{W}_a + \mathcal{W}_b \cap \mathcal{V}$$

Exercise 18.5   Suppose $\mathcal{V}$, $\mathcal{W} \subset X$ are subspaces. Show that there exists an F such that

$$(A + BF)\mathcal{V} \subset \mathcal{V}, \quad (A + BF)\mathcal{W} \subset \mathcal{W}$$

if and only if

$$A\mathcal{V} \subset \mathcal{V} + \mathcal{B}, \quad A\mathcal{W} \subset \mathcal{W} + \mathcal{B}$$

$$A(\mathcal{V} \cap \mathcal{W}) \subset \mathcal{V} \cap \mathcal{W} + \mathcal{B}$$

Exercise 18.6   If $\hat{\mathcal{B}} \subset \mathcal{B}$, prove that

$$<A \,|\, \mathcal{B} \cap <A \,|\, \hat{\mathcal{B}}>> = <A \,|\, \hat{\mathcal{B}}>$$

If

$$<A \,|\, \mathcal{B} \cap <A \,|\, c>> = <A \,|\, c>$$

prove that there exists an $m \times m$ matrix G such that

$$<A \,|\, Im \,[BG\,]> = <A \,|\, c>$$

Exercise 18.7   For the state equation in Example 18.15, characterize the following subspaces of $X = R^4$ in terms of the standard basis:
(a) all controllability subspaces,
(b) all controlled invariant subspaces.
Formulate a reasonable definition of stabilizability subspace, and similarly characterize all stabilizability subspaces.

Exercise 18.8   Show that $<A \,|\, \mathcal{B}>$ is precisely the set of states that can be reached from the zero initial state in finite time with a continuous input signal.

Exercise 18.9   Prove that the linear state equation

$$\dot{x}(t) = Ax(t) + Bu(t)$$

$$y(t) = Cx(t)$$

with *rank C* $= p$ is *output controllable* in the sense of Exercise 9.10 if and only if

$$C<A \,|\, \mathcal{B}> = \mathcal{Y}$$

Exercise 18.10   Show that the closed-loop state equation

$$\dot{x}(t) = (A + BF)x(t)$$

$$y(t) = Cx(t)$$

is observable for all gain matrices F if and only if the only controlled invariant subspace contained in $Ker\,[C\,]$ for the open-loop state equation is 0.

Exercise 18.11   Suppose $\mathcal{R}$ is a controllability subspace for

$$\dot{x}(t) = Ax(t) + Bu(t)$$

and, in terms of the columns of B,

$$\mathcal{B} \cap \mathcal{R} = Im[B_1] + \cdots + Im[B_q]$$

Suppose the columns of the $n \times n$ matrix P form a basis for X that is adapted to the nested set of subspaces

$$\mathcal{B} \cap \mathcal{R} \subset \mathcal{R} \subset <A|\mathcal{B}> \subset X$$

Using the state variable change $z(t) = P^{-1}x(t)$, what structural features does the resulting state equation have? (Note that there is no state feedback involved in this question.)

Exercise 18.12   Suppose $\mathcal{K} \subset R^n$ is a subspace, and $z(t)$ is a continuously-differentiable $n \times 1$ function of time that satisfies $z(t) \in \mathcal{K}$ for all $t \geq 0$. Show that $\dot{z}(t) \in \mathcal{K}$ for all $t \geq 0$.

Exercise 18.13   Consider a linear state equation

$$\dot{x}(t) = Ax(t) + Bu(t)$$

and suppose $z(t)$ is a continuously-differentiable $n \times 1$ function satisfying $z(t) \in \mathcal{B} \cap A^{-1}\mathcal{B}$ for all $t \geq 0$. Show that there exists a continuous input signal such that with $x(0) = z(0)$ the solution of the state equation is $x(t) = z(t)$ for $t \geq 0$. (*Hint*: Use Exercise 18.12.)

NOTES

Note 18.1   Though often viewed by beginners as the system theory from another galaxy, the geometric approach arose on Earth in the late 1960's in independent work reported in the papers

G. Basile, G. Marro, "Controlled and conditioned invariant subspaces in linear system theory," *Journal of Optimization Theory and Applications,* Vol. 3, No. 5, pp. 306–315, 1969

W.M. Wonham, A.S. Morse, "Decoupling and pole assignment in linear multivariable systems: A geometric approach," *SIAM Journal on Control and Optimization,* Vol. 8, No. 1, pp. 1–18, 1970

In the latter paper controlled invariant subspaces are called (A, B)-*invariant subspaces,* a term that has fallen somewhat out of favor in recent years. In the former paper a dual notion is presented that recalls Definition 18.30. A subspace $\mathcal{V} \subset X$ is called a *conditioned invariant subspace* for the usual linear state equation if there exists an $n \times p$ matrix H such that

$$(A + HC)\mathcal{V} \subset \mathcal{V}$$

This construct provides the basis for a geometric development of state observers and other notions related to dynamic compensators. See also

W.M. Wonham, "Dynamic observers — geometric theory," *IEEE Transactions on Automatic Control,* Vol. 15, No. 2, pp. 258–259, 1970

Note 18.2   For further study of the geometric theory, a standard reference is

W.M. Wonham, *Linear Multivariable Control: A Geometric Approach,* Third Edition, Springer-Verlag, New York, 1985

This book makes use of linear-algebra concepts at a more advanced level than our introductory treatment. For example dual spaces and factor spaces play an important role in further developments. More than this, the purist prefers to keep the proofs coordinate free, rather than adopt a particularly convenient basis as we have so often done. Satisfying this preference requires more sophisticated proof technique in many instances.

Note 18.3   From a Laplace-transform viewpoint, the various subspaces introduced in this chapter can be characterized in terms of rational solutions to polynomial equations. Thus the geometric theory makes contact with polynomial fraction descriptions. As a start, consult

M.L.J. Hautus, "(A, B)-invariant and stabilizability subspaces, a frequency domain description," *Automatica,* Vol. 16, pp. 703–707, 1980

Note 18.4   Eigenvalue assignment properties of nested collections of controlled invariant subspaces are neatly discussed in

J.M. Schumacher, "A complement on pole placement," *IEEE Transactions on Automatic Control,* Vol. 25, No. 2, pp. 281–282, 1980

Eigenvalue assignment using friends of a specified controlled invariant subspace $\mathcal{V}$ will be an important issue in Chapter 19, and it might not be surprising that the *largest* controllability subspace contained in $\mathcal{V}$ plays a major role. Geometric interpretations of various concepts of system zeros, including transmission zeros discussed in Chapter 17, are presented in

H. Aling, J.M. Schumacher, "A nine-fold canonical decomposition for linear systems," *International Journal of Control,* Vol. 39, No. 4, pp. 779–805, 1984

This leads to a geometry-based refinement of the Canonical Structure theorem described in Note 10.2.

Note 18.5   A subspace $S \subset X$ is called a *stabilizability subspace* for (3) if S is a controlled invariant subspace for (3), and there is a friend F of S such that the component of

$$\dot{x}(t) = (A + BF)x(t)$$

corresponding to S is exponentially stable. Characterizations of stabilizability subspaces and applications to control problems are discussed in the paper by Hautus cited in Note 18.3. In Lemma 3.2 of

J.M. Schumacher, "Regulator synthesis using (C, A, B)-pairs," *IEEE Transactions on Automatic Control,* Vol. 27, No. 6, pp. 1211 -1221, 1982

a characterization of stabilizable subspaces, there called *inner stabilizable subspaces,* is given that is a geometric cousin of the rank condition in Exercise 14.8.

Note 18.6   An approximation notion related to invariant subspaces is introduced in the papers

J.C. Willems, "Almost invariant subspaces: An approach to high-gain feedback design — Part I: Almost controlled invariant subspaces," *IEEE Transactions on Automatic Control,* Vol. 26, No. 1, pp. 235–252, 1981; "Part II: Almost conditionally invariant subspaces," *IEEE Transactions on Automatic Control,* Vol. 27, No. 5, pp. 1071–1085, 1981

Loosely speaking, for an initial state in an almost controlled invariant subspace there are input signals such that the state trajectory remains as close as desired to that subspace. This so-called *almost* geometric theory can be applied to many of the same control problems as the basic geometric theory, including the problems addressed in Chapter 19. Consult

R. Marino, W. Respondek, A.J. Van der Schaft, "Direct approach to almost disturbance and almost input-output decoupling," *International Journal of Control,* Vol. 48, No. 1, pp. 353–383, 1986

19

APPLICATIONS OF GEOMETRIC THEORY

In this chapter we apply the geometric theory for a time-invariant linear state equation (often called the *plant* or *open-loop state equation* in the context of feedback)

$$\dot{x}(t) = Ax(t) + Bu(t)$$
$$y(t) = Cx(t) \tag{1}$$

to linear control problems involving rejection of unknown disturbance signals, and isolation of specified entries of the output signal from specified input-signal entries. In both problems the control objective can be phrased in terms of invariant subspaces for the closed-loop state equation. Thus the geometric theory is a natural tool.

New features of the subspaces introduced in Chapter 18 are required by the development. These include notions of maximal controlled-invariant and controllability subspaces contained in a specified subspace, and methods for their calculation.

Disturbance Decoupling

A disturbance input can be added to (1) to obtain the linear state equation

$$\dot{x}(t) = Ax(t) + Bu(t) + Ew(t)$$
$$y(t) = Cx(t) \tag{2}$$

We suppose that $w(t)$ is a $q \times 1$ signal that is unknown, but continuous in keeping with the usual default, and E is an $n \times q$ coefficient matrix that describes the way the disturbance affects the plant. All other dimensions, assumptions, and notations from Chapter 18 are preserved. Of course the various geometric constructs are unchanged by adding the disturbance input. That is, invariant subspaces for A and controlled invariant subspaces with regard to the plant input are the same for (2) as for (1).

The control objective is to choose time-invariant linear state feedback

$$u(t) = Fx(t) + Gv(t)$$

so that, regardless of the reference input $v(t)$ and initial state x_o, the output signal of the closed-loop state equation

$$\dot{x}(t) = (A + BF)x(t) + BGv(t) + Ew(t) , \quad x(0) = x_o$$

$$y(t) = Cx(t) \tag{3}$$

is uninfluenced by $w(t)$. Of course the component of $y(t)$ due to $w(t)$ is independent of the initial state, so we assume $x_o = 0$. Then, representing the solution of (3) in terms of Laplace transforms, a compact way of posing the problem is to require that F be chosen so that the transfer function from disturbance signal to output signal is zero:

$$C(sI - A - BF)^{-1}E = 0 \tag{4}$$

When this condition is satisfied the closed-loop state equation is said to be *disturbance decoupled*. Note that no stability requirement is imposed on the closed-loop state equation — a deficiency addressed in the sequel.

The choice of reference-input gain G plays no role in disturbance decoupling. Furthermore it is clear from (4) that the objective is attained precisely when F is such that

$$<A+BF \,|\, Im\,[E\,]> \subset Ker\,[C\,]$$

In words, the disturbance decoupling problem is solvable if and only if there exists a state feedback gain F such that the smallest $(A + BF)$ invariant subspace containing $Im\,[E\,]$ is a subspace of $Ker\,[C\,]$. This can be rephrased in terms of the plant as follows. The disturbance decoupling problem is solvable if and only if there exists a controlled invariant subspace $\mathcal{V} \subset Ker\,[C\,]$ for (2) with the property that $Im\,[E\,] \subset \mathcal{V}$. To turn this statement into a checkable necessary and sufficient condition for solvability of the disturbance decoupling problem, we need to develop a notion of the largest controlled invariant subspace for (1) that is contained in a specified subspace of X, in this instance the subspace $Ker\,[C\,]$.

Suppose $\mathcal{K} \subset X$ is a subspace. By definition a *maximal* controlled invariant subspace contained in $\mathcal{K}$ for (1) contains every other controlled invariant subspace contained in $\mathcal{K}$ for (1). The first task is to show existence of such a maximal controlled invariant subspace, denoted by $\mathcal{V}^*$. (The dependence on $\mathcal{K}$ is left understood.) Then the relevance of $\mathcal{V}^*$ to the disturbance decoupling problem is shown, and the computation of $\mathcal{V}^*$ is addressed.

19.1 Theorem   Suppose $\mathcal{K} \subset X$ is a subspace. Then there exists a unique maximal controlled invariant subspace $\mathcal{V}^*$ contained in $\mathcal{K}$ for (1).

Proof   The key to the proof is to show that a sum of controlled invariant subspaces contained in $\mathcal{K}$ also is a controlled invariant subspace contained in $\mathcal{K}$. First note that there is at least one controlled invariant subspace contained in $\mathcal{K}$, namely the subspace 0, so our argument is not vacuous. If $\mathcal{V}_a$ and $\mathcal{V}_b$ are any two controlled invariant subspaces contained in $\mathcal{K}$, then

$$A\mathcal{V}_a \subset \mathcal{V}_a + \mathcal{B}, \quad A\mathcal{V}_b \subset \mathcal{V}_b + \mathcal{B}$$

Also $\mathcal{V}_a + \mathcal{V}_b \subset \mathcal{K}$, and

$$A(\mathcal{V}_a + \mathcal{V}_b) = A\mathcal{V}_a + A\mathcal{V}_b \subset \mathcal{V}_a + \mathcal{V}_b + \mathcal{B}$$

That is, by Theorem 18.19, $\mathcal{V}_a + \mathcal{V}_b$ is a controlled invariant subspace contained in $\mathcal{K}$.

Forming the sum of all controlled invariant subspaces contained in $\mathcal{K}$, and using the finite dimensionality of $\mathcal{K}$, a simple argument shows that there is a controlled invariant subspace contained in $\mathcal{K}$ of largest dimension, say $\mathcal{V}^*$. To show $\mathcal{V}^*$ is maximal, if $\mathcal{V} \subset \mathcal{K}$ is another controlled invariant subspace for (1), then so is $\mathcal{V} + \mathcal{V}^*$. But then

$$\dim \mathcal{V}^* \leq \dim (\mathcal{V} + \mathcal{V}^*) \leq \dim \mathcal{V}^*$$

and this inequality shows that $\mathcal{V} \subset \mathcal{V}^*$. Therefore $\mathcal{V}^*$ is a maximal controlled invariant subspace contained in $\mathcal{K}$. To show uniqueness simply argue that two maximal controlled invariant subspaces contained in $\mathcal{K}$ for (1) must contain each other, and thus are identical.
□□□

Returning to the disturbance decoupling problem, the basic solvability condition is straightforward to establish in terms of $\mathcal{V}^*$.

19.2 Theorem   There exists a state feedback gain F that solves the disturbance decoupling problem for the plant (2) if and only if

$$Im\,[E] \subset \mathcal{V}^* \tag{5}$$

where $\mathcal{V}^*$ is the maximal controlled invariant subspace contained in $Ker\,[C]$ for (2).

Proof   If (5) holds, then choosing any friend F of $\mathcal{V}^*$ we have, since $\mathcal{V}^*$ is an invariant subspace for $A + BF$,

$$\int_0^t e^{(A+BF)(t-\sigma)} Ew(\sigma)\, d\sigma \in \mathcal{V}^*, \quad t \geq 0$$

for any disturbance signal. Since $\mathcal{V}^* \subset Ker\,[C]$,

$$C \int_0^t e^{(A+BF)(t-\sigma)} Ew(\sigma)\, d\sigma = 0, \quad t \geq 0$$

again for any disturbance signal, and taking the Laplace transform gives (4).

Conversely if (4) holds, then

$$Ce^{(A+BF)t}E = 0, \quad t \geq 0 \tag{6}$$

and therefore

$$CE = C(A+BF)E = \cdots = C(A+BF)^{n-1}E = 0$$

This implies that $<A+BF \,|\, Im \,[E]>$, an invariant subspace for $A + BF$, is contained in $Ker \,[C]$. Since $\mathcal{V}^*$ is the maximal controlled invariant subspace contained in $Ker \,[C]$, we have

$$Im \,[E] \subset \; <A+BF \,|\, Im \,[E]> \; \subset \mathcal{V}^*$$

□□□

      Application of the solvability condition in (5) requires computation of the maximal controlled invariant subspace $\mathcal{V}^*$ contained in a specified subspace $\mathcal{K}$. This is addressed in two steps: first a conceptual algorithm is established, and then, at the end of the chapter, a matrix algorithm that implements the conceptual algorithm is presented. Roughly speaking the conceptual algorithm generates a nested set of decreasing-dimension subspaces, beginning with $\mathcal{K}$, that yields $\mathcal{V}^*$ in a finite number of steps. Then the matrix algorithm provides a method for calculating bases for these subspaces.

      Once the computation of $\mathcal{V}^*$ is settled, the first part of the proof of Theorem 19.2 shows that any friend of $\mathcal{V}^*$ specifies a state feedback that achieves disturbance decoupling. The construction of such a friend is easily lifted from the proof of Theorem 18.19. Let $v_1, \ldots, v_n$ be a basis for X adapted to $\mathcal{V}^*$, so that $v_1, \ldots, v_\nu$ is a basis for $\mathcal{V}^*$. Since $A\mathcal{V}^* \subset \mathcal{V}^* + \mathcal{B}$, for $k = 1, \ldots, \nu$, we can solve for $w_k \in \mathcal{V}^*$ and $u_k \in \mathcal{U}$, the input space, such that $Av_k = w_k - Bu_k$. Then with arbitrary $m \times 1$ vectors $u_{\nu+1}, \ldots, u_n$, set

$$F = \begin{bmatrix} u_1 & \cdots & u_n \end{bmatrix} \begin{bmatrix} v_1 & \cdots & v_n \end{bmatrix}^{-1}$$

If $\mathcal{V}$ is any controlled invariant subspace with $Im \,[E] \subset \mathcal{V} \subset \mathcal{V}^* \subset Ker \,[C]$, then the first part of the proof of Theorem 19.2 also shows that any friend of $\mathcal{V}$ achieves disturbance decoupling. Furthermore the construction of a friend of $\mathcal{V}$ proceeds just as above.

19.3 Theorem   Suppose $\mathcal{K} \subset X$ is a subspace, and define a sequence of subspaces of $\mathcal{K}$ by

$$\mathcal{V}^0 = \mathcal{K}$$

$$\mathcal{V}^k = \mathcal{K} \cap A^{-1}(\mathcal{B} + \mathcal{V}^{k-1}), \quad k = 1, 2, \cdots \tag{7}$$

Then $\mathcal{V}^n$ is the maximal controlled invariant subspace contained in $\mathcal{K}$ for (1), that is,

$$\mathcal{V}^n = \mathcal{V}^* \tag{8}$$

    Proof   First we show by induction that $\mathcal{V}^k \subset \mathcal{V}^{k-1}$, $k = 0, 1, \cdots$. Obviously $\mathcal{V}^1 \subset \mathcal{V}^0$. Supposing that $K \geq 2$ is such that $\mathcal{V}^K \subset \mathcal{V}^{K-1}$,

$$\mathcal{V}^{K+1} = \mathcal{K} \cap A^{-1}(\mathcal{B} + \mathcal{V}^K)$$

$$\subset \mathcal{K} \cap A^{-1}(\mathcal{B} + \mathcal{V}^{K-1})$$

$$= \mathcal{V}^K$$

and the induction is complete.

It follows that $dim\ \mathcal{V}^k \leq dim\ \mathcal{V}^{k-1}$, $k = 0, 1, \cdots$. Furthermore, if $\mathcal{V}^k = \mathcal{V}^{k-1}$ for some value of k, then

$$\mathcal{V}^{k+1} = \mathcal{K} \cap A^{-1}(\mathcal{B} + \mathcal{V}^k)$$
$$= \mathcal{K} \cap A^{-1}(\mathcal{B} + \mathcal{V}^{k-1})$$
$$= \mathcal{V}^k = \mathcal{V}^{k-1} \quad (9)$$

This implies that $\mathcal{V}^{k+j} = \mathcal{V}^{k-1}$ for all $j = 1, 2, \cdots$. Therefore at each iteration the dimension of the generated subspace must decrease, or the algorithm effectively terminates. Since $dim\ \mathcal{V}^0 \leq n$, the dimension can decrease for at most n iterations, and thus $\mathcal{V}^{n+j} = \mathcal{V}^n$ for all $j = 1, 2, \cdots$. Now

$$\mathcal{V}^n = \mathcal{V}^{n+1} = \mathcal{K} \cap A^{-1}(\mathcal{V}^n + \mathcal{B})$$

and this implies $\mathcal{V}^n \subset A^{-1}(\mathcal{V}^n + \mathcal{B})$ and $\mathcal{V}^n \subset \mathcal{K}$. Equivalently $A\mathcal{V}^n \subset \mathcal{V}^n + \mathcal{B}$ and $\mathcal{V}^n \subset \mathcal{K}$, and therefore $\mathcal{V}^n$ is a controlled invariant subspace contained in $\mathcal{K}$.

Finally, to show that $\mathcal{V}^n$ is maximal, suppose $\mathcal{V}$ is any controlled invariant subspace contained in $\mathcal{K}$. By definition $\mathcal{V} \subset \mathcal{V}^0$, and if we assume $\mathcal{V} \subset \mathcal{V}^K$ then an induction argument can be completed as follows. By Theorem 18.19,

$$A\mathcal{V} \subset \mathcal{V} + \mathcal{B} \subset \mathcal{V}^K + \mathcal{B}$$

that is,

$$\mathcal{V} \subset A^{-1}(\mathcal{V}^K + \mathcal{B})$$

Therefore

$$\mathcal{V} \subset \mathcal{K} \cap A^{-1}(\mathcal{V}^K + \mathcal{B}) = \mathcal{V}^{K+1}$$

This induction proves that $\mathcal{V} \subset \mathcal{V}^k$ for all $k = 0, 1, \cdots$, and thus $\mathcal{V} \subset \mathcal{V}^n$. Therefore $\mathcal{V}^n = \mathcal{V}^*$, the maximal controlled invariant subspace contained in $\mathcal{K}$.
□□□

The algorithm in (7) can be sharpened in a couple of respects. It is obvious from the proof that $\mathcal{V}^*$ is obtained in at most n steps — the n is chosen here only for simplicity of notation. Also, because of the containment relationship of the iterates, the general step of the algorithm can be recast as

$$\mathcal{V}^k = \mathcal{V}^{k-1} \cap A^{-1}(\mathcal{V}^{k-1} + \mathcal{B}) \quad (10)$$

19.4 Example For the linear state equation (2), suppose $\mathcal{V}^*$ is the maximal controlled invariant subspace contained in $Ker\,[C]$, with the dimension of $\mathcal{V}^*$ denoted ν, and $Im\,[E] \subset \mathcal{V}^*$. Then for any friend F^a of $\mathcal{V}^*$, consider the corresponding state feedback for (3):

$$u(t) = F^a x(t) + v(t)$$

The closed-loop state equation, after a state variable change $z(t) = P^{-1}x(t)$, where the columns of P comprise a basis for X adapted to $\mathcal{V}^*$, can be written as

$$\begin{bmatrix} \dot{z}_a(t) \\ \dot{z}_b(t) \end{bmatrix} = \begin{bmatrix} \hat{A}_{11} & \hat{A}_{12} \\ 0 & \hat{A}_{22} \end{bmatrix} \begin{bmatrix} z_a(t) \\ z_b(t) \end{bmatrix} + \begin{bmatrix} \hat{B}_{11} \\ \hat{B}_{21} \end{bmatrix} v(t) + \begin{bmatrix} \hat{E}_{11} \\ 0_{(n-v) \times q} \end{bmatrix} w(t)$$

$$y(t) = \begin{bmatrix} 0_{p \times v} & \hat{C}_{12} \end{bmatrix} \begin{bmatrix} z_a(t) \\ z_b(t) \end{bmatrix} \tag{11}$$

From the form of the coefficient matrices, and especially from the diagram in Figure 19.5, it is clear that (11) is disturbance decoupled. And it is straightforward to verify (in terms of the state variable $z(t)$) that

$$F = F^a + \begin{bmatrix} 0_{m \times v} & F_{12}^b \end{bmatrix} P^{-1}$$

also is a friend of $\mathcal{V}^*$, for any $m \times (n-v)$ matrix F_{12}^b. This suggests that there is flexibility to achieve goals for the closed-loop state equation in addition to disturbance decoupling. Moreover if $\mathcal{V} \subset \mathcal{V}^*$ is a smaller-dimension controlled invariant subspace contained in $Ker[C]$ with $Im[E] \subset \mathcal{V}$, then this analysis can be repeated for $\mathcal{V}$. Greater flexibility is obtained since the size of F_{12}^b will be larger.

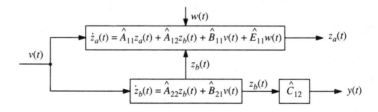

19.5 Figure   Structure of the disturbance-decoupled state equation (11).

Disturbance Decoupling with Eigenvalue Assignment

Disturbance decoupling alone is a limited objective, and next we consider the problem of simultaneously achieving eigenvalue assignment for the closed-loop state equation. (The intermediate problem of disturbance decoupling with exponential stability is discussed in Note 19.1.) The proof of Theorem 19.2 shows that if $\mathcal{V}$ is a controlled invariant subspace such that $Im[E] \subset \mathcal{V} \subset Ker[C]$, then any friend of $\mathcal{V}$ can be used to achieve disturbance decoupling. Thus we need to consider eigenvalue assignment for the closed-loop state equation using friends of $\mathcal{V}$ as feedback gains. Not surprisingly, in view of Theorem 18.26, this involves certain controllability subspaces for the plant. A solvability condition can be given in terms of a maximal controllability subspace, and therefore we first consider the existence and conceptual computation of maximal controllability subspaces. Fortunately good use can be made of the computation for maximal controlled invariant subspaces. The star notation for maximality is continued for controllability subspaces.

19.6 Theorem   Suppose $\mathcal{K} \subset \mathcal{X}$ is a subspace, $\mathcal{V}^*$ is the maximal controlled invariant subspace contained in $\mathcal{K}$ for (1), and F is a friend of $\mathcal{V}^*$. Then

$$\mathcal{R}^* = <A+BF \mid \mathcal{B} \cap \mathcal{V}^*> \tag{12}$$

is the unique maximal controllability subspace contained in $\mathcal{K}$ for (1).

Proof   Let F be a friend of $\mathcal{V}^*$ and, as in the proof of Theorem 18.23, compute an $m \times m$ matrix G such that $Im\,[BG] = \mathcal{B} \cap \mathcal{V}^*$. Then let

$$\mathcal{R} = <A+BF \mid \mathcal{B} \cap \mathcal{V}^*> = <A+BF \mid Im\,[BG]> \tag{13}$$

Clearly $\mathcal{R}$ is a controllability subspace, $\mathcal{R} \subset \mathcal{V}^* \subset \mathcal{K}$, and by definition F also is a friend of $\mathcal{R}$. We next show that if F^b is any other friend of $\mathcal{V}^*$, then F^b is a friend of $\mathcal{R}$. That is,

$$<A+BF^b \mid \mathcal{B} \cap \mathcal{V}^*> = \mathcal{R} \tag{14}$$

To show the left side is contained in the right side, we use induction. Of course $\mathcal{B} \cap \mathcal{V}^* \subset \mathcal{R}$, and if $(A + BF^b)^K (\mathcal{B} \cap \mathcal{V}^*) \subset \mathcal{R}$, then

$$(A + BF^b)^{K+1}(\mathcal{B} \cap \mathcal{V}^*) = (A + BF^b)[(A + BF^b)^K(\mathcal{B} \cap \mathcal{V}^*)]$$

$$\subset (A + BF^b)\mathcal{R}$$

$$\subset (A + BF)\mathcal{R} + B(F^b - F)\mathcal{R} \tag{15}$$

Since F is a friend of $\mathcal{R}$, $(A + BF)\mathcal{R} \subset \mathcal{R}$. To show $B(F^b - F)\mathcal{R} \subset \mathcal{R}$, note that Theorem 18.20 implies $B(F^b - F)\mathcal{V}^* \subset \mathcal{V}^*$ since both F and F^b are friends of $\mathcal{V}^*$. Obviously $B(F^b - F)\mathcal{V}^* \subset \mathcal{B}$, so we have

$$B(F^b - F)\mathcal{V}^* \subset \mathcal{B} \cap \mathcal{V}^* \subset \mathcal{R}$$

Therefore

$$B(F^b - F)\mathcal{R} \subset \mathcal{R}$$

and (15) gives

$$(A + BF^b)^{K+1}(\mathcal{B} \cap \mathcal{V}^*) \subset \mathcal{R}$$

This completes the induction proof that

$$<A+BF^b \mid \mathcal{B} \cap \mathcal{V}^*> \subset \mathcal{R}$$

The reverse inclusion is obtained by an exactly analogous induction argument. Thus (14) is verified, and any friend of $\mathcal{V}^*$ is a friend of $\mathcal{R}$. (In particular this guarantees that (12) is well defined — any friend F of $\mathcal{V}^*$ can be used.)

To show $\mathcal{R}$ is maximal suppose $\mathcal{R}_a$ is any other controllability subspace contained in $\mathcal{K}$ for (1). Then by Theorem 18.23 there exists an F^a such that

$$\mathcal{R}_a = <A+BF^a \mid \mathcal{B} \cap \mathcal{R}_a>$$

Furthermore since $\mathcal{R}_a$ also is a controlled invariant subspace contained in $\mathcal{K}$ for (1), $\mathcal{R}_a \subset \mathcal{V}^*$. To prove that $\mathcal{R}_a \subset \mathcal{R}$ involves finding a common friend of these two controllability subspaces, but by the first part of the proof we need only compute a common friend F^c for $\mathcal{R}_a$ and $\mathcal{V}^*$.

Select a basis $p_1, \ldots, p_n$ for $\mathcal{X}$ such that $p_1, \ldots, p_\rho$ is a basis for $\mathcal{R}_a$, and $p_1, \ldots, p_\nu$ is a basis for $\mathcal{V}^*$. Then the property $A\mathcal{V}^* \subset \mathcal{V}^* + \mathcal{B}$ implies in particular that there exist $v_{\rho+1}, \ldots, v_\nu \in \mathcal{V}^*$ and $u_{\rho+1}, \ldots, u_\nu \in \mathcal{U}$ such that

$$Ap_j = v_j - Bu_j, \quad j = \rho+1, \ldots, \nu$$

Choosing

$$F^c = \begin{bmatrix} F^a p_1 & \cdots & F^a p_\rho & u_{\rho+1} & \cdots & u_\nu & 0_{m \times (n-\nu)} \end{bmatrix} \begin{bmatrix} p_1 & \cdots & p_n \end{bmatrix}^{-1}$$

it follows that

$$(A + BF^c)p_j = \begin{cases} (A + BF^a)p_j \in \mathcal{R}_a, & j = 1, \ldots, \rho \\ v_j \in \mathcal{V}^*, & j = \rho+1, \ldots, \nu \\ 0, & j = \nu+1, \ldots, n \end{cases} \tag{16}$$

This shows F^c is a friend of both $\mathcal{R}_a$ and $\mathcal{V}^*$.

Since F^c is a friend of $\mathcal{R}_a$ and $\mathcal{V}^*$, and hence $\mathcal{R}$, from $\mathcal{R}_a \subset \mathcal{V}^*$ we have

$$\mathcal{R}_a = <A+BF^c \,|\, \mathcal{B} \cap \mathcal{R}_a>$$
$$\subset <A+BF^c \,|\, \mathcal{B} \cap \mathcal{V}^*>$$
$$= \mathcal{R}$$

Therefore $\mathcal{R}$ in (13) is a maximal controllability subspace contained in $\mathcal{K}$ for (1). Finally uniqueness is obvious, since any two such subspaces must contain each other. □□□

The conceptual computation of $\mathcal{R}^*$ suggested by Theorem 19.6 involves first computing $\mathcal{V}^*$. Then, as discussed in Chapter 18, a friend F of $\mathcal{V}^*$ can be computed, from which it is straightforward to compute $\mathcal{R}^* = <A+BF \,|\, \mathcal{B} \cap \mathcal{V}^*>$. In addition the proof of Theorem 19.6 provides a theoretical result that deserves display.

19.7 Corollary   With $\mathcal{R}^* \subset \mathcal{V}^* \subset \mathcal{K}$ as in Theorem 19.6, if F is a friend of $\mathcal{V}^*$, then F is a friend of $\mathcal{R}^*$.

19.8 Example   It is interesting to explore the structure that can be induced in a closed-loop state equation via these geometric constructions. Suppose that $\mathcal{V}$ is a controlled invariant subspace for the state equation (1), and $\mathcal{R}^*$ is the maximal controllability subspace contained in $\mathcal{V}$. Supposing that F^a is a friend of $\mathcal{V}$, Corollary 19.7 gives that F^a is a friend of $\mathcal{R}^*$ via the device of viewing $\mathcal{V}$ as the maximal controlled invariant subspace contained in $\mathcal{V}$ for (1). Furthermore, suppose $q = dim \ \mathcal{B} \cap \mathcal{R}^*$, and let $G = [\ G_1 \ \ G_2\ ]$ be an invertible $m \times m$ matrix with $m \times q$ partition G_1 such that

$$Im\,[BG_1] = \mathcal{B} \cap \mathcal{R}^*$$

Now for the closed-loop state equation

$$\dot{x}(t) = (A + BF^a)x(t) + BGv(t) \tag{17}$$

consider a change of state variables using a basis adapted to the nested set of subspaces $\mathcal{B} \cap \mathcal{R}^*, \mathcal{R}^*$, and $\mathcal{V}$. Specifically let $p_1, \ldots, \ p_q$ be a basis for $\mathcal{B} \cap \mathcal{R}^*, p_1, \ldots, \ p_\rho$ be a basis for $\mathcal{R}^*, p_1, \ldots, \ p_\nu$ be a basis for $\mathcal{V}$, and $p_1, \ldots, \ p_n$ be a basis for X, with $0 < q < \rho < \nu < n$ to avoid vacuity. Then with

$$z(t) = [\ p_1 \ \ \cdots \ \ p_n\ ]^{-1}x(t)$$

the closed-loop state equation (17) can be written in the partitioned form

$$\dot{z}(t) = \begin{bmatrix} \hat{A}_{11} & \hat{A}_{12} & \hat{A}_{13} \\ 0 & \hat{A}_{22} & \hat{A}_{23} \\ 0 & 0 & \hat{A}_{33} \end{bmatrix} z(t) + \begin{bmatrix} \hat{B}_{11} & \hat{B}_{12} \\ 0 & \hat{B}_{22} \\ 0 & \hat{B}_{32} \end{bmatrix} v(t)$$

Here $\hat{A}_{11}$ is $\rho \times \rho$, $\hat{B}_{11}$ is $\rho \times q$, $\hat{B}_{12}$ is $\rho \times (m-q)$, $\hat{A}_{22}$ is $(\nu-\rho) \times (\nu-\rho)$, $\hat{B}_{22}$ is $(\nu-\rho) \times (m-q)$, $\hat{A}_{33}$ is $(n-\nu) \times (n-\nu)$, and $\hat{B}_{32}$ is $(n-\nu) \times (m-q)$.

Consider next the state feedback gain

$$F = F^a + GF^bP^{-1}$$

where F^b has the partitioned form

$$F^b = \begin{bmatrix} F_{11}^b & 0 & 0 \\ 0 & 0 & F_{23}^b \end{bmatrix}$$

The resulting closed-loop state equation

$$\dot{x}(t) = (A + BF)x(t) + BGv(t)$$

after the same state variable change is given by

$$\dot{z}(t) = \begin{bmatrix} \hat{A}_{11}+\hat{B}_{11}F_{11}^b & \hat{A}_{12} & \hat{A}_{13}+\hat{B}_{12}F_{23}^b \\ 0 & \hat{A}_{22} & \hat{A}_{23}+\hat{B}_{22}F_{23}^b \\ 0 & 0 & \hat{A}_{33}+\hat{B}_{32}F_{23}^b \end{bmatrix} z(t) + \begin{bmatrix} \hat{B}_{11} & \hat{B}_{12} \\ 0 & \hat{B}_{22} \\ 0 & \hat{B}_{32} \end{bmatrix} v(t) \tag{18}$$

In this set of coordinates it is apparent that F is a friend of $\mathcal{V}$, and a friend of $\mathcal{R}^*$. The characteristic polynomial of the closed-loop state equation is

$$det\,(\lambda I - \hat{A}_{11} - \hat{B}_{11}F_{11}^b)\,det\,(\lambda I - \hat{A}_{22})\,det\,(\lambda I - \hat{A}_{33} - \hat{B}_{32}F_{23}^b)$$

and under a controllability hypothesis, F_{11}^b and F_{23}^b can be chosen to obtain desired coefficients for the associated polynomial factors. However the characteristic polynomial of $\hat{A}_{22}$ remains fixed. Of course we have used a special choice of F^b to arrive at this conclusion. In particular the zero blocks in the bottom block row of F^b preserve the block-upper-triangular structure of $P^{-1}(A + BF)P$, thus displaying the eigenvalues of $A + BF$. The zero blocks in the top row of F^b are not critical — entries there do not affect eigenvalues. Using a more abstract analysis it can be shown that the characteristic polynomial of $\hat{A}_{22}$ remains fixed for *every* friend F of $\mathcal{V}$.
□□□

With all this machinery established, we are ready to prove a basic solvability condition for the disturbance decoupling problem with eigenvalue assignment. The particular choice of basis in Example 19.8 provides the key to an elementary treatment, though in more detail than is needed. Moreover the conditions we present as sufficient conditions can be shown to be both necessary and sufficient. In the notation of Example 19.8, necessity requires a proof that the eigenvalues of $\hat{A}_{22}$ in (18) are fixed for every friend of $\mathcal{V}$.

19.9 Lemma   Suppose the plant (1) is controllable, $\mathcal{V}$ is a v-dimensional controlled invariant subspace, $v \geq 1$, and $\mathcal{R}^*$ is the maximal controllability subspace contained in $\mathcal{V}$. If $\mathcal{R}^* = \mathcal{V}$, then for any degree-v polynomial $p_v(\lambda)$ and any degree-$(n-v)$ polynomial $p_{n-v}(\lambda)$ there exists a friend F of $\mathcal{V}$, such that

$$\det (\lambda I - A - BF) = p_v(\lambda)p_{n-v}(\lambda) \tag{19}$$

Proof   Given $p_v(\lambda)$ and $p_{n-v}(\lambda)$, first select a friend F^a of $\mathcal{V} = \mathcal{R}^*$ so that the state feedback

$$u(t) = F^a x(t) + v(t)$$

applied to (1) yields, by Theorem 18.26, the characteristic polynomial $p_v(\lambda)$ for the component of the closed-loop state equation corresponding to $\mathcal{R}^*$. Applying a state variable change $z(t) = P^{-1}x(t)$, where the columns of P form a basis for X adapted to $\mathcal{R}^* = \mathcal{V}$, gives the closed-loop state equation in partitioned form,

$$\dot{z}(t) = \begin{bmatrix} \hat{A}_{11} & \hat{A}_{12} \\ 0 & \hat{A}_{22} \end{bmatrix} z(t) + \begin{bmatrix} \hat{B}_{11} \\ \hat{B}_{21} \end{bmatrix} v(t) \tag{20}$$

where $\det (\lambda I - \hat{A}_{11}) = p_v(\lambda)$. Now consider, in place of F^a, a feedback gain of the form

$$F = F^a + \begin{bmatrix} 0 & F_{12}^b \end{bmatrix} P^{-1}$$

This new feedback gain is easily shown to be a friend of $\mathcal{V} = \mathcal{R}^*$ that gives the closed-loop state equation, in terms of the state variable $z(t)$,

$$\dot{z}(t) = \begin{bmatrix} \hat{A}_{11} & \hat{A}_{12} + \hat{B}_{11}F_{12}^b \\ 0 & \hat{A}_{22} + \hat{B}_{21}F_{12}^b \end{bmatrix} z(t) + \begin{bmatrix} \hat{B}_{11} \\ \hat{B}_{21} \end{bmatrix} v(t)$$

The characteristic polynomial of this closed-loop state equation is

$$p_v(\lambda) \det (\lambda I - \hat{A}_{22} - \hat{B}_{21}F_{12}^b)$$

By hypothesis the plant is controllable, and therefore the second component state equation in (20) is controllable. Thus F_{12}^b can be chosen to obtain the characteristic polynomial factor

$$\det (\lambda I - \hat{A}_{22} - \hat{B}_{21}F_{12}^b) = p_{n-v}(\lambda)$$

□□□

    The reason for the factored characteristic polynomial in Lemma 19.9, and the next result, is subtle. But the issue should become apparent on considering an example where $n = 2$, $v = 1$, and the specified characteristic polynomial is λ^2+1.

19.10 Theorem   Suppose the plant (2) is controllable, and $\mathcal{R}^*$ of dimension $\rho \geq 1$ is the maximal controllability subspace contained in $Ker\,[C]$. Given any degree-ρ polynomial $p_\rho(\lambda)$ and any degree-$(n-\rho)$ polynomial $p_{n-\rho}(\lambda)$, there exists a state feedback gain F such that the closed-loop state equation (3) is disturbance decoupled and has characteristic polynomial $p_\rho(\lambda)p_{n-\rho}(\lambda)$ if

$$Im\,[E] \subset \mathcal{R}^* \tag{21}$$

    Proof   Viewing $\mathcal{V} = \mathcal{R}^*$ as a controlled invariant subspace contained in $Ker\,[C]$, since $Im\,[E] \subset \mathcal{V}$ the first part of the proof of Theorem 19.2 shows that for any state feedback gain F that is a friend of $\mathcal{V}$ the closed-loop state equation is disturbance decoupled. Then Lemma 19.9 gives that a friend of $\mathcal{V}$ can be selected such that the characteristic polynomial of the disturbance-decoupled, closed-loop state equation is $p_\rho(\lambda)p_{n-\rho}(\lambda)$.

Noninteracting Control

    The noninteracting control problem is treated in Chapter 14 for time-varying linear state equations with $p = m$, and then specialized to the time-invariant case. Here we reformulate the time-invariant problem in a geometric setting and assume $p \geq m$ so that the objective in general involves scalar input components and blocks of output components. It is convenient to adjust notation by partitioning the output matrix C to write the plant in the form

$$\dot{x}(t) = Ax(t) + Bu(t)$$

$$y_j(t) = C_j x(t), \quad j = 1, \ldots, m \tag{22}$$

where C_j is a $p_j \times n$ matrix, and $p_1 + \cdots + p_m = p$. With G_i denoting the ith-column of the $m \times m$ matrix G, linear state feedback can be written as

$$u(t) = Fx(t) + \sum_{i=1}^{m} G_i v_i(t)$$

The resulting closed-loop state equation is

$$\dot{x}(t) = (A + BF)x(t) + \sum_{i=1}^{m} BG_i v_i(t)$$

$$y_j(t) = C_j x(t) , \quad j = 1, \ldots, m \tag{23}$$

a notation that focuses attention on the scalar components of the input signal and the $p_j \times 1$ vector partitions of the output signal.

The objectives for the closed-loop state equation involve only input-output behavior, and so zero initial state is assumed. The first objective is that for $i \neq j$ the jth output (block) $y_j(t)$ should be uninfluenced by the ith input $v_i(t)$. This is called *noninteracting control,* and in terms of the component closed-loop transfer functions,

$$Y_j(s) = C_j(sI - A - BF)^{-1} BG_i \, V_i(s) , \quad i, j = 1, \ldots, m$$

the first objective is, simply, $Y_j(s)/V_i(s) = 0$, for $i \neq j$. The second objective is that the closed-loop state equation be output controllable in the sense of Exercise 9.10. This imposes the requirement that the jth-output block *is* influenced by the jth-input. For example, from the solution of Exercise 9.11, if $p_1 = \cdots = p_m = 1$, then the output controllability requirement is that each scalar transfer function $Y_j(s)/V_j(s)$ be a nonzero rational function of s.

It is straightforward to translate this problem into geometric terms. For any F and G, the controllable subspace of the closed-loop state equation corresponding to the ith-input is $<A+BF \,|\, Im\,[BG_i]>$. Then the noninteraction requirement is equivalent to existence of feedback gains F and G such that

$$<A+BF \,|\, Im\,[BG_i]> \subset Ker\,[C_j] , \quad j \neq i$$

Stated another way, noninteraction is equivalent to existence of F and G such that

$$<A+BF \,|\, Im\,[BG_i]> \subset \mathcal{K}_i , \quad i = 1, \ldots, m$$

where

$$\mathcal{K}_i = \bigcap_{\substack{j=1 \\ j \neq i}}^{m} Ker\,[C_j] , \quad i = 1, \ldots, m \tag{24}$$

Also, by Exercise 18.9, the output controllability requirement can be written as

$$C_i <A+BF \,|\, Im\,[BG_i]> = \mathcal{Y}_i , \quad i = 1, \ldots, m$$

where $\mathcal{Y}_i = Im\,[C_i]$.

It is traditional, though not necessary, to further rephrase the noninteracting control problem in terms of controllability subspaces characterized as in Theorem 18.23, so that G is implicit. This focuses attention more directly on geometric aspects, and leads to the following problem statement. Compute an $m \times n$ matrix F and controllability subspaces $\mathcal{R}_1, \ldots, \mathcal{R}_m$ such that

$$\mathcal{R}_i = <A+BF \,|\, \mathcal{B} \cap \mathcal{R}_i >$$

$$\mathcal{R}_i \subset \mathcal{K}_i$$

$$C_i \mathcal{R}_i = \mathcal{Y}_i \tag{25}$$

for $i = 1, \ldots, m$. The key issue is existence of a single F that is a friend of all the controllability subspaces $\mathcal{R}_1, \ldots, \mathcal{R}_m$. Controllability subspaces that have a common friend are called *compatible,* and this terminology is applied also to controlled invariant subspaces.

Conditions for solvability of the noninteracting control problem can be presented either in terms of maximal controlled invariant subspaces, or maximal controllability subspaces. Because an input gain G is involved, we use controllability subspaces for congeniality with basic definitions of the subspaces. To rule out trivially unsolvable problems, and thus obtain a compact condition that is necessary as well as sufficient, familiar assumptions are adopted. (See Exercise 19.12.) These assumptions have the added benefit of harmony with existence of a state feedback with invertible G that solves the noninteracting control problem — a desirable feature in typical situations.

19.11 Theorem   Suppose the plant (22) is controllable, with *rank* $B = m$ and *rank* $C = p$. Then there exist feedback gains F and invertible G that solve the noninteracting control problem if and only if

$$\mathcal{B} = \mathcal{B} \cap \mathcal{R}_1{}^* + \cdots + \mathcal{B} \cap \mathcal{R}_m{}^* \tag{26}$$

where, for $i = 1, \ldots, m$, $\mathcal{R}_i{}^*$ is the maximal controllability subspace contained in $\mathcal{K}_i$ for (22).

Proof   To show (26) is a necessary condition, suppose F and invertible G are such that the closed-loop state equation (23) is noninteracting. Then the controllability subspaces

$$\mathcal{R}_i = Im\,[BG_i] + (A + BF)Im\,[BG_i] + \cdots + (A + BF)^{n-1} Im\,[BG_i]$$

satisfy

$$\mathcal{R}_i \subset \mathcal{K}_i, \quad i = 1, \ldots, m$$

and, of course, $\mathcal{R}_i \subset \mathcal{R}_i{}^*$. Therefore $Im\,[BG_i] \subset \mathcal{R}_i{}^*$, and since $Im\,[BG_i] \subset \mathcal{B}$,

$$Im\,[BG_i] \subset \mathcal{B} \cap \mathcal{R}_i{}^*, \quad i = 1, \ldots, m$$

Using the invertibility of G,

$$\mathcal{B} = Im\,[BG_1] + \cdots + Im\,[BG_m]$$
$$\subset \mathcal{B} \cap \mathcal{R}_1{}^* + \cdots + \mathcal{B} \cap \mathcal{R}_m{}^* \tag{27}$$

Since the reverse inclusion is obvious, we have established (26).

It is a much more intricate task to prove that (26) is a sufficient condition for solvability of the noninteracting control problem. For convenience we divide the proof and state two lemmas. The first presents a refinement of (26), and the second proves compatibility of a certain set of controlled invariant subspaces as an intermediate step in proving compatibility of $\mathcal{R}_1{}^*, \ldots, \mathcal{R}_m{}^*$.

19.12 Lemma   Under the hypotheses of Theorem 19.11, if (26) holds, then

$$\sum_{j=1}^{m} \mathcal{R}_j^* = X \tag{28}$$

$$\dim \; \mathcal{B} \cap \mathcal{R}_j^* = 1 \;, \quad j = 1, \ldots, m \tag{29}$$

$$\mathcal{B} = \mathcal{B} \cap \mathcal{R}_1^* \; \oplus \; \cdots \; \oplus \; \mathcal{B} \cap \mathcal{R}_m^* \tag{30}$$

Proof   Since a sum of controlled invariant subspaces is a controlled invariant subspace,

$$\sum_{j=1}^{m} \mathcal{R}_j^*$$

is a controlled invariant subspace that, by (26), contains $\mathcal{B}$. But $<A|\mathcal{B}>$ is the minimal controlled invariant subspace that contains $\mathcal{B}$, and the controllability hypothesis and Corollary 18.7 therefore give (28).

Next we show that $\mathcal{B} \cap \mathcal{R}_1^*$ has dimension one. Let

$$\gamma_1 = \dim \; \mathcal{B} \cap \mathcal{R}_1^*$$

$$\gamma_i = \dim \left(\sum_{j=1}^{i} \mathcal{B} \cap \mathcal{R}_j^* \right) - \dim \left(\sum_{j=1}^{i-1} \mathcal{B} \cap \mathcal{R}_j^* \right), \quad i = 2, \ldots, m \tag{31}$$

These obviously are nonnegative integers, and the following contradiction argument proves that $\gamma_1, \ldots, \gamma_m \geq 1$. If $\gamma_i = 0$ for some value of i, then

$$\mathcal{B} \cap \mathcal{R}_i^* \subset \sum_{j=1}^{i-1} \mathcal{B} \cap \mathcal{R}_j^*$$

$$\subset \sum_{\substack{j=1 \\ j \neq i}}^{m} \mathcal{B} \cap \mathcal{R}_j^* \tag{32}$$

Setting

$$\tilde{\mathcal{R}}_i = \sum_{\substack{j=1 \\ j \neq i}}^{m} \mathcal{R}_j^*$$

(32) together with (26) gives that $\mathcal{B} \subset \tilde{\mathcal{R}}_i$. Thus $\tilde{\mathcal{R}}_i$ is a controlled invariant subspace that contains $\mathcal{B}$, and, summoning Corollary 18.7 again, $\tilde{\mathcal{R}}_i = X$. By the definition of $\mathcal{R}_1^*, \ldots, \mathcal{R}_m^*$, $\tilde{\mathcal{R}}_i \subset Ker[C_i]$, which implies $Ker[C_i] = X$, and this contradicts the assumption *rank C = p*.

Having established that $\gamma_1, \ldots, \gamma_m \geq 1$, we further observe from (26) and (31) that

$$\gamma_1 + \cdots + \gamma_m = \dim \; \mathcal{B} = m$$

An immediate consequence is

$$\gamma_1 = \cdots = \gamma_m = 1$$

Of course this shows $dim \; \mathcal{B} \cap \mathcal{R}_1{}^* = 1$.

To establish (29) for any other value of j, simply reverse the roles of $\mathcal{B} \cap \mathcal{R}_j{}^*$ and $\mathcal{B} \cap \mathcal{R}_1{}^*$ in the definition of integers $\gamma_1, \ldots, \gamma_m$, and apply the same argument. Finally, (30) holds as a consequence of (26), (29) and $dim \; \mathcal{B} = m$.

19.13 Lemma   Under the hypotheses of Theorem 19.11, suppose (26) holds. Let $\mathcal{V}_i{}^*$ denote the maximal controlled invariant subspace contained in $\mathcal{K}_i$, $i = 1, \ldots, m$. Then the subspaces defined by

$$\tilde{\mathcal{V}_i} = \sum_{\substack{j=1 \\ j \neq i}}^{m} \mathcal{V}_j{}^* \; , \quad i = 1, \ldots, m \tag{33}$$

are compatible controlled invariant subspaces.

Proof   The calculation

$$A \tilde{\mathcal{V}_i} = \sum_{\substack{j=1 \\ j \neq i}}^{m} A \mathcal{V}_j{}^*$$

$$\subset \sum_{\substack{j=1 \\ j \neq i}}^{m} (\mathcal{V}_j{}^* + \mathcal{B})$$

$$= \tilde{\mathcal{V}_i} + \mathcal{B}$$

proves that $\tilde{\mathcal{V}_1}, \ldots, \tilde{\mathcal{V}_m}$ are controlled invariant subspaces. Using (26), and the fact that $\mathcal{R}_i{}^* \subset \mathcal{V}_i{}^*$,

$$A \tilde{\mathcal{V}_i} \subset \tilde{\mathcal{V}_i} + \mathcal{B} \cap \mathcal{R}_1{}^* \oplus \cdots \oplus \mathcal{B} \cap \mathcal{R}_m{}^*$$

$$= \tilde{\mathcal{V}_i} + \mathcal{B} \cap \mathcal{R}_i{}^* \; , \quad i = 1, \ldots, m \tag{34}$$

By (29) we can choose $n \times 1$ vectors $\tilde{B}_1, \ldots, \tilde{B}_m$ such that

$$Im \, [\tilde{B}_i] = \mathcal{B} \cap \mathcal{R}_i{}^* \; , \quad i = 1, \ldots, m$$

Then from (34),

$$A \tilde{\mathcal{V}_i} \subset \tilde{\mathcal{V}_i} + Im \, [\tilde{B}_i] \; , \quad i = 1, \ldots, m$$

and, calling on Theorem 18.19, there exist $1 \times n$ matrices $\tilde{F}_1, \ldots, \tilde{F}_m$ such that

$$(A + \tilde{B}_i \tilde{F}_i) \tilde{\mathcal{V}_i} \subset \tilde{\mathcal{V}_i} \; , \quad i = 1, \ldots, m$$

From this data a common friend F for $\tilde{\mathcal{V}_1}, \ldots, \tilde{\mathcal{V}_m}$ can be constructed. Let $v_1, \ldots, v_n$ be a basis for X. Since $Im \, [\tilde{B}_i] \subset \mathcal{B}$, there exist $m \times 1$ vectors $u_1, \ldots, u_n$ such that

$$Bu_k = \sum_{j=1}^{m} \widetilde{B}_j \widetilde{F}_j v_k , \quad k = 1, \dots, n$$

Let

$$F = \begin{bmatrix} u_1 & \cdots & u_n \end{bmatrix} \begin{bmatrix} v_1 & \cdots & v_n \end{bmatrix}^{-1} \tag{35}$$

so that

$$BFv_k = B \begin{bmatrix} u_1 & \cdots & u_n \end{bmatrix} e_k$$

$$= \sum_{j=1}^{m} \widetilde{B}_j \widetilde{F}_j v_k , \quad k = 1, \dots, n$$

Since any vector in $\widetilde{\mathcal{V}}_i$ can be written as a linear combination of $v_1, \dots, v_n$,

$$(A + BF)\widetilde{\mathcal{V}}_i = \left(A + \widetilde{B}_i \widetilde{F}_i + \sum_{\substack{j=1 \\ j \neq i}}^{m} \widetilde{B}_j \widetilde{F}_j \right) \widetilde{\mathcal{V}}_i$$

$$\subset (A + \widetilde{B}_i \widetilde{F}_i)\widetilde{\mathcal{V}}_i + \sum_{\substack{j=1 \\ j \neq i}}^{m} \mathcal{B} \cap \mathcal{R}_j{}^*$$

$$\subset \widetilde{\mathcal{V}}_i + \sum_{\substack{j=1 \\ j \neq i}}^{m} \mathcal{R}_j{}^*$$

$$= \widetilde{\mathcal{V}}_i , \quad i = 1, \dots, m \tag{36}$$

Therefore the controlled invariant subspaces $\widetilde{\mathcal{V}}_1, \dots, \widetilde{\mathcal{V}}_m$ are compatible, with common friend F given by (35).
□□□

Returning to the sufficiency proof for Theorem 19.11, we now show that (26) implies existence of F and invertible G such that $\mathcal{R}_1{}^*, \dots, \mathcal{R}_m{}^*$ satisfy the conditions in (25). The major effort involves proving that $\mathcal{R}_1{}^*, \dots, \mathcal{R}_m{}^*$ are compatible. To this end we use Lemma 19.13 and show that F in (35) satisfies

$$(A + BF)\mathcal{V}_i{}^* \subset \mathcal{V}_i{}^* , \quad i = 1, \dots, m$$

Then it follows from Corollary 19.7 that F is a common friend of $\mathcal{R}_1{}^*, \dots, \mathcal{R}_m{}^*$. In other words, we show that compatibility of $\widetilde{\mathcal{V}}_1, \dots, \widetilde{\mathcal{V}}_m$ implies compatibility of $\mathcal{R}_1{}^*, \dots, \mathcal{R}_m{}^*$.
Let

$$\mathcal{V}_i = \bigcap_{\substack{j=1 \\ j \neq i}}^{m} \widetilde{\mathcal{V}}_j , \quad i = 1, \dots, m \tag{37}$$

Since each $\widetilde{\mathcal{V}}_j$ is an invariant subspace for $(A + BF)$, it is easy to show that $\mathcal{V}_1, \ldots, \mathcal{V}_m$ also are invariant subspaces for $(A + BF)$. We next prove that $\mathcal{V}_i = \mathcal{V}_i^*$, $i = 1, \ldots, m$, a step that brings us close to the end.

From the definition of $\widetilde{\mathcal{V}}_j$ in (33), $\mathcal{V}_i^* \subset \widetilde{\mathcal{V}}_j$ for all $i \neq j$. Then, from the definition of $\mathcal{V}_i$ in (37), $\mathcal{V}_i^* \subset \mathcal{V}_i$, $i = 1, \ldots, m$. To show the reverse containment matters must be written out in detail. From (33) and (37)

$$\mathcal{V}_i = \bigcap_{\substack{j=1 \\ j \neq i}}^{m} \sum_{\substack{k=1 \\ k \neq j}}^{m} \mathcal{V}_k^* , \quad i = 1, \ldots, m$$

Since

$$\mathcal{V}_k^* \subset \mathcal{K}_k = \bigcap_{\substack{l=1 \\ l \neq k}}^{m} Ker[C_l] , \quad k = 1, \ldots, m$$

it follows that

$$\mathcal{V}_i \subset \bigcap_{\substack{j=1 \\ j \neq i}}^{m} \sum_{\substack{k=1 \\ k \neq j}}^{m} \bigcap_{\substack{l=1 \\ l \neq k}}^{m} Ker[C_l] \tag{38}$$

Noting that $Ker[C_j]$ is common to each intersection in the sum of intersections

$$\sum_{\substack{k=1 \\ k \neq j}}^{m} \bigcap_{\substack{l=1 \\ l \neq k}}^{m} Ker[C_l]$$

we can apply the first part of Exercise 18.4 (after easy generalization to sums of more than two intersections) to obtain

$$\sum_{\substack{k=1 \\ k \neq j}}^{m} \bigcap_{\substack{l=1 \\ l \neq k}}^{m} Ker[C_l] \subset Ker[C_j] \cap \sum_{\substack{k=1 \\ k \neq j}}^{m} \bigcap_{\substack{l=1 \\ l \neq k, j}}^{m} Ker[C_l]$$

This gives, from (38),

$$\mathcal{V}_i \subset \bigcap_{\substack{j=1 \\ j \neq i}}^{m} \left[Ker[C_j] \cap \sum_{\substack{k=1 \\ k \neq j}}^{m} \bigcap_{\substack{l=1 \\ l \neq k, j}}^{m} Ker[C_l] \right]$$

$$\subset \bigcap_{\substack{j=1 \\ j \neq i}}^{m} Ker[C_j] = \mathcal{K}_i , \quad i = 1, \ldots, m \tag{39}$$

Therefore $\mathcal{V}_i \subset \mathcal{V}_i^*$, $i = 1, \ldots, m$, by maximality of each $\mathcal{V}_i^*$, and we have shown $\mathcal{V}_i = \mathcal{V}_i^*$, $i = 1, \ldots, m$.

With the argument above we have compatibility of $\mathcal{V}_1^*, \ldots, \mathcal{V}_m^*$, hence compatibility of $\mathcal{R}_1^*, \ldots, \mathcal{R}_m^*$. Lemma 19.13 provides a construction for a common friend F, and it remains only to determine the invertible gain G. From (29) we can compute $m \times 1$ vectors $G_1, \ldots, G_m$ such that

$$Im[BG_i] = B \cap R_i^* , \quad i = 1, \ldots, m \tag{40}$$

then

$$R_i^* = <A + BF \,|\, Im[BG_i]> , \quad i = 1, \ldots, m$$

and it is immediate from (30) that G is invertible.

We conclude the proof that $R_1^*, \ldots, R_m^*$ satisfy the geometric conditions in (25) by demonstrating output controllability for the closed-loop state equation. Using (28) and the inclusion $\tilde{R}_i \subset Ker[C_i]$ noted in the proof of Lemma 19.12 yields

$$R_i^* + Ker[C_i] = X, \quad 1 = 1, \ldots, m$$

But then

$$C_i R_i^* = C_i \left[R_i^* + Ker[C_i] \right] = C_i X = Y_i , \quad 1 = 1, \ldots, m$$

and the proof is complete.
□□□

After a blizzard of subspaces, and before a matrix-computation procedure for V^*, and hence R^*, it might be helpful to work a simple problem freestyle from the basic theory.

19.14 Example   Consider $X = R^3$ with the standard basis e_1, e_2, e_3, and a linear plant specified by

$$A = \begin{bmatrix} 1 & 0 & 0 \\ 2 & 3 & 4 \\ 0 & 0 & 5 \end{bmatrix}, \quad B = \begin{bmatrix} 0 & 1 \\ 0 & 0 \\ 2 & 0 \end{bmatrix}, \quad C = \begin{bmatrix} 1 & 0 & 0 \\ 0 & 0 & 2 \end{bmatrix} \tag{41}$$

The assumptions of Theorem 19.11 are satisfied, and the main task in ascertaining solvability of the noninteracting control problem is to compute R_1^* and R_2^*, the maximal controllability subspaces contained in $Ker[C_2]$ and $Ker[C_1]$, respectively.

Retracing the approach described immediately above Corollary 19.7, we first compute V_1^* and V_2^*, the maximal controlled invariant subspaces contained in $Ker[C_2]$ and $Ker[C_1]$, respectively. Since B is spanned by e_1, e_3, and $Ker[C_2]$ is spanned by e_1, e_2, written

$$B = \text{span } \{e_1, e_3\}$$

$$Ker[C_2] = \text{span } \{e_1, e_2\}$$

the algorithm in Theorem 19.3 gives

$$V_1^0 = \text{span } \{e_1, e_2\}$$

$$V_1^1 = \left[\text{span } \{e_1, e_2\} \right] \cap A^{-1} \left[\text{span } \{e_1, e_3\} + \text{span } \{e_1, e_2\} \right]$$

Thus

$$V_1^* = \text{span } \{e_1, e_2\}$$

Friends of V_1^* can be characterized via the condition $(A + BF)V_1^* \subset V_1^*$. That is, writing

$$F = \begin{bmatrix} f_{11} & f_{12} & f_{13} \\ f_{21} & f_{22} & f_{23} \end{bmatrix}$$

we consider

$$\begin{bmatrix} 1+f_{21} & f_{22} & f_{23} \\ 2 & 3 & 4 \\ 2f_{11} & 2f_{12} & 5+2f_{13} \end{bmatrix} \text{ span } \{e_1, e_2\} \subset \text{span } \{e_1, e_2\} \tag{42}$$

This gives that F is a friend of V_1^* if and only if $f_{11} = f_{12} = 0$. The simplest friend of V_1^* is $F = 0$, and since $B \cap V_1^* = e_1$,

$$R_1^* = <A + BF \mid B \cap V_1^* >$$

$$= \text{span } \{e_1\} + A \text{ span } \{e_1\} + A^2 \text{ span } \{e_1\}$$

$$= \text{span } \{e_1, e_2\}$$

$$= V_1^*$$

A similar calculation gives that

$$R_2^* = V_2^* = \text{span } \{e_2, e_3\}$$

and F is a friend of V_2^* if and only if $f_{22} = f_{23} = 0$.

Applying the solvability condition (26),

$$B \cap R_1^* + B \cap R_2^* = \text{span } \{e_1\} + \text{span } \{e_3\} = B$$

and noninteracting control is feasible. Using (40) immediately gives the reference-input gain

$$G = \begin{bmatrix} 0 & 1 \\ 1 & 0 \end{bmatrix} \tag{43}$$

A gain F provides noninteracting control if and only if it is a common friend of R_1^* and R_2^*. Therefore the class of state-feedback gains for noninteracting control is described by

$$F = \begin{bmatrix} 0 & 0 & f_{13} \\ f_{21} & 0 & 0 \end{bmatrix} \tag{44}$$

where f_{13} and f_{21} are arbitrary.

A straightforward calculation shows that $A + BF$ has a fixed eigenvalue at 3 for any F of the form (44). Therefore noninteracting control and exponential stability cannot be achieved simultaneously by static state feedback in this example.

Maximal Controlled Invariant Subspace Computation

There are two main steps needed to translate the conceptual algorithm for V^* in Theorem 19.3 into a numerical algorithm. First is the computation of a basis for the intersection of two subspaces from the subspace bases. Second, and less easy, we need a method to compute a basis for the inverse image of a subspace under a linear map. But a preliminary result converts this second step into two simpler computations. The proof uses the basic linear-algebra fact that if H is an $n \times q$ matrix,

$$R^n = Im[H] \oplus Ker[H^T] \tag{45}$$

19.15 Lemma   Suppose A is an $n \times n$ matrix, and H is an $n \times q$ matrix. If L is a maximal rank $n \times l$ matrix such that $L^T H = 0$, then $A^{-1}Im[H] = Ker[L^T A]$.

Proof   If $x \in A^{-1}Im[H]$, then there exists a vector $y \in Im[H]$ such that $Ax = y$. Since y can be written as a linear combination of the columns of H, the definition of L gives

$$0 = L^T y = L^T Ax$$

That is, $x \in Ker[L^T A]$.

On the other hand suppose $x \in Ker[L^T A]$. Letting $y = Ax$ again, by (45) there exist unique $n \times 1$ vectors $y_a \in Im[H]$ and $y_b \in Ker[H^T]$ such that $y = y_a + y_b$. Then

$$0 = L^T y = L^T y_a + L^T y_b = L^T y_b$$

Furthermore $H^T y_b = 0$ gives $y_b^T H = 0$, and it follows from the maximal rank property of L that y_b^T must be a linear combination of the rows of L^T. If the coefficients in this linear combination are $\alpha_1, \ldots, \alpha_l$, then

$$y_b^T y_b = [\, \alpha_1 \quad \cdots \quad \alpha_l \,]L^T y_b = 0 \tag{46}$$

Thus $y_b = 0$ and we have shown that $y = y_a \in Im[H]$. Therefore $x \in A^{-1}Im[H]$.
□□□

Given A, B, and a subspace $\mathcal{K} \subset X$, the following sequence of matrix computations delivers a basis for the maximal controlled invariant subspace $V^* \subset \mathcal{K}$. We assume that $\mathcal{K}$ is specified as the image of an n-row, full-column-rank matrix V_0; in other words, the columns of V_0 form a basis for $\mathcal{K}$. Each step of the matrix algorithm implements a portion of the conceptual algorithm in Theorem 19.3, as indicated by parenthetical comments.

19.16 Algorithm

(i) With $Im[V_0] = \mathcal{K} = V^0$, compute a maximal-rank matrix L_0 such that $L_0^T V_0 = 0$. (By Lemma 19.15 with $A = I$, this gives $V^0 = Ker[L_0^T]$.)

(ii) Construct a matrix $\hat{V}_0$ by deleting linearly dependent columns from the partitioned matrix $[\, B \quad V_0 \,]$. (Then $Im[\hat{V}_0] = \mathcal{B} + V^0$.)

(iii) Compute a maximal-rank matrix L_1 such that $L_1^T \hat{V}_0 = 0$. (Then by Lemma 19.15, $Ker[L_1^T A] = A^{-1}(\mathcal{B} + \mathcal{V}^0)$.)

(iv) Compute a maximal rank matrix V_1 such that

$$\begin{bmatrix} L_0^T \\ L_1^T A \end{bmatrix} V_1 = 0 \qquad\qquad (47)$$

(Thus $Im[V_1] = \mathcal{V}^0 \cap A^{-1}(\mathcal{B} + \mathcal{V}^0)$.)

(v) Continue by iterating the previous three steps.
□□□

Specifically the algorithm continues by deleting linearly dependent columns from $[B \quad V_1]$ to form $\hat{V}_1$, computing a maximal rank L_2 such that $L_2^T \hat{V}_1 = 0$, and then computing a maximal rank V_2 such that

$$\begin{bmatrix} L_0^T \\ L_2^T A \end{bmatrix} V_2 = 0 \qquad\qquad (60)$$

Then $\mathcal{V}^2 = Im[V_2]$, and so on. Repeating this until the first step k, where $rank\ V_{k+1} = rank\ V_k$, $k \le n$ guaranteed, gives $\mathcal{V}^* = Im[V_k]$.

EXERCISES

Exercise 19.1   With a basis for $X = R^n$ fixed, and $S \subset X$ a subspace, let

$$S^\perp = \left\{ z \in X \mid z^T x = 0 \text{ for all } x \in S \right\}$$

(Note that this definition is not coordinate free.) If $W \subset X$ is another subspace, show that

$$[W + S]^\perp = W^\perp \cap S^\perp$$

If A is an $n \times n$ matrix, show that

$$(A^T S)^\perp = A^{-1} S^\perp$$

If C is $p \times n$, show that

$$(Ker[C])^\perp = Im[C^T]$$

Exercise 19.2   Corresponding to the linear state equation

$$\dot{x}(t) = Ax(t) + Bu(t)$$

suppose $\mathcal{K} \subset X$ is a specified subspace. Define the corresponding sequence of subspaces (see Exercise 19.1 for definitions)

$$\mathcal{W}^0 = \mathcal{K}^\perp$$
$$\mathcal{W}^k = \mathcal{W}^{k-1} + A^T(\mathcal{W}^{k-1} \cap \mathcal{B}^\perp), \quad k = 1, 2, \cdots$$

Show that the maximal controlled invariant subspace contained in $\mathcal{K}$ is given by

$$\mathcal{V}^* = (\mathcal{W}^n)^\perp$$

(*Hint*: Compare this algorithm with the algorithm for $\mathcal{V}^*$, and use Exercise 19.1 to show that $(\mathcal{W}^k)^\perp = \mathcal{V}^k$ for $k = 0, 1, \cdots$.)

Exercise 19.3   For a single-output linear state equation

$$\dot{x}(t) = Ax(t) + Bu(t)$$

$$y(t) = cx(t)$$

suppose κ is a finite positive integer such that

$$cA^j B = 0 , \quad j = 0, \ldots, \kappa{-}2 ; \quad cA^{\kappa-1}B \neq 0$$

Show that the maximal controlled invariant subspace contained in $Ker[c]$ is

$$\mathcal{V}^* = \bigcap_{k=0}^{\kappa-1} Ker[cA^k]$$

(*Hint*: Use the algorithm in Exercise 19.2 to compute $\mathcal{V}^*$.)

Exercise 19.4   Suppose $\mathcal{V}^*$ is the maximal controlled invariant subspace contained in $\mathcal{K} \subset \mathcal{X}$. Define a corresponding sequence of subspaces by

$$\mathcal{R}^0 = 0$$

$$\mathcal{R}^k = \mathcal{V}^* \cap (A\mathcal{R}^{k-1} + \mathcal{B}) , \quad k = 1, 2, \cdots$$

Show that $\mathcal{R}^n = \mathcal{R}^*$, the maximal controllability subspace contained in $\mathcal{K}$. *Hint*: Using Exercise 18.4, show that if F is a friend of $\mathcal{V}^*$, then

$$\mathcal{R}^k = \sum_{j=1}^{k} (A + BF)^{j-1}(\mathcal{B} \cap \mathcal{V}^*)$$

Exercise 19.5   For the linear state equation

$$\dot{x}(t) = Ax(t) + Bu(t)$$

$$y(t) = Cx(t)$$

denote the jth-row of C by C_j. If $\mathcal{V}^*$ is the maximal controlled invariant subspace contained in $Ker[C]$, and $\mathcal{V}_j^*$ is the maximal controlled invariant subspace contained in $Ker[C_j]$, $j = 1, \ldots, p$, show that

$$\mathcal{V}^* \subset \bigcap_{j=1}^{p} \mathcal{V}_j^*$$

Exercise 19.6   Corresponding to the linear state equation

$$\dot{x}(t) = Ax(t) + Bu(t)$$

show that there exists a unique maximal subspace $\mathcal{Z}^*$ among all subspaces that satisfy

$$A\mathcal{Z} + \mathcal{Z} \subset \mathcal{B}$$

Furthermore, show that

$$Z^* = B \cap A^{-1}B$$

(This relates to *perfect tracking* as explored in Exercise 18.13.)

Exercise 19.7   Suppose that the disturbance input $w(t)$ to the plant

$$\dot{x}(t) = Ax(t) + Bu(t) + Ew(t)$$

$$y(t) = Cx(t)$$

is measurable. Show that the disturbance decoupling problem is solvable with state/disturbance feedback of the form

$$u(t) = Fx(t) + Kw(t) + Gv(t)$$

if and only if

$$Im[E] \subset V^* + B$$

where V^* is the maximal controlled invariant subspace contained in $Ker[C]$.

Exercise 19.8   Corresponding to the linear state equation

$$\dot{x}(t) = Ax(t) + Bu(t)$$

suppose $K \subset X$ is a subspace, V^* is the maximal controlled invariant subspace contained in K, and R^* is the maximal controllability subspace contained in K. Show that

$$B \cap V^* = B \cap R^*$$

Use this fact to restate Theorem 19.11.

Exercise 19.9   If the conditions in Theorem 19.11 for existence of a solution of the noninteracting control problem are satisfied, show that there is no other set of controllability subspaces $R_i \subset K_i$, $i = 1, \ldots, m$, such that

$$B = B \cap R_1 + \cdots + B \cap R_m$$

That is, $R_1^*, \ldots, R_m^*$ provide the only solution of (26).

Exercise 19.10   Consider the additional hypothesis $p = n$ for Theorem 19.11 (so that C is invertible). Show that then (26) can be replaced by the equivalent condition

$$R_i^* + Ker[C_i] = X, \quad i = 1, \ldots, m$$

Exercise 19.11   Consider a linear state equation with $m = 2$ that satisfies the conditions for noninteracting control in Theorem 19.11. For the noninteracting closed-loop state equation

$$\dot{x}(t) = (A + BF)x(t) + BG_1v_1(t) + BG_2v_2(t)$$

$$y_1(t) = C_1x(t)$$

$$y_2(t) = C_2x(t)$$

consider a state variable change adapted to the nested set of subspaces

$$\text{span } \{p_{n-q}, \ldots, p_n\} = \mathcal{R}_1{}^* \cap \mathcal{R}_2{}^*$$

$$\text{span } \{p_1, \ldots, p_r; p_{n-q}, \ldots, p_n\} = \mathcal{R}_1{}^*$$

$$\text{span } \{p_1, \ldots, p_n\} = \mathcal{R}_1{}^* + \mathcal{R}_2{}^* = X$$

What is the partitioned form of the closed-loop state equation in the new coordinates?

Exercise 19.12   Justify the assumptions *rank B = m* and *rank C = p* in Theorem 19.11 by providing simple examples with $m = p = 2$ to show that removal of either assumption admits obviously unsolvable problems.

NOTES

Note 19.1   Further developments on disturbance decoupling, including refinements of the basic problem studied here and output-feedback solutions, can be found in

S.P. Bhattacharyya, "Disturbance rejection in linear systems," *International Journal of Systems Science,* Vol. 5, pp. 633–637, 1974

J.C. Willems, C. Commault, "Disturbance decoupling by measurement feedback with stability or pole placement," *SIAM Journal of Control and Optimization,* Vol. 19, pp. 490–504, 1981

We have not discussed the problem of disturbance decoupling with stability, where eigenvalue assignment is not required. But it should be no surprise that this problem involves the stabilizability condition in Theorem 18.28 and the condition $Im[E] \subset S^*$, where S^* is the maximal stabilizability subspace contained in $Ker[C]$. For further information see the references in Note 18.5.

Note 19.2   Numerical aspects of the computation of maximal controlled invariant subspaces are discussed in the papers

B.C. Moore, A.J. Laub, "Computation of supremal (A, B)-invariant and (A, B)-controllability subspaces," *IEEE Transactions on Automatic Control,* Vol. AC-23, No. 5, pp. 783–792, 1978

A. Linnemann, "Numerical aspects of disturbance decoupling by measurement feedback," *IEEE Transactions on Automatic Control,* Vol. AC-32, No. 10, pp. 922–926, 1987

The *singular values* of a matrix A are the nonnegative square roots of the eigenvalues of A^TA. The associated *singular value decomposition* provides efficient methods for calculating sums of subspaces, inverse images, and so on. For an introduction see

V.C. Klema, A.J. Laub, "The singular value decomposition: its computation and some applications," *IEEE Transactions on Automatic Control,* Vol. 25, No. 2, pp. 164–176, 1980

Note 19.3   The noninteracting control problem, also known simply as the *decoupling* problem, has a rich history. Early geometric work is surveyed in the paper

A.S. Morse, W.M. Wonham, "Status of noninteracting control," *IEEE Transactions on Automatic Control,* Vol. AC-16, No. 6, pp. 568–581, 1971

The proof of Theorem 19.11 follows the broad outlines of the development in

A.S. Morse, W.M. Wonham, "Decoupling and pole assignment by dynamic compensation," *SIAM Journal on Control and Optimization,* Vol. 8, No. 3, pp. 317–337, 1970 with refinements deduced from the treatment of a nonlinear noninteracting control problem in

H. Nijmeijer, J.M. Schumacher, "The regular local noninteracting control problem for nonlinear control systems," *SIAM Journal on Control and Optimization,* Vol. 24, No. 6, pp. 1232–1245, 1986

Independent early work on the geometric approach to noninteracting control for linear systems is reported in

G. Basile, G. Marro, "A state space approach to noninteracting controls," *Ricerche di Automatica,* Vol. 1, No. 1, pp. 68–77, 1970

Fundamental papers on algebraic approaches to noninteracting control include

P.L. Falb, W.A. Wolovich, "Decoupling in the design and synthesis of multivariable control systems," *IEEE Transactions on Automatic Control,* Vol. AC-12, No. 6, pp. 651–659, 1967

E.G. Gilbert, "The decoupling of multivariable systems by state feedback," *SIAM Journal on Control and Optimization,* Vol. 7, No. 1, pp. 50–63 , 1969

L.M. Silverman, H.J. Payne, "Input-output structure of linear systems with application to the decoupling problem," *SIAM Journal on Control and Optimization,* Vol. 9, No. 2, pp. 199–233, 1971

Note 19.4 The important problem of using static state feedback to simultaneously achieve noninteracting control and exponential stability for the closed-loop state equation is neglected in our introductory treatment. Conditions under which this can be achieved are established via algebraic arguments for the case $m = p$ in the paper by Gilbert cited in Note 19.3. For more general linear plants, geometric conditions are derived in

J.W. Grizzle, A. Isidori, "Block noninteracting control with stability via static state feedback," *Mathematics of Control, Signals, and Systems,* Vol. 2, No. 4, pp. 315–342, 1989

These authors begin with an alternate geometric formulation of the noninteracting control problem that involves controlled invariant subspaces containing $Im\,[BG_i]$, and contained in $Ker\,[C_i]$. This leads to a different solvability condition that is of independent interest.

If dynamic state feedback is permitted, then solvability of the noninteracting control problem with static state feedback implies solvability of the problem with exponential stability via dynamic state feedback. See the papers by Morse and Wonham cited in Note 19.3.

Note 19.5 Another control problem that has been treated extensively via the geometric approach is the *servomechanism* or *output regulation* problem. This involves stabilizing the closed-loop system while achieving asymptotic tracking of any reference input generated by a specified, exogenous linear system, and asymptotic rejection of any disturbance signal generated by another specified, exogenous linear system. The servomechanism problem treated algebraically in Chapter 14 is an example where the (unmentioned) exogenous systems are simply integrators. Consult the geometric treatment in

B.A. Francis, "The linear multivariable regulator problem," *SIAM Journal on Control and Optimization,* Vol. 15, No. 3, pp. 486–505, 1977

a paper than contains references to a variety of other approaches. Other problems involving dynamic state feedback, observers, and dynamic output feedback can be treated from a geometric viewpoint. See the citations in Note 18.1, and

W.M. Wonham, *Linear Multivariable Control: A Geometric Approach,* Third Edition, Springer-Verlag, New York, 1985

Note 19.6   Geometric methods are prominent in nonlinear system and control theory. Entries to the literature are provided by the books

A. Isidori, *Nonlinear Control Systems,* Second Edition, Springer-Verlag, Berlin, 1989

H. Nijmeijer, A.J. van der Schaft, *Nonlinear Dynamical Control Systems,* Springer-Verlag, New York, 1990

INDEX

Author Index

A

Ackermann, J., 236
Aeyels, D., 136
Ailon, A., 136
Aling, H., 321
Anderson, B.D.O., 112, 178, 192, 256, 295
Antoulas, A.C., 178
Apostlol, T.M., 65
Arbib, M.A., 157, 178, 256
Ascher, U.M., 49

B

Baratchart, L., 136
Barnett, S., 279
Basile, G., 320, 347
Bass, R.W., 235
Belevitch, V., 214
Bellman, R., 96, 123
Bentsman, J., 123
Berlinski, D.J., 35
Bernstein, D.S., 82
Bhattacharyya, S.P., 257, 346
Bittanti, S., 136
Blair, W.B., 65
Blomberg, H., 279
Boley, D., 157

B (continued, right column)

Bongiorno, J.J., 256
Brockett, R.W., 17, 49, 135, 235
Bruni, C., 157, 178
Brunovsky, P., 136, 237
Bucy, R.S., 18, 215
Byrnes, C.I., 237

C

Callier, F.M., 295
Champetier, C., 237
Chen, C.T., 135, 295
Cheng, V.H.L., 235
Christov, N.N., 18
Chua, L.O., 34
Colaneri, P., 136
Commault, C., 346
Coppel, W.A., 97, 122

D

D'Alessandro, P., 156
Damen, A.A.H., 178
D'Angelo, H., 83
Davison, E.J., 236, 257
DeCarlo, R.A., 18
Delchamps, D.F., 17, 156, 279

Subject Index